SPACE EXPLORATION

ISSN 1551-210X

SPACE EXPLORATION

Kim Masters Evans

INFORMATION PLUS® REFERENCE SERIES
Formerly Published by Information Plus, Wylie, Texas

Farmington Hills, Mich • San Francisco • New York • Waterville, Maine
Meriden, Conn • Mason, Ohio • Chicago

Space Exploration

Kim Masters Evans

Kepos Media, Inc.: Steven Long and Janice Jorgensen, Series Editors

Project Editors: Laura Avery, Tracie Moy

Rights Acquisition and Management: Ashley M. Maynard, Carissa Poweleit

Composition: Evi Abou-El-Seoud, Mary Beth Trimper

Manufacturing: Rita Wimberley

Cover photograph: iurii/Shutterstock.com.

Gale
27500 Drake Rd.
Farmington Hills, MI 48331-3535

ISBN-13: 978-0-7876-5103-9 (set)
ISBN-13: 978-1-57302-704-5

ISSN 1551-210X

This title is also available as an e-book.
ISBN-13: 978-1-57302-710-6 (set)
Contact your Gale sales representative for ordering information.

Printed in the United States of America
1 2 3 4 5 20 19 18 17 16

TABLE OF CONTENTS

PREFACE

Space Exploration is part of the *Information Plus Reference Series*. The purpose of each volume of the series is to present the latest facts on a topic of pressing concern in modern American life. These topics include the most controversial and studied social issues of the 21st century: abortion, capital punishment, care for the elderly, crime, health care, the environment, immigration, race and ethnicity, social welfare, women, youth, and many more. Although this series is written especially for high school and undergraduate students, it is an excellent resource for anyone in need of factual information on current affairs.

By presenting the facts, it is the intention of Gale, Cengage Learning, to provide its readers with everything they need to reach an informed opinion on current issues. To that end, there is a particular emphasis in this series on the presentation of scientific studies, surveys, and statistics. These data are generally presented in the form of tables, charts, and other graphics placed within the text of each book. Every graphic is directly referred to and carefully explained in the text. The source of each graphic is presented within the graphic itself. The data used in these graphics are drawn from the most reputable and reliable sources, such as from the various branches of the U.S. government and from private organizations and associations. Every effort has been made to secure the most recent information available. Readers should bear in mind that many major studies take years to conduct and that additional years often pass before the data from these studies are made available to the public. Therefore, in many cases the most recent information available in 2016 is dated from 2013 or 2014. Older statistics are sometimes presented as well, if they are landmark studies or of particular interest and no more-recent information exists.

Although statistics are a major focus of the *Information Plus Reference Series*, they are by no means its only content. Each book also presents the widely held positions and important ideas that shape how the book's subject is discussed in the United States. These positions are explained in detail and, where possible, in the words of their proponents. Some of the other material to be found in these books includes historical background, descriptions of major events related to the subject, relevant laws and court cases, and examples of how these issues play out in American life. Some books also feature primary documents or have pro and con debate sections that provide the words and opinions of prominent Americans on both sides of a controversial topic. All material is presented in an evenhanded and unbiased manner; readers will never be encouraged to accept one view of an issue over another.

HOW TO USE THIS BOOK

The achievements of the National Aeronautics and Space Administration (NASA) and its counterparts in other nations are widely admired. Nevertheless, humankind's space exploration efforts have been expensive, and include some disturbing failures and tragic accidents. As a result, space exploration is at the center of numerous controversies. This volume presents the facts on space exploration's successes and failures, and the questions that surround them. How can tragedies such as the *Columbia* disaster happen? Are manned missions necessary or are robotic probes more cost effective? What do we gain from space exploration? Where should our priorities lie? Should we continue to explore space at all?

Space Exploration consists of nine chapters and three appendixes. Each chapter is devoted to a particular aspect of space exploration. For a summary of the information that is covered in each chapter, please see the synopses that are provided in the Table of Contents. Chapters generally begin with an overview of the basic facts and background information on the chapter's topic, then proceed to examine subtopics of particular interest. For example,

Chapter 1: Introduction to Space Exploration provides historical information about some of the key scientists and innovators whose discoveries made space travel possible. It also describes the importance of the cold war, which was a time of tense relations between the United States and the Soviet Union following World War II. The two superpowers developed robust space programs as part of a race to land humans on the moon. They also invested time and resources in missile programs and deploying satellites for navigation, communications, and other commercial purposes. The chapter briefly summarizes the major developments in human and robotic spaceflight that have occurred through the early 21st century. It ends with a reminder about the dangers of space exploration and a mention of the lives that have been lost as humans have pursued perilous journeys off Earth. Readers can find their way through a chapter by looking for the section and subsection headings, which are clearly set off from the text. They can also refer to the book's extensive Index if they already know what they are looking for.

Statistical Information

The tables and figures featured throughout *Space Exploration* will be of particular use to readers in learning about this issue. The tables and figures represent an extensive collection of the most recent and important statistics on space exploration, as well as related issues—for example, graphics depict breakdowns of NASA's budget; public opinion about NASA and space exploration in general; a listing of planetary bodies with known or suspected watery oceans; and diagrams of dozens of spacecraft and other equipment. Gale, Cengage Learning, believes that making this information available to readers is the most important way to fulfill the goal of this book: to help readers understand the issues and controversies surrounding space exploration and reach their own conclusions about them.

Each table or figure has a unique identifier appearing above it, for ease of identification and reference. Titles for the tables and figures explain their purpose. At the end of each table or figure, the original source of the data is provided.

To help readers understand these often complicated statistics, all tables and figures are explained in the text. References in the text direct readers to the relevant statistics. Furthermore, the contents of all tables and figures are fully indexed. Please see the opening section of the Index at the back of this volume for a description of how to find tables and figures within it.

Appendixes

Besides the main body text and images, *Space Exploration* has three appendixes. The first is the Important Names and Addresses directory. Here, readers will find contact information for a number of government and private organizations that can provide further information on aspects of space exploration. The second appendix is the Resources section, which can also assist readers in conducting their own research. In this section, the author and editors of *Space Exploration* describe some of the sources that were most useful during the compilation of this book. The final appendix is the detailed Index. It has been greatly expanded from previous editions and should make it even easier to find specific topics in this book.

COMMENTS AND SUGGESTIONS

The editors of the *Information Plus Reference Series* welcome your feedback on *Space Exploration*. Please direct all correspondence to:

Editors
Information Plus Reference Series
27500 Drake Rd.
Farmington Hills, MI 48331-3535

CHAPTER 1
INTRODUCTION TO SPACE EXPLORATION

Mankind will migrate into space, and will cross the airless Saharas which separate planet from planet and sun from sun.

—Winwood Reade, *The Martyrdom of Man* (1874)

Humans have always been explorers. When ancient humans stumbled across unknown lands or seas, they were compelled to explore them. They were driven by a desire to dare and conquer new frontiers and by a thirst for knowledge, wealth, and prestige. These are the same motivations that drove people of the 20th century to venture into space.

The very idea of space exploration has a sense of mystery and excitement about it. Americans call their space explorers astronauts. The term *astronaut* is a combination of two Greek words: *astron* (star) and *nautes* (sailor). Thus, astronauts are those who sail among the stars. This romantic imagery adds to the allure of space travel.

The truth is that space holds many dangers to humans. Potentially harmful radiation flows in the form of cosmic rays from deep space and electromagnetic waves that emanate from the sun and other stars. Tiny bits of rock and ice hurtle around in space at high velocities, like miniature missiles.

Space is not readily accessible. It takes a tremendous amount of power and thrust to hurl something off the surface of Earth. It is a fight against the force of Earth's gravity and the heavy drag of an air-filled atmosphere.

Getting into space is not easy, and getting back to Earth safely is even more challenging. Returning to Earth from space requires conquering another mighty force: friction. Any object that penetrates Earth's atmosphere from space encounters layers and layers of dense air molecules. Traveling at high speed and rubbing against these molecules produces a fiery blaze that can rip apart most objects.

It was not until the 1950s that the proper combination of skills and technology existed to overcome the obstacles of space travel. The political climate was also just right. Two rich and powerful nations (the Soviet Union and the United States) devoted their resources to besting one another in space instead of on the battlefield. It was this spirit of competition that pushed humans off the planet and onto the moon in 1969.

Once that race was over, space priorities changed. In the 21st century computerized machines do most of the exploring. They investigate planets, asteroids, comets, and the sun. Human explorers stay much closer to Earth. They visit and live aboard a space station that is in orbit a couple hundred miles above the planet. On Earth people dream of longer journeys because the vast majority of space is still an unknown sea that is just waiting to be explored.

THE MECHANICS OF SPACEFLIGHT

The mechanics of spaceflight were explored by the British physicist Isaac Newton (1642–1727). Newton first unraveled the mysteries of gravity on Earth and then extended his findings into space. He was the first to explain how a satellite (an orbiting body) could be put into orbit around Earth.

Newton's thought experiment, as it was called, proposed a cannon atop a tall mountain as a theoretical means for putting an object into orbit around Earth. The cannon shoots out projectiles one after another, using more gunpowder with each successive firing. Newton said each projectile would travel farther horizontally than the previous one before falling. Finally, there would be a projectile shot with enough gunpowder that it would travel very far horizontally and when it began to fall toward Earth, its path would have the same curvature as Earth's surface. The projectile would not fall to Earth's surface but continue to circle around the planet. It would be another two centuries before Newton's theory could be proven experimentally.

PIONEERS OF ROCKET SCIENCE

There are four men in history who are considered to be the founders of modern rocket science: Konstantin Tsiolkovsky of Russia, Hermann Oberth of Austria-Hungary, Robert H. Goddard of the United States, and Wernher von Braun of Germany. All four were working on rocket science during the early years of the 20th century. Although they were scattered around the world, they reached similar conclusions at about the same time.

Konstantin Tsiolkovsky

Konstantin Tsiolkovsky (1857–1935), a Russian schoolteacher, studied the theoretical concepts of rocket flight, such as gravity effects, escape velocity, and fuel needs. He developed a simple mathematical equation relating the initial velocity of a rocket to the final velocity, the starting and ending mass of the rocket, and the velocity of the rocket exhaust gases. Tsiolkovsky's equation became a fundamental concept of rocket science and is still taught in the 21st century.

In 1895 Tsiolkovsky wrote *Grezy o Zemle i Nebe* (*Dreams of the Earth and Sky*). The book described how a satellite could be launched into an orbit around Earth. Later publications included "Exploration of the Universe with Reaction Machines" and "Research into Interplanetary Space by Means of Rocket Power," both published in 1903. Figure 1.1 shows some of Tsiolkovsky's designs for liquid-propelled rockets. Two decades later he wrote *Plan of Space Exploration* (1926) and *The Space Rocket Trains* (1929).

FIGURE 1.1

Tsiolkovsky's rocket designs, 1903

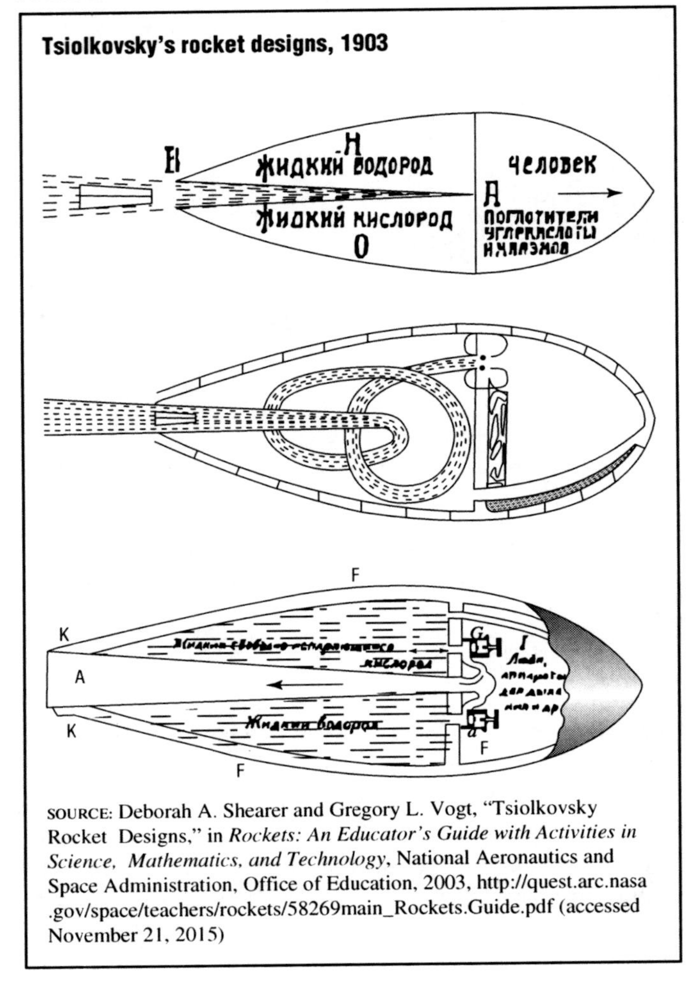

SOURCE: Deborah A. Shearer and Gregory L. Vogt, "Tsiolkovsky Rocket Designs," in *Rockets: An Educator's Guide with Activities in Science, Mathematics, and Technology*, National Aeronautics and Space Administration, Office of Education, 2003, http://quest.arc.nasa.gov/space/teachers/rockets/58269main_Rockets.Guide.pdf (accessed November 21, 2015)

Tsiolkovsky believed that rockets launched into space would have to include multiple stages. That is, instead of having one big cylinder loaded with fuel, the fuel would be divided up among smaller rocket stages linked together. As each stage used up its fuel, it could be jettisoned away so that the remainder did not have to carry dead weight. Tsiolkovsky reasoned that this was the only way for the mass of a rocket to be reduced as its fuel supply was depleted.

In 1911 Tsiolkovsky predicted in a letter, "Mankind will not remain on Earth forever, but in its quest for light and space will at first timidly penetrate beyond the confines of the atmosphere, and later will conquer for itself all the space near the Sun."

Hermann Oberth

Hermann Oberth (1894–1989) was born in Transylvania, Romania, which was part of the Austro-Hungarian Empire. As a teenager, Oberth studied mathematics and began developing sophisticated rocket theories. He studied medicine and physics at the University of Munich. During the 1920s he wrote two important books: *Die Rakete zu den Planetenräumen* (*The Rocket into Interplanetary Space*) and *Wege zur Raumschiffahrt* (*Methods of Achieving Space Flight*).

In 1923 Oberth predicted that rockets "can be built so powerfully that they could be capable of carrying a man aloft." He proposed bullet-shaped rockets for manned missions to Mars and an Earth-orbiting space station for refueling rockets. Like his Russian counterpart, Oberth advocated multistage rockets that were fueled by liquid propellants. He also inspired the German rocket club known as Verein für Raumschiffahrt (VfR; Society for Spaceship Travel). The VfR put Oberth's theories into practice by building and launching rockets that were based on his designs.

During the 1930s the Nazi government put Oberth and other VfR members to work developing rockets for warfare, rather than for space flight.

Robert H. Goddard

Robert H. Goddard (1882–1945) was an American physicist born in Worcester, Massachusetts. In a speech made in 1904, he said, "It is difficult to say what is impossible, for the dream of yesterday is the hope of today and the reality of tomorrow," and Goddard spent the rest of his life making rocket flight a reality. After graduating from college, he taught physics at Clark University in his hometown. He also spent time on a relative's farm experimenting with explosive rocket propellants. Unlike his Russian and Austro-Hungarian counterparts, Goddard's rocket science was more experimental than theoretical. In all, he was granted 70 patents for his inventions. The first two came in 1914 for a liquid-fuel gun rocket and a multistage step rocket. He is believed to be the first person to prove experimentally that a rocket can provide thrust in a vacuum.

Much of Goddard's research was funded by the Smithsonian Institution and the Guggenheim Foundation. The U.S. government showed little interest in rocket science except for its possible use in warfare. Late in World War I (1914–1918) Goddard presented the military with the concept for a new rocket weapon, later called the bazooka. After the war Goddard worked part time as a weapons consultant to the U.S. armed forces.

In 1920 the Smithsonian published Goddard's famous paper "A Method of Attaining Extreme Altitudes," in which he described how a rocket could be sent to the moon. The idea was greeted with skepticism from scientists and derision from the media. The *New York Times* published a scornful editorial that ridiculed Goddard for this fanciful notion. Stung by the criticism, Goddard spent the rest of his life avoiding publicity. His low-profile approach kept his work from being well known for many years. The one person who did take a keen interest in him was Charles A. Lindbergh Jr. (1902–1974), who in 1927 had become the first aviator to fly nonstop from New York to Paris. Lindbergh played a key role in securing funding for Goddard's rocket research from the Guggenheim Foundation.

FIGURE 1.2

Goddard's rocket design, 1926

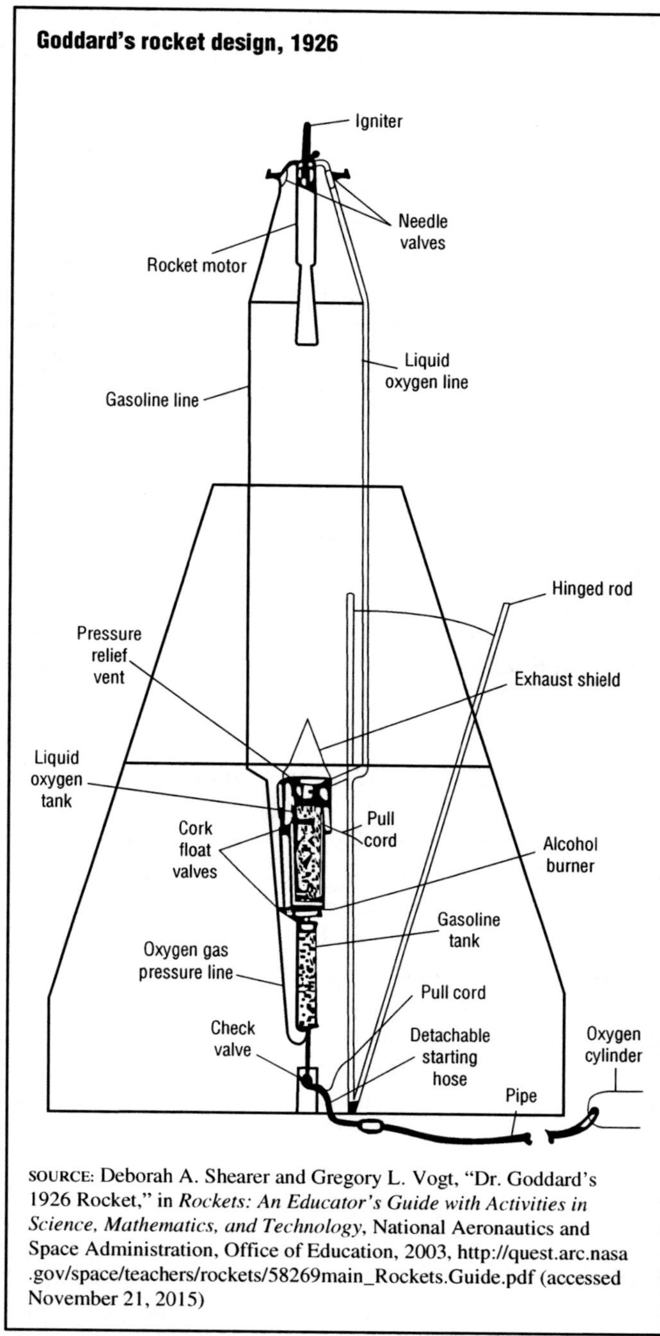

SOURCE: Deborah A. Shearer and Gregory L. Vogt, "Dr. Goddard's 1926 Rocket," in *Rockets: An Educator's Guide with Activities in Science, Mathematics, and Technology*, National Aeronautics and Space Administration, Office of Education, 2003, http://quest.arc.nasa.gov/space/teachers/rockets/58269main_Rockets.Guide.pdf (accessed November 21, 2015)

On March 16, 1926, Goddard achieved the first-known successful flight of a liquid-propelled rocket. (See Figure 1.2.) Throughout the next decade he labored quietly in the desert near Roswell, New Mexico, developing increasingly more powerful rockets. In 1935 Goddard launched the first supersonic liquid-fuel rocket. (Supersonic means faster than the speed of sound. Sound waves travel at about 700 miles per hour [1,130 kph], depending on the air temperature.) A year later Goddard's achievements finally received recognition when the Smithsonian published another of his papers, "Liquid Propellant Rocket Development."

He continued his work until 1941, when the United States entered World War II (1939–1945). Until his death in 1945, Goddard worked with the military to develop rocket applications for aircraft. Three decades later the *New York Times* finally issued an apology for its 1920 editorial about him. The date was July 17, 1969, and three American astronauts were on their way to the moon. The newspaper admitted that Goddard had been right after all.

Wernher von Braun

World War II ushered in the rocket age. The Nazi government of Germany was eager to use rockets against its enemies. The great talents and minds of the VfR were directed to forget about space travel and concentrate on warfare. During the early 1940s Germany developed the most sophisticated rocket program in the world. At its helm was the brilliant young Wernher von Braun (1912–1977).

Von Braun had been Oberth's assistant during the 1930s and an active member of the VfR. He was put in charge of developing a rocket weapon to terrorize the British population. Von Braun's team included Oberth and hundreds of people who worked on the remote island of Peenemünde. They developed the rocket-powered Vergeltungswaffens (weapons of vengeance), which were called V weapons for short.

There were two series of V weapons. The V-1 carried 1 ton (0.9 t) of explosives and traveled at a top speed of about 400 miles per hour (645 kph). This was slow enough that British gunners could blow apart the V-1s as they descended through the air. Thousands of V-1s were launched against Great Britain, but roughly half never impacted the ground.

Far more lethal was the V-2. This was truly a rocket with a top speed around 2,000 miles per hour (3,200 kph). The V-2s traveled far too fast to be shot down, and they terrified the British public. In "V-2: Hitler's Last Weapon of Terror" (BBC.co.uk, September 7, 2004), Paul Rincon reports that about 1,300 V-2s were launched against Great Britain during World War II, killing more than 2,700 people.

The first V-2 rocket fell on London in September 1944. However, the tide had already turned against Germany. By early 1945 the country was being invaded by the Soviets from the east and the Americans, British, and other Allies from the west. To be in position to surrender to U.S. forces, von Braun moved his team near the German-Swiss border. A negotiated surrender was arranged in which von Braun turned over himself, people on his team, and vital plans, drawings, rocket parts, and documents. In exchange, the U.S. Army agreed to transport the team to the United States and fund its work on a U.S. rocket program. The army called the agreement Operation Paperclip. It had no way of knowing that this move was going to put Americans on the moon.

THE EDGE OF SPACE?

Space begins at the edge of Earth's atmosphere, just beyond the protective blanket of gases and heat that surrounds the planet. This blanket is thick and dense near the surface and light and wispy farther away from the planet. At the dawn of the modern space age during the 1950s a debate began about the altitude above Earth at which space begins. According to the American space historian Dennis R. Jenkins, in "A Word about the Definition of Space" (March 1, 2008, http://www.nasa.gov/centers/dryden/news/X-Press/stories/2005/102105_Schneider.html), U.S. engineers who were engaged in designing space-faring craft during the 1950s decided to designate 50 miles (80.1 km) as the space boundary. Jenkins notes that atmospheric gases at this altitude are so low in density that "traditional aerodynamic control surfaces would be rendered largely useless." The U.S. military embraced the 50-mile boundary and has since used it to distinguish between air flight and spaceflight.

However, another altitude for the Earth-space boundary also came into use during the 1950s: the so-called Kármán Line, named after the Hungarian American aeronautical engineer Theodore von Kármán (1881–1963). Jenkins reports that von Kármán calculated the altitude above which a vehicle must fly faster than orbital velocity to stay aloft. (A vehicle at orbital velocity achieves a balance between the inertia of its own movement and the force of Earth's gravity, allowing the vehicle to circle Earth.) Although this altitude varies somewhat based on several factors, von Kármán found it to be in the neighborhood of 62 miles (100 km). He suggested the 62-mile boundary to the Fédération Aéronautique Internationale (FAI). Jenkins notes that the FAI is "the internationally recognized body" for recording air and spaceflight feats.

According to Jenkins, the Kármán Line is also commonly used in international treaties that are related to spaceflight. However, he points out that the U.S. government has never "officially adopted" a specific altitude for the boundary between air and space.

THE X-SERIES

Even before World War II ended the United States began developing rocket-powered planes. In 1943 the National Advisory Committee for Aeronautics (which had been established in 1915 to oversee the development of aircraft design and flight theory) initiated a research program in conjunction with the air force and the navy. Because the planes were experimental, they were given the name X-aircraft. In 1944 a company called Bell Aircraft began work on the XS-1, with the "S" standing for supersonic. Later, the "S" was dropped, and the plane became the X-1.

On October 14, 1947, Captain Chuck Yeager (1923–) of the U.S. Air Force flew the X-1 at the speed of sound, which is known as Mach 1. The X-1 was only the first of many high-performance planes that were tested in the program. Eventually, X-planes flew at hypersonic speeds, that is, speeds greater than Mach 5 (five times the speed of sound). In "X-15: Hypersonic Research at the Edge of Space" (February 24, 2000, http://history.nasa.gov/x15/cover.html), the National Aeronautics and Space Administration (NASA) explains that the X-15 was a rocket-fueled plane tested during the late 1950s and early 1960s. It was taken up to an altitude of approximately 45,000 feet (13,700 m) by a B-52 airplane and released. A rocket engine was then fired to propel the X-15 to incredible speeds and heights. On November 9, 1961, an X-15 flew at Mach 6.04, the fastest suborbital speed ever reached. On August 22, 1963, an X-15 soared across the boundary into space to an altitude of 67 miles (108 km). This record would remain unbroken for more than four decades.

The early X-series were high-speed, high-altitude planes unlike any ever built before. Most of them were tested over desolate desert areas near Muroc, California. Daring young test pilots flew the planes. However, this was a very dangerous profession. Many pilots were killed or seriously injured while testing the experimental planes. The pilots who survived became the first men considered for the nation's astronaut program.

It should be noted that the development of X-series planes still continued in early 2016 by various U.S. government entities, including NASA. However, not all X-series planes relate to space travel. For example, in "NASA's Electric-Propulsion Wing Test Helps Shape Next X-Plane" (AviationWeek.com, September 4, 2015), Graham Warwick describes work by NASA on an X-series airplane propelled by electric motors.

A COLD WAR IN SPACE BEGINS

The term *cold war* is used to describe U.S. relations with the Soviet Union from the end of World War II to 1991, when the Soviet Union collapsed. During this period—which is chiefly marked by a mutual mistrust and rivalry that led to a buildup of arms—both nations developed extensive nuclear weapons programs. Each thought the other was militarily aggressive, deceitful, and dangerous. Each feared the other wanted to take over the world. This paranoia was in full force when space exploration began.

Ballistic Missiles

Following World War II both the United States and the Soviet Union began researching the feasibility of attaching warheads to long-range rockets that were capable of traveling halfway around the world. These weapons were eventually called intercontinental ballistic missiles (ICBMs). They could be equipped with conventional or nuclear warheads. The United States had introduced the nuclear warfare age by dropping two atomic bombs on Japan to end World War II in August 1945.

By the early 1950s the U.S. Air Force was actively testing three different ICBMs under the Navaho, Snark, and Atlas Programs. This work was highly classified as a matter of national security. The United States and the Soviet Union both engaged in massive spying campaigns throughout the cold war. In 1955 American spies brought word that the Soviets were close to completing ICBMs capable of reaching U.S. cities.

The Soviet rocket work was spearheaded by Sergei Korolev (1906–1966). He oversaw the development of the R-7, the world's first ICBM, and is considered to be the father of the Soviet space program.

The International Geophysical Year

In 1952 a group of American scientists proposed that the International Council of Scientific Unions (ICSU) should sponsor a worldwide research program to learn more about Earth's polar regions. Eventually, the project was expanded to include the entire planet and the space around it. The ICSU decided to hold the project between July 1957 and December 1958 and call it the International Geophysical Year (IGY). Geophysics is a branch of earth science that focuses on physical processes and phenomena in Earth and its vicinity.

The IGY period was selected to coincide with an expected phase of heightened solar activity. Approximately every 11 years the sun undergoes a one- to two-year period of extra radioactive and magnetic activity. This is called the solar maximum. The ICSU hoped that rocket technology would progress enough to put satellites in Earth orbit during the next solar maximum and collect data on this phenomenon.

Sixty-seven countries participated in various capacities in the IGY project. The American delegation to the ICSU was led by the National Academy of Sciences, which consisted of a team of scientists from businesses, universities, and private and military research laboratories to conduct American activities during the IGY. Many of these activities would later be taken over by the National Oceanic and Atmospheric Administration (NOAA). According to NOAA, in "Rockets, Radar, and Computers: The International Geophysical Year" (July 19, 2012, http://celebrating200years.noaa.gov/magazine/igy/welcome.html), the IGY was a "resounding success" from a scientific standpoint. Rockets and balloons carrying sophisticated monitoring equipment provided new data about Earth's magnetic field, upper atmosphere, oceans, and weather. However, all the scientific accomplishments of the IGY would be overshadowed by one monumental event: the launch of a Soviet satellite named *Sputnik*.

Sputnik 1

Throughout the mid-1950s the United States worked unsuccessfully to construct a science satellite for the IGY. This work proceeded separately from ICBM development. However, at the time only the military had the expertise and resources to build rockets that were capable of leaving Earth's atmosphere. The U.S. Navy was charged with developing a rocket that was capable of carrying a package of scientific instruments into Earth orbit. In 1957 testing of the U.S. IGY satellite was still ongoing and proceeding poorly when the United States received shocking news.

On the evening of October 4, 1957, the Soviet Union news service announced that the nation had successfully launched the first-ever artificial satellite into Earth orbit. It was called *Sputnik*, which means "companion" in

English. The word also translates as "satellite," because a satellite is Earth's companion in an astronomical sense. Launched atop an R-7 Semiorka rocket, the satellite weighed 184 pounds (83 kg) on Earth, was about the size of a basketball, and orbited Earth every 98 minutes.

A Secret Surprise

The launch announcement of *Sputnik 1* was both a disappointment and a surprise to American scientists. They knew their Soviet counterparts were working on a science satellite for the IGY, but they had no idea the Soviets had progressed so quickly. The American scientists had openly shared information about their research during ICSU meetings. By contrast, the Soviet government forbade its scientists from disclosing any details about their work. *Sputnik 1* had been developed and launched in near total secrecy. According to Hugh Sidey, in "The Space Race Lifts Off" (Time.com, March 31, 2003), Lloyd Berkner, the president of the ICSU, learned about the launch while at a dinner party at the Soviet embassy in Washington, D.C., when a reporter from the *New York Times* whispered the news to him.

The American public was even more shocked by the announcement. Millions went outside in the darkness to look for the satellite in the night sky. Witnesses said it was a tiny twinkling pinpoint of light that moved steadily across the horizon. The satellite continuously broadcast radio signals that were picked up by ham radio operators all over the world. Ham radio is communication using short-wave radio signals on small amateur stations.

The *Sputnik 1* signals were another unpleasant surprise for American scientists. It had been universally agreed that IGY satellites would broadcast radio signals at a frequency of 108 megahertz. The United States had already built a satellite tracking system that was designed for this frequency. However, *Sputnik 1* transmitted at much lower frequencies, ensuring that U.S. scientists would not be able to pick up its data.

Sputnik 2—A Dog in Space

The success of *Sputnik 1* caught the United States off guard and unprepared. For the first time, the American public realized that the Soviets probably had the capability of launching long-range nuclear missiles against the United States. A month later there was even further dismay when the Soviets launched a second satellite.

Sputnik 2 was much larger than its predecessor and carried a live dog, a husky-mix named Laika, into orbit. It was a one-way trip for her because the Soviet scientists had not yet worked out how to bring the spacecraft safely back to Earth. At the time the Soviet news agency bragged that Laika survived for a week aboard the spacecraft. Decades later scientists admitted that Laika died only hours after launch, when she panicked and overheated in her tiny cabin.

The United States Reacts

The American public was scared by the size of *Sputnik 2*, which weighed more than 1,000 pounds (450 kg) on Earth. Furthermore, it was common knowledge that the United States did not have a rocket capable of carrying that much weight into space. There was an uproar in the media, and politicians demanded to know how the Soviet Union had gotten so far ahead of the United States in space technology. President Dwight D. Eisenhower (1890–1969) charged the U.S. military to do whatever it took to put a satellite in space.

Because the navy's efforts to build a satellite had proved unsuccessful, the U.S. military decided to turn to von Braun and his team of rocket scientists, who were working for the army. On January 31, 1958, the first American satellite soared into orbit. It was named *Explorer 1* and rode atop a Jupiter-C rocket developed by the von Braun team at Huntsville, Alabama. (See Figure 1.3.)

On May 15, 1958, the Soviets countered with *Sputnik 3*, which was a miniature physics laboratory sent into orbit to collect scientific data.

Later that year, in October 1958, the United States formed NASA to oversee the nation's space endeavors. Although it was a civilian agency charged with operating peaceful missions in space, NASA would rely heavily on military resources to achieve its goals.

FIRST HUMAN IN SPACE

On April 12, 1961, the Soviet cosmonaut Yuri Gagarin (1934–1968) became the first human to travel beyond Earth's atmosphere, enter the frontier of space, and return safely to Earth. Gagarin was born in a village near Gzhatsk (now Gagarin) in central Russia. He grew up in a peasant family, dreaming of becoming a pilot. Before being recruited to be a cosmonaut (a Russian astronaut), Gagarin was serving as a lieutenant in the Soviet air force.

The article "Gagarin: Son of a Peasant, Star of Space" (BBC.co.uk, April 1, 1998) reports that his flight took him roughly 186 miles (300 km) above Earth and that he spent 1 hour and 49 minutes circling the planet, completing one entire orbit and part of another one. His cramped spacecraft was equipped with a radio for communicating with ground control. Looking down at the planet beneath him, he said, "The Earth is blue. How wonderful. It is amazing."

The weightlessness (the feeling of no forces on oneself) bestowed by space travel had always been a worry for scientists. There is a common misconception among

FIGURE 1.3

***Explorer I* in space**

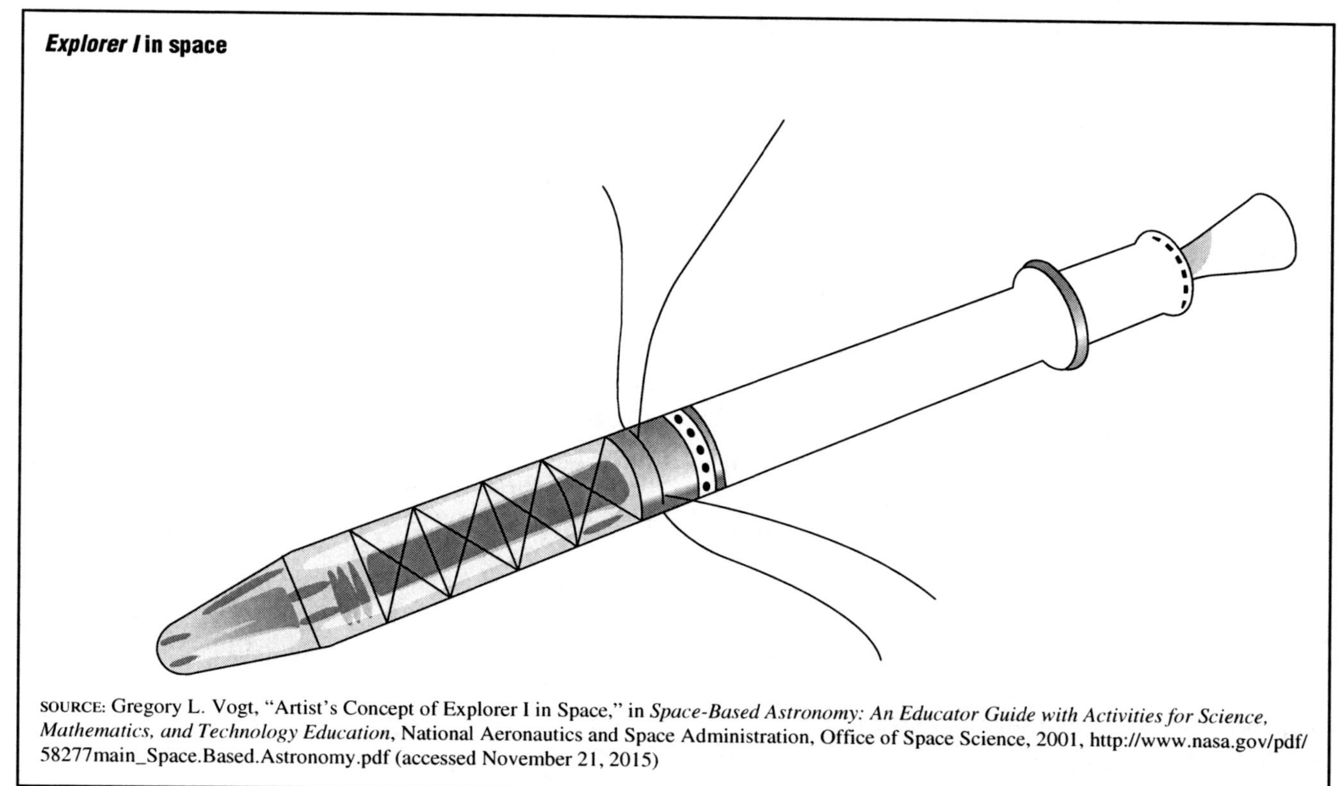

SOURCE: Gregory L. Vogt, "Artist's Concept of Explorer I in Space," in *Space-Based Astronomy: An Educator Guide with Activities for Science, Mathematics, and Technology Education*, National Aeronautics and Space Administration, Office of Space Science, 2001, http://www.nasa.gov/pdf/58277main_Space.Based.Astronomy.pdf (accessed November 21, 2015)

the public that there is no gravity in space. This is not true. Actually, the force of gravity remains strong for great distances around Earth. People who leave Earth's atmosphere experience the sensation of weightlessness because they are essentially in free fall toward Earth throughout their trip.

At the time of Gagarin's flight, scientists were not certain how the human body would react to weightlessness. His spacecraft included a computerized automatic pilot, in case Gagarin lost consciousness or was unable to move. This fear proved to be unfounded. The mission showed that humans can function in weightlessness.

Gagarin returned to Earth safely. He ejected from his spacecraft somewhere over Russia and parachuted to the ground. He became a national hero and an international sensation. His picture was on the front page of every major newspaper in the world.

The scientific teams in the United States were impressed with Gagarin's accomplishment but also envious of it. In "Yuri Gagarin: First Man in Space" (April 13, 2011, http://www.nasa.gov/mission_pages/shuttle/sts1/gagarin_anniversary.html), NASA notes that an agency spokesperson congratulated the Soviets for their achievement and summed up the U.S. space program with these glum words: "So close, but yet so far." In Huntsville, Alabama, von Braun was more blunt, saying, "To keep up, [the] U.S.A. must run like hell."

RACE TO THE MOON

On May 5, 1961, the American astronaut Alan Shepard (1923–1998) soared to an altitude of 116 miles (187 km) in the spaceship *Freedom 7*. He spent 15 minutes and 28 seconds in a suborbital flight. Suborbital means unable to attain one revolution in orbit. In other words, a suborbital flight does not complete an entire revolution around Earth. Shepard's flight was much shorter in distance and time than Gagarin's flight had been. A few months later Gherman Stepanovich Titov (1935–2000), the second cosmonaut in space, completed 17 and a half orbits around Earth. NASA knew it would be a year or more before it could accomplish a similar feat.

The United States was tired of coming in second place. Because there was no way to beat the Soviets at the orbital space race, President John F. Kennedy (1917–1963) decided to start a new race. His advisers recommended that the United States should put a manned spacecraft in orbit around the moon or even land a man on the moon. Either one would require the development of a huge new rocket to supply the lifting power that was needed to boost a spaceship out of Earth orbit. Neither the Soviets nor the Americans had such a rocket.

On May 25, 1961, President Kennedy revealed his decision to the world in the speech "Special Message to the Congress on Urgent National Needs" (http://www.presidency.ucsb.edu/ws/index.php?pid=8151&st=&st1=).

His words ignited the biggest race in human history: "First, I believe that this nation should commit itself to achieving the goal, before this decade is out, of landing a man on the moon and returning him safely to the earth. No single space project in this period will be more impressive to mankind, or more important for the long-range exploration of space; and none will be so difficult or expensive to accomplish."

AIMING FOR DRY SEAS

Suddenly, all eyes were on the moon. Earth's closest neighbor had been a subject of fascination since the first humans gazed up at the night sky.

Most of the features on the moon were named during the 1600s by the Italian astronomer Giambattista Riccioli (1598–1671). Riccioli was a priest, a member of the Roman Catholic order the Society of Jesus, which is devoted to missionary and educational work. At the request of the church, Riccioli devoted his life to astronomy and telescopic studies. He published a detailed lunar map that he developed with Francesco Maria Grimaldi (1618–1663), a fellow Jesuit and Italian physicist. This map featured Latin names for lunar features, elevations and depressions were named after famous astronomers and philosophers, and large dark flat areas that looked like bodies of water were named oceans or seas.

Four hundred years later humans on opposite sides of Earth took aim at these features. During the early and mid-1960s NASA and the Soviets sent dozens of photographic probes to take pictures of the moon. Some probes proved successful, and some did not. Four NASA probes crashed into the moon, but they had beamed back valuable photographs before impacting the lunar surface. In February 1966 the Soviet probe *Luna 9* softly set down in the Ocean of Storms, the largest of the lunar "seas." Four months later NASA's *Surveyor 1* probe landed nearby.

Both countries needed lunar data to support their efforts to send humans to the moon. During this time Soviet officials did not even acknowledge that they had a manned lunar program. Those in the U.S. program suspected that they did but could not be sure. It was not until years later, when the Russian diplomat Sergei Leskov recounted the story in "How We Didn't Get to the Moon" (*Izvestiya*, August 18, 1989), that the United States learned how determined the Soviet Union was at trying to beat the Americans to the moon.

HARD WORK

The U.S. effort to put men on the moon consisted of three phases:

- Mercury—short-duration suborbital and orbital missions consisting of one astronaut
- Gemini—longer-duration orbital missions consisting of two astronauts including extravehicular activity (space walking) and docking of spacecraft in space
- Apollo—manned missions to the moon involving three astronauts; a lunar module containing two astronauts would land on the moon, enabling them to explore the moon's surface, while a third astronaut remained in lunar orbit

Shepard's historic 1961 flight was considered the first Mercury mission. Over the next two years five more successful Mercury flights were conducted. In 1965 a series of 10 manned Gemini missions began. They were completed near the end of 1966.

Soon after it started, it became apparent that the moon program was going to be expensive. On September 12, 1962, President Kennedy (http://er.jsc.nasa.gov/seh/ricetalk.htm) reinforced his commitment to the project during a speech at Rice University in Houston, Texas. Kennedy said, "We choose to go to the moon. We choose to go to the moon in this decade and do the other things, not because they are easy, but because they are hard, because that goal will serve to organize and measure the best of our energies and skills, because that challenge is one that we are willing to accept, one we are unwilling to postpone, and one which we intend to win."

ROCKETS ARE KEY

One key goal for the United States and the Soviet Union was the development of a large and powerful rocket—a so-called superbooster. NASA called its superbooster a Saturn rocket, and the Soviets named their rocket the N-1.

Development of the Saturn rocket series began in 1961 under the direction of von Braun. He had actually been pitching the idea to the military for several years. Before the Apollo Program, NASA used relatively small rockets that were capable of lifting only a few hundred to a few thousand pounds into Earth orbit.

The Scout rocket was used to launch small satellites and probes weighing up to 300 pounds (136 kg) on Earth. It was devised by combining aspects of rockets that were used by the armed forces (the navy's Polaris and Vanguard rockets and the army's Sergeant rockets). The Thor, Atlas, and Titan series evolved from air force rockets that were first developed as ICBMs. During the mid-1960s the military replaced most of its Atlas rockets with Minuteman missiles. Modified Atlas rockets were used to launch satellites and for Project Mercury. The Titan II was used during Project Gemini.

The Saturn series evolved from von Braun's Jupiter series. Legend has it that the Saturn series got its name because it was one step beyond the Jupiter series, just as Saturn is the next planet beyond Jupiter in the solar system. The Saturn V, with a height of 364 feet (111 m), was at the time the largest rocket ever

FIGURE 1.4

Saturn V rocket being transported to launch pad

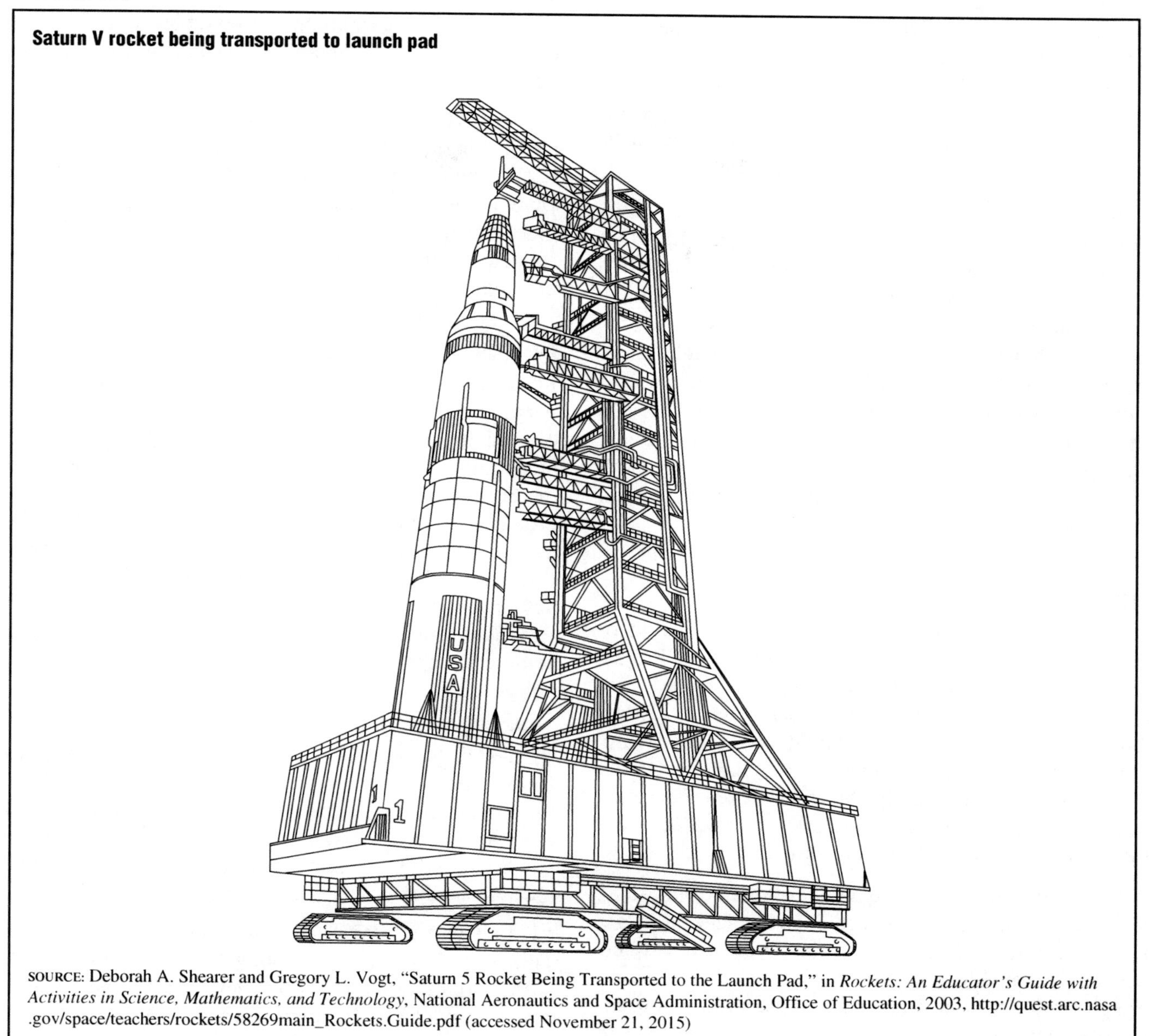

SOURCE: Deborah A. Shearer and Gregory L. Vogt, "Saturn 5 Rocket Being Transported to the Launch Pad," in *Rockets: An Educator's Guide with Activities in Science, Mathematics, and Technology*, National Aeronautics and Space Administration, Office of Education, 2003, http://quest.arc.nasa.gov/space/teachers/rockets/58269main_Rockets.Guide.pdf (accessed November 21, 2015)

built. (See Figure 1.4.) It had to be to push the 100-ton (90.7-t) *Apollo* spacecraft toward the moon.

APOLLO: TRAGEDY AND TRIUMPH

The Soviet space program continued to flourish. In September 1968 an unmanned probe called *Zond 5* became the first spacecraft to travel around the moon and return to Earth. The pressure was on NASA to speed up the Apollo Program.

On January 27, 1967, three American astronauts—Virgil I. Grissom (1926–1967), Edward Higgins White (1930–1967), and Roger B. Chaffee (1935–1967)—were killed when a flash fire raced through their capsule during a routine practice drill. They were the first human casualties of the space program to die while training for a scheduled NASA mission. This mission was called Apollo 1.

The next manned Apollo mission was launched on October 11, 1968. Apollo 7 successfully conducted a flight test and returned to Earth. It was followed in rapid succession by the more ambitious missions of Apollo 8, Apollo 9, and Apollo 10, each of which tested a lunar or command module in lunar or Earth orbit. The mission to set humans on the moon was named Apollo 11, and it was scheduled for July 1969.

By this time the Soviets had desperately tried to get their own manned lunar program going. However, the N-1 rocket kept failing its launch tests. The Soviets realized that it would not be ready before the Apollo 11 launch. Still hoping to steal some of the thunder from the Americans, the Soviets launched the robotic probe *Luna 15* to the moon. It was designed to gather samples from the lunar surface and return to Earth before the Apollo 11 expedition. Launched on July 13, 1969, *Luna 15* completed 52 moon orbits before it crashed into the lunar surface on July 21, 1969, and was lost.

Meanwhile, on July 20, 1969, *Apollo 11* set down safely on the moon near the Sea of Tranquility. Late that evening the astronaut Neil Armstrong (1930–2012) stepped out of the spacecraft to become the first human to stand on the moon. Approximately half a billion people on Earth watched the historic event on television. Four days later the *Apollo 11* crew returned to Earth to a heroes' welcome. There were six more Apollo missions to the moon (five that successfully landed on the moon, with *Apollo 13* forced to return home after it was damaged by a violent explosion on the way to the moon) before the program ended in 1972.

THE RIGHT STUFF

NASA's space program introduced a new kind of hero to American culture: the astronaut. When Project Mercury began, NASA selected seven men to be astronauts: Grissom, Shepard, Scott Carpenter (1925–2013), Gordon Cooper (1927–2004), John Glenn (1921–), Wally Schirra (1923–2007), and Donald K. Slayton (1924–1993). They were called the Mercury Seven. The men were all successful military test pilots known for their bravery and professional piloting skills.

The men had to pass strenuous batteries of physical, mental, and medical tests to become astronauts and begin their training to go into space. In the eyes of the American public, the Mercury Seven captured the bold and daring spirit of famous flyers such as Manfred von Richthofen (1882–1918), a German World War I ace, and Lindbergh. They were instant superstars and began receiving thousands of fan letters. Once NASA realized the great popularity of the astronauts, it used them as goodwill ambassadors for the agency. The astronauts traveled around the country speaking to civic groups and clubs to elicit public support for the space program.

NASA scientists originally envisioned astronauts as mere guinea pigs for space experiments. They were intended to be passive passengers covered with medical sensors and sealed inside space capsules that were completely controlled by operators on the ground through onboard computers. The astronauts rebelled at this notion and insisted on many changes, including installation of windows and manual piloting controls on the space capsules. When Project Gemini began, NASA selected nine more astronaut candidates and soon dozens after that. NASA indicates in "Chronology of U.S. Astronaut Missions (1961–1972)" (June 29, 2011, http://nssdc.gsfc.nasa.gov/planetary/chrono_astronaut.html) that by the end of the Apollo Program, 34 American astronauts had traveled into space.

In 1979 the story of the original Mercury Seven was profiled in the book *The Right Stuff* by Tom Wolfe. In the book, Wolfe describes the tremendous pressures that were put on the first astronauts during the space program, their dedication to serving their country, and how they reacted to fame and glory. In 1983 the book was made into a popular movie of the same name.

DÉTENTE IN SPACE

During the early years of space flight, U.S. relations with the Soviet Union were at their worst. John Pike et al. explain in "R-46" (July 29, 2000, http://www.fas.org/nuke/guide/russia/icbm/r-46.htm) that only months after the Soviets put their first cosmonauts in space the Soviet premier Nikita Khrushchev (1894–1971) made the veiled threat, "We placed Gagarin and Titov in space, and we can replace them with other loads that can be directed to any place on Earth." The meaning was clear to the American public: the Soviet Union's powerful rockets could carry nuclear warheads just as easily as they carried humans.

Détente is a French word that means a relaxation of strained relations. The United States and the Soviet Union occasionally enjoyed periods of détente during the cold war, particularly in their space activities. In 1965 a joint project was undertaken in which American and Soviet scientists shared information they had learned about space biology and medicine.

In October 1967 the two countries negotiated the Treaty on Principles Governing the Activities of States in the Exploration and Use of Outer Space, Including the Moon and Other Celestial Bodies (2016, http://www.unoosa.org/oosa/en/ourwork/spacelaw/treaties/introouterspacetreaty.html), which is more commonly known as the Outer Space Treaty. The treaty provides a basic framework for the activities that are and are not allowed in space and during space travel. The main principles are:

- Nations cannot place nuclear weapons or other weapons of mass destruction in Earth orbit or elsewhere in space.
- Outer space is open to all humankind and all nations for exploration and use.
- Outer space cannot be appropriated or claimed for ownership by any nation.
- Celestial bodies can only be used for peaceful purposes.
- Nations cannot contaminate outer space or celestial bodies.

- Astronauts are "envoys of mankind."
- Nations are responsible for all their national space activities whether conducted by governmental agencies or nongovernmental organizations.
- Nations are liable for any damage caused by objects they put into space.

The Outer Space Treaty was signed on January 27, 1967, by the United States, the Soviet Union, and the United Kingdom. Over the following decades it would be signed by more than 100 nations.

In 1969 NASA proposed the development of American and Soviet spacecraft that could dock with each other in space for future missions of mutual interest. In July 1975 the docking procedure proved to be successful during the Apollo-Soyuz Test Rendezvous and Docking Test Project. The mission was largely symbolic, and many people considered it wasted money that could have been spent on space exploration.

Near the end of the Apollo Program the two countries agreed to a number of cooperative projects including the sharing of lunar samples, weather satellite data, and space medical data.

HUMAN SPACEFLIGHT SINCE THE APOLLO ERA

In the minds of most Americans, the space race was over the day *Apollo 11* set down on the moon. Although NASA carried out six more Apollo missions, public interest and political support for them faded quickly. Neither the U.S. nor the Soviet government was interested in racing to somewhere else in space. Both governments decided to concentrate on putting crewed scientific space stations in low Earth orbit (LEO).

LEO is approximately 100 to 1,200 miles (160 to 1,900 km) above Earth's surface. Below this altitude air drag from Earth's atmosphere is still dense enough to pull spacecraft downward quickly. Beyond LEO lies a thick region of radiation that is known as the inner Van Allen radiation belt. This region poses a hazard to human life and to sensitive electronic equipment. Spacecraft in LEO travel at about 17,000 miles per hour (27,400 kph) and orbit Earth once every 90 minutes or so.

Since the end of the Apollo era no humans have ventured beyond LEO. During the 1970s and 1980s the Soviet Union put eight space stations into LEO or just below it. The most famous was *Mir*. NASA also had an orbiting space station called *Skylab*. These crewed projects evolved into the *International Space Station* (*ISS*), which currently orbits Earth in LEO and is described in detail in Chapter 5. The *ISS* is a collaboration of more than a dozen countries that are led by the United States, Russia, Canada, and Japan. The program has allowed humans from many different nations to travel into space.

By the early 1990s the Soviet Union had politically disintegrated. The United States' former archenemy splintered into dozens of individual republics, and the cold war came to an end. The Russian Federation took over the space program that was established by the Soviet Union. China, which does not participate in the *ISS*, launched its first taikonaut (Chinese astronaut) into orbit in 2003. Its space program, as well as that of Russia, Japan, France, and the European Space Agency (a confederation of European nations), is described in Chapter 3.

Between 1981 and 2011 NASA flew astronauts to and from LEO aboard reusable space planes called space shuttles. In total, 135 launches took place. Chapter 2 discusses the development, operation, and termination of this important program, which wrought both fantastic results and horrific tragedies. NASA's next-generation space transportation system was still being developed in early 2016. Its evolution has been hindered by the deep economic recession that engulfed the United States between 2007 and 2009 and by subsequent efforts to dramatically cut government spending. In a strange historical irony, after the space shuttle era ended the United States became dependent on Russia, its old space rival, to transport American astronauts to and from the *ISS*.

The Commercial Space Act of 1998 encouraged NASA to facilitate the participation of the private sector in the operation, use, and servicing of the *ISS*. This act received little attention until 2004, when the decision was made to retire the space shuttle fleet. Since then, the U.S. government has actively funded private industry involvement in NASA's space missions, particularly to and from the *ISS*. Chapter 4 describes the private spacecraft that have been developed to transport cargo to the orbiting station and the ones under development to fly humans there.

For more than three decades the only way for humans to access space was through government-operated space programs. This all changed on June 21, 2004, when *SpaceShipOne* carried Mike Melvill (1940–) to an altitude of 62 miles (100 km), which is considered to be the boundary of space by the FAI. This was the first privately built and financed craft to fly a human into space. As is explained in Chapter 4, its success offers tantalizing prospects for private individuals to travel into space (however briefly) as space tourists.

SPACE-AGE SCIENCE FICTION

Science fiction is a category of literature in which an imaginative story is told that incorporates at least some scientific principles to give it a sense of authenticity and believability. It is a mixture of science and imagination.

The advent of the space age introduced a wealth of information to science-fiction authors.

One of the most innovative of these authors was Gene Roddenberry (1921–1991). During the mid-1960s he created the television show *Star Trek*. This was a futuristic tale about a mixed crew of humans and aliens that explored the galaxy in the starship *Enterprise* during the 23rd century. The television show was not popular during its original run, but over the next few decades it developed a loyal fan base and spawned a number of movies, including *Star Trek* (2009), *Star Trek into Darkness* (2013), and *Star Trek Beyond* (2016).

In 1974 thousands of *Star Trek* fans sent letters to NASA requesting that one of the newly developed space shuttles be named *Enterprise*. NASA gave this name to the prototype shuttle model that was used for flight testing.

Another notable science-fiction work of the 1960s was the film *2001: A Space Odyssey* (1968), which was based on the short story "The Sentinel" (1948) by Arthur C. Clarke (1917–2008). Astronauts exploring the moon find a mysterious artifact. Believing that it came from Jupiter, they set off for that planet on an amazing spacecraft. The ship is equipped with a supercomputer named HAL that malfunctions and turns against the human crew. The film featured little dialogue, but it became a hit for its very imaginative plot and spectacular views of futuristic space travel.

In 1977 the science-fiction film *Star Wars* debuted and became one of the most popular movies of all time. Set "a long time ago in a galaxy far, far away," the film tells the story of an adventurous young man who leaves his home world to join a band of rebels fighting against a tyrannical empire. The movie was renowned for its story, characters, adventure, and special effects. The *Star Wars* franchise went on to include six more highly successful films and a book series.

Hollywood movies featuring hostile space aliens invading Earth were a staple of pop culture during the 1950s. Such films captured the paranoia and fear that Americans felt about the communist threat from the Soviet Union. Beginning in the 1970s a gentler viewpoint of aliens emerged in movies such as *Close Encounters of the Third Kind* (1977), *ET: The Extra-Terrestrial* (1982), *Cocoon* (1985), and *Contact* (1997). Science-fiction films released in the 21st century have included both murderous and friendly aliens. In addition, there has been an emphasis on more realistic space experiences in movies such as *Gravity* (2013), *Interstellar* (2014), and *The Martian* (2015).

ROBOTIC SPACECRAFT

Space programs that use human explorers are expensive. It is cheaper to build and send mechanized (robotic) spacecraft to do the exploring. These explorers are discussed in detail in Chapters 6, 7, and 8. The crude lunar probes that were launched at the dawn of the space age have evolved into highly sophisticated robotic spacecraft that visit planets, comets, and asteroids throughout the solar system and travel far beyond. Some have even taken samples of extraterrestrial bodies and returned those samples to Earth.

Not all space exploration requires long-distance travel. Advances in computers and telescopes have allowed scientists to do a lot of exploring with robotic spacecraft that are stationed nearby Earth. Dozens of these high-tech machines take photographs, measure radiation waves, and collect data on solar and galactic phenomena.

Since the 1950s hundreds of robotic spacecraft have been launched into space. However, the vast majority were not designed to explore space, but to exploit their position in space to provide practical benefits on Earth. These spacecraft are commonly called application satellites.

Application Satellites

Application satellites serve as tools of earth science or for navigation, communication, or other commercial purposes. They would not be possible without the technology of the space age. Most satellites include two major components: a bus (or platform) that provides the infrastructure and utilities (such as energy and propulsion) for the satellite and a payload, which is the scientific or commercial equipment that collects data or performs some other activity. Satellites have limited power supplies, and eventually the bus and/or payload quit functioning. Inactive or "dead" satellites are satellites that have lost their ability to function. However, they can continue to orbit Earth for many years because they were placed in very high orbits above Earth so that atmospheric drag at such high altitudes is negligible. Over the decades the space around Earth has become littered with inactive satellites that pose a collision danger to active satellites and other spacecraft. This problem is discussed at length in Chapter 2.

On April 1, 1960, NASA launched the first successful meteorological satellite *TIROS 1* (Television Infrared Observation Satellite). The satellite was equipped with television cameras to film cloud cover around Earth. Over the next few decades weather satellites grew increasingly more sophisticated in their capabilities. Other application satellites perform various duties for earth scientists, such as mapping oceans and land masses or measuring the heat and moisture content of Earth's surface.

On August 12, 1960, NASA launched *Echo 1A*, its first communications satellite. It was a large metallic

FIGURE 1.5

***TDRS-L* spacecraft**

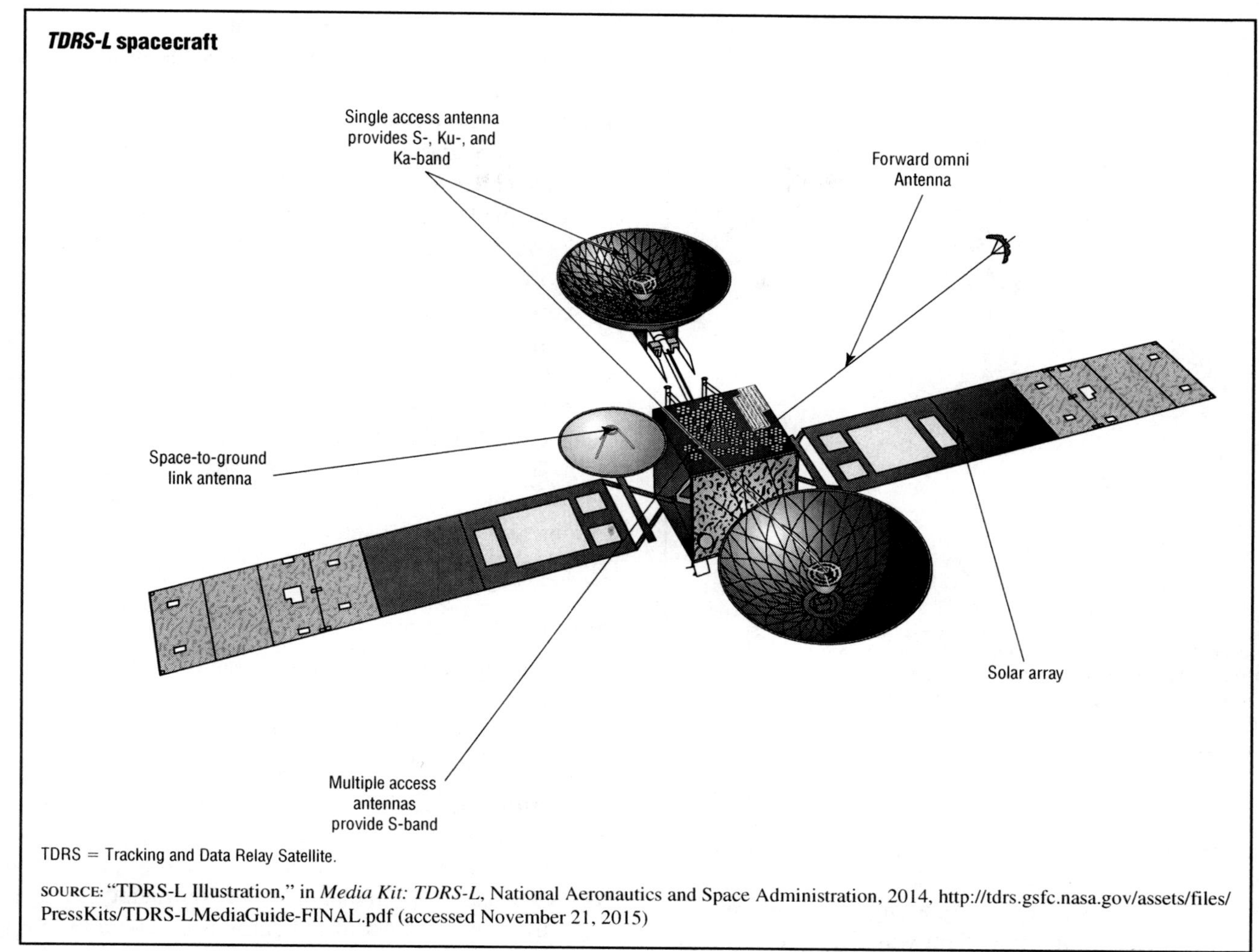

TDRS = Tracking and Data Relay Satellite.

SOURCE: "TDRS-L Illustration," in *Media Kit: TDRS-L*, National Aeronautics and Space Administration, 2014, http://tdrs.gsfc.nasa.gov/assets/files/PressKits/TDRS-LMediaGuide-FINAL.pdf (accessed November 21, 2015)

sphere that reflected radio signals. NASA maintains a whole series of communications satellites in Earth orbit that allow the agency to communicate with astronauts and relay data to robotic spacecraft during missions. They are part of the agency's Tracking and Data Relay Satellite (TDRS) fleet. Figure 1.5 shows the *TDRS-L* satellite, which was launched by NASA in January 2014.

Many communications satellites are placed in Earth orbit 22,230 miles (35,780 km) from the planet's surface. At this distance they are anchored in place by Earth's gravity and are in synch with its revolution rate. In other words, they move around Earth at the same speed that it revolves around its axis. This is called a geosynchronous orbit. Some satellites are placed in a geosynchronous orbit directly above Earth's equator and appear from Earth to hover in space at the exact same location all the time. They are in a geostationary orbit.

Navigation is the act of determining one's position relative to other locations. Before the invention of satellites, navigational signals were transmitted by land-based systems using antennas. These antennas sent low-frequency radio signals that traveled along Earth's surface or reflected off the ionosphere to reach their target receptor. The ionosphere is a layer of atmosphere that begins about 30 miles (48 km) above Earth's surface. Atmospheric gases undergo electrical and chemical changes within the ionosphere. This is what gives it reflective properties.

During the 1970s the U.S. military developed a space-based navigational system called the Global Positioning System (GPS). This system relies on a series of satellites in Earth orbit to handle signal transmissions. (See Figure 1.6.) During the 1980s GPS was made available for international civil use. Over the next two decades it became one of the most popular navigational tools in the world. Figure 1.7 illustrates how a GPS device on Earth determines its location based on satellite radio signals.

Commercial Launch Services

In 1962 Congress passed the Communications Satellite Act, opening the door for commercial use of satellites in space. For decades these satellites could only be launched by national space agencies (such as NASA). In 1980 Arianespace (a subsidiary of the European Space

FIGURE 1.6

Global Positioning System satellites

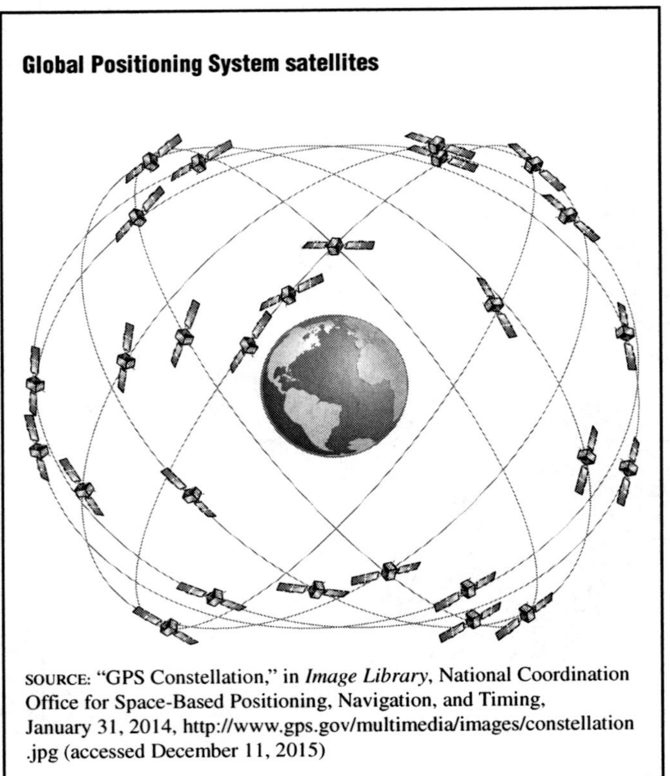

SOURCE: "GPS Constellation," in *Image Library*, National Coordination Office for Space-Based Positioning, Navigation, and Timing, January 31, 2014, http://www.gps.gov/multimedia/images/constellation.jpg (accessed December 11, 2015)

Agency) became the world's first commercial space transportation company. Arianespace began offering satellite launches using Ariane rockets at its spaceport in French Guiana (a small country along the northern coast of South America). Its first client was a U.S. telecommunications company.

In 1984 the Commercial Space Launch Act was passed in the United States. The act granted power to the U.S. private sector to develop and provide satellite launching, reentry, and associated services, which are described in Chapter 4.

SPACE LAUNCHES

Only a handful of launches take place each year to support human spaceflight programs that are operated by the United States, Russia, and China. The vast majority of launches take place to put uncrewed commercial, science, and military satellites into Earth orbit. The science missions are largely devoted to earth science, studying Earth's weather, climate patterns, atmospheric conditions, and so on. Military satellites perform reconnaissance (spying) from space or support the communications and navigation needs of armed forces around the world.

According to the Federal Aviation Administration, in *The Annual Compendium of Commercial Space Transportation: 2014* (February 2015, http://www.faa.gov/about/office_org/headquarters_offices/ast/media/FAA_Annual_Compendium_2014.pdf), there were 92 attempted orbital launches in 2014. All but three were deemed completely successful. Some launch events included multiple payloads. As a result, more than 150 payloads were launched into orbit in 2014. Most of them were devoted to communications, development, and remote-sensing purposes. Table 1.1 provides a breakdown of the 92 orbital launches by country. Russia conducted the most launches (32), followed by the United States (23), and China (16). Figure 1.8 breaks down the launches by country and launch vehicle. Many of the launch vehicles are described in detail in subsequent chapters. Table 1.2 provides information about the 23 U.S. orbital launches attempted in 2014. Most (14) of the launches put payloads into orbit for commercial clients, such as telecommunications companies, or for NASA. Nine of the launches were conducted for national security purposes by the U.S. Air Force or the National Reconnaissance Office, an agency that designs and launches satellites that are devoted to gathering intelligence (information).

SPACE CASUALTIES

Exploration has always been dangerous. Many ancient explorers died during their journeys across deserts, seas, mountains, and jungles. Space exploration has its own casualties.

During the earliest days of space travel dozens of animals were sacrificed for the space program. The United States sent a variety of small animals, most notably primates, up in rockets to test the safety of space flight for humans. Few of the animals survived the flight or the examination afterward. Some of the so-called astro-monkeys and astro-chimps that died were named Able, Albert, Bonny, Goliath, Gordo, and Scatback. The Soviets preferred to use dogs to test their spacecraft. Dogs named Bars, Laika, Lisichka, Mushka, and Pchelka died as a result.

Space programs in both countries have suffered human losses throughout the years as described in subsequent chapters. In January 2004 the NASA administrator Sean O'Keefe (1956–) announced that the last Thursday in January will become a day of remembrance for the lives that have been lost in the U.S. space program. Each year on this day, NASA employees observe a moment of silence and flags are flown at half-staff to honor the dead.

Like all journeys of discovery, space exploration is a bold and perilous undertaking. Major sacrifices have been made to move humankind closer to the stars. On September 13, 1962, President Kennedy (http://www.fordham.edu/halsall/mod/1962JFK-space.html) aptly described the combination of fear, hope, and yearning that characterizes every journey into space: "As we set sail, we ask God's blessing on the most hazardous and dangerous and greatest adventure on which man has ever embarked."

FIGURE 1.7

Operation of the Global Positioning System

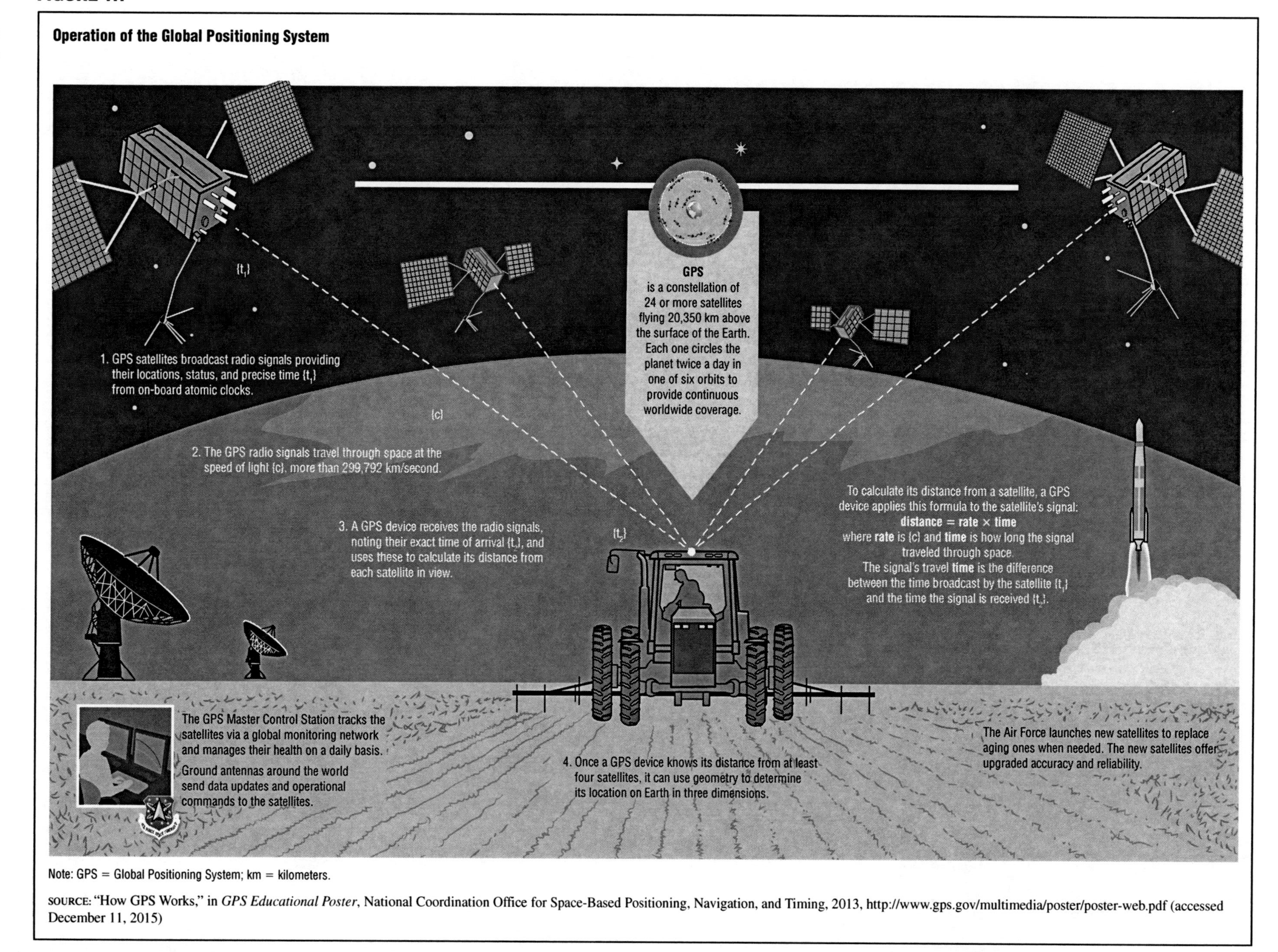

Note: GPS = Global Positioning System; km = kilometers.

SOURCE: "How GPS Works," in *GPS Educational Poster*, National Coordination Office for Space-Based Positioning, Navigation, and Timing, 2013, http://www.gps.gov/multimedia/poster/poster-web.pdf (accessed December 11, 2015)

TABLE 1.1

Worldwide orbital launches by country and launch type, 2014

Country/region	Commercial launches	Non-commercial launches	Launches
United States	11	12	23
Russia	4	28	32
Europe	6	5	11
China	0	16	16
Japan	0	4	4
India	1	3	4
Israel	0	1	1
Multinational	1	0	1
Total	**23**	**69**	**92**

SOURCE: "Table 1. 2014 Worldwide Orbital Launch Events," in *The Annual Compendium of Commercial Space Transportation: 2014*, Federal Aviation Administration, February 2015, http://www.faa.gov/about/office_org/headquarters_offices/ast/media/FAA_Annual_Compendium_2014.pdf (accessed November 21, 2015)

FIGURE 1.8

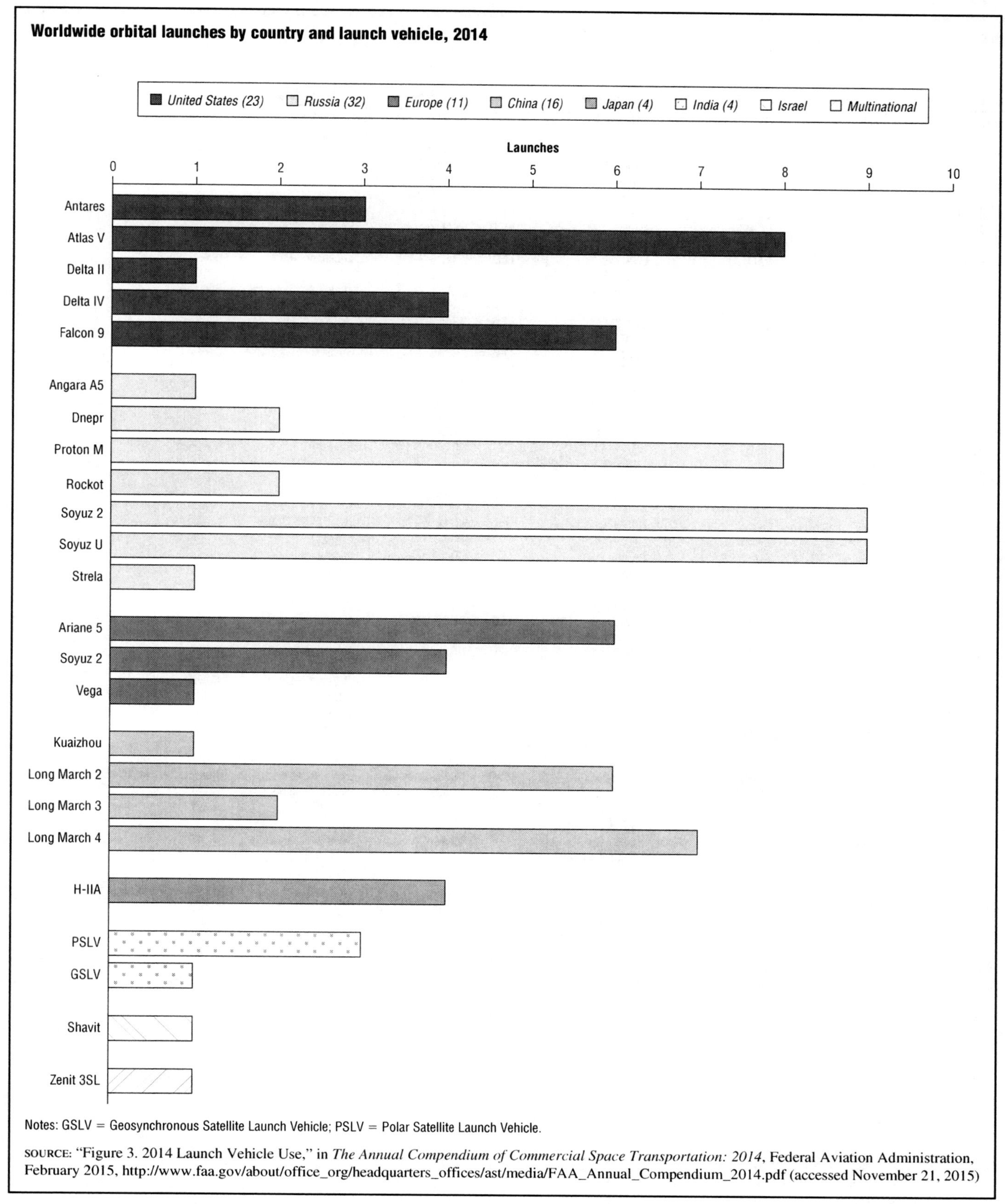

Notes: GSLV = Geosynchronous Satellite Launch Vehicle; PSLV = Polar Satellite Launch Vehicle.

SOURCE: "Figure 3. 2014 Launch Vehicle Use," in *The Annual Compendium of Commercial Space Transportation: 2014*, Federal Aviation Administration, February 2015, http://www.faa.gov/about/office_org/headquarters_offices/ast/media/FAA_Annual_Compendium_2014.pdf (accessed November 21, 2015)

TABLE 1.2

U.S. orbital launches, 2014

Date			Vehicle	Site		Payload(s)	Orbit	Operator	Manufacturer	Use	Comm'l price	L	M
6-Jan-14	a	b	Falcon 9	CCAFS	c	Thaicom 6	GEO	Thaicom	Orbital Sciences Corp.	Communications	$61.2M	S	S
9-Jan-14	a	b	Antares 120	MARS		Orb 1 (plus 33 satellites internally for ISS deployment)	LEO	Orbital Sciences Corp.	Orbital Sciences Corp.	Cargo	$80M	S	S
23-Jan-14			Atlas V 401	CCAFS		TDRS L	GEO	NASA	Boeing	Communications		S	S
20-Feb-14			Delta IV Medium+(4,2)	CCAFS		Navstar GPS 2F-05	MEO	USAF	Boeing	Navigation		S	S
3-Apr-14			Atlas V 401	VAFB		USA 249 (DMSP-5D3-F19)	LEO	USAF	Lockheed Martin	Meteorological		S	S
3-Apr-14			Soyuz 2.1a	Guiana Space Center		Sentinel 1A	SSO	ESA	Thales Alenia Space	Scientific		S	S
10-Apr-14			Atlas V 541	CCAFS		USA 250 (NRO L-67)	TBD	NRO	Classified	Classified		S	S
					c	Spx 3	LEO	SpaceX	SpaceX	Cargo			S
						KickSat	LEO	Cornell University	Cornell University	Communications			F
18-Apr-14	a	b	Falcon 9	CCAFS		ALL-STAR THEIA	LEO	CoSGC	CoSGC	Development	$61.2M	S	S
						SporeSat 1	LEO	Purdue University	Purdue University	Scientific			S
						TestSat-Lite	LEO	Taylor University	Taylor University	Development			S
						PhoneSat 2.5	LEO	NASA	NASA	Development			S
17-May-14			Delta IV Medium+(4,2)	CCAFS		USA 251 (Navstar GPS 2F-6)	MEO	USAF	Boeing	Navigation		S	S
22-May-14			Atlas V 401	CCAFS		USA 252	TBD	NRO	Classified	Classified		S	S
26-May-14	a	b	Zenit 3SL	Odyssey Platform	c	Eutelsat 3B	GEO	Eutelsat	Airbus	Communications	$95M	S	S
2-Jul-14			Delta II 7320-10	VAFB		OCO 2	SSO	NASA	Orbital Sciences Corp.	Remote Sensing		S	S
13-Jul-14	a	b	Antares 120	MARS	c	Orb 2 (plus 33 satellites internally for ISS deployment)	LEO	Orbital Sciences Corp.	Orbital Sciences Corp.	Cargo	$80M	S	S
14-Jul-14	a	b	Falcon 9	CCAFS	c	ORBCOMM 2 F3, F4, F6, F7, F9, F11	LEO	ORBCOMM	Sierra Nevada Corp.	Communications	$61.2M	S	S
28-Jul-14			Delta IV Medium+(4,2)	CCAFS		USA 253 (GSSAP 1)	GEO	USAF	Orbital Sciences Corp.	IMINT			S
						USA 254 (GSSAP 2)	GEO	USAF	Orbital Sciences Corp.	IMINT		S	S
						USA 255 (ANGELS)	GEO	AFRL	Lockheed Martin	Development			S
2-Aug-14			Atlas V 401	CCAFS		USA 256 (Navstar GPS 2F-7)	MEO	USAF	Boeing	Navigation		S	S
5-Aug-14	a	b	Falcon 9	CCAFS	c	AsiaSat 8	GEO	AsiaSat	Space Systems Loral	Communications	$61.2M	S	S
13-Aug-14	a	b	Atlas V 401	VAFB	c	WorldView 3	SSO	DigitalGlobe	Ball Aerospace	Remote Sensing	$150M	S	S
7-Sep-14	a	b	Falcon 9	CCAFS	c	AsiaSat 6	GEO	AsiaSat	Space Systems Loral	Communications	$61.2M	S	S
17-Sep-14			Atlas V 401	CCAFS		USA 257 (CLIO)	TBD	NRO	Classified	Classified		S	S
21-Sep-14	a	b	Falcon 9	CCAFS	c	Spx 4 (plus 1 satellite internally for ISS deployment)	LEO	SpaceX	SpaceX	Cargo	$61.2M	S	S
28-Oct-14	a	b	Antares 120	MARS	c	Orb 3 (plus 29 satellites internally for ISS deployment)	LEO	Orbital Sciences Corp.	Orbital Sciences Corp.	Cargo	$80M	F	F
29-Oct-14			Soyuz 2.1a	Baikonur		Progress M25	LEO	Roscosmos	RSC Energia	Cargo		S	S
29-Oct-14			Atlas V 401	CCAFS		USA 258 (Navstar GPS 2F-8)	MEO	USAF	Boeing	Navigation		S	S
5-Dec-14	a	b	Delta IV Heavy	CCAFS		EFT 1	LEO	NASA	Lockheed Martin	Development	$350M	S	S

TABLE 1.2

U.S. orbital launches, 2014 [CONTINUED]

[a]Denotes commercial launch, defined as a launch that is internationally competed or FAA-licensed, or privately financed launch activity. For multiple manifested launches, certain secondary payloads whose launches were commercially procured may also constitute a commercial launch.
[b]Denotes FAA-licensed launch.
[c]Denotes a commercial payload, defined as a spacecraft that serves a commercial function or is operated by a commercial entity.
L and M refer to the outcome of the Launch and Mission: S = Success, P = Partial Success, F = Failure.
Notes: All prices are estimates. All launch dates are based on local time at the launch site.
CCAFS = Cape Canaveral Air Force Station. GEO = geostationary orbit. GPS = global positioning system. ISS = International Space Station. LEO = low Earth orbit.
MARS = Mid-Atlantic Regional Spaceport (Virginia). MEO = medium Earth orbit. NASA = National Aeronautics and Space Administration. NRO = National Reconnaissance Office.
OCO = Orbiting Carbon Observatory. Spx 4 = SpaceX CRS-4 (Commercial Resupply Services). SSO = Sun-synchronous orbit. TBD = To Be Determined. USAF = U.S. Air Force.
VAFB = Vandenberg Air Force Base.

SOURCE: Adapted from "2014 Worldwide Orbital Launch Events," in *The Annual Compendium of Commercial Space Transportation: 2014*, Federal Aviation Administration, February 2015, http://www.faa.gov/about/office_org/headquarters_offices/ast/media/FAA_Annual_Compendium_2014.pdf (accessed November 21, 2015)

CHAPTER 2

SPACE ORGANIZATIONS PART 1: THE NATIONAL AERONAUTICS AND SPACE ADMINISTRATION

It is the policy of the United States that activities in space should be devoted to peaceful purposes for the benefit of all mankind.

—National Aeronautics and Space Act of 1958

Once it became obvious that space exploration was an achievable reality, it became a national priority for rich and powerful countries. Following World War II (1939–1945), there were only two superpowers in the world—the United States and the Soviet Union—and they considered each other enemies.

Both superpowers had military, scientific, and political reasons to pursue space travel. The military saw outer space not only as a potential battlefield but also as an opportunity to spy on enemies on the other side of the world. Scientists valued space travel for another reason. They wanted to gather data from space to help them unravel the mysteries of the universe. From a political standpoint, a successful space program was a source of national pride and a symbol of national superiority. This motivation above all others drove space exploration during the latter half of the 20th century.

The Soviet Union's space program was under the control of the military. By contrast, the United States split its space program into two parts. The U.S. military was given control over space projects related to national security, and a new civilian agency called the National Aeronautics and Space Administration (NASA) was formed within the government in 1958 to oversee nonmilitary space programs.

Throughout its history, NASA has been associated with spectacular feats and horrific disasters in space exploration. It has received great praise for its successes and harsh criticism for its failures. Space travel is an expensive enterprise and, as a government agency, NASA is bound by federal budget constraints. This budget rises and falls according to the political climate. U.S. presidents set space goals, but Congress sets NASA's budget.

In 1961 President John F. Kennedy (1917–1963) charged NASA with the monumental task of putting a man on the moon before the end of the decade. Congress allocated billions of dollars to NASA, and this goal was accomplished. Later presidents also set grand goals for the agency, but they proved difficult to achieve. Every major endeavor went over budget and fell behind schedule, the public seemed to lose interest in space travel, and Congress lacked the political motivation to dramatically increase NASA's funding. Figure 2.1 shows the agency's budget authority in at-that-time dollars and in 2015 dollars (i.e., assuming that the dollar had the same buying power as it did in 2015). NASA's budget soared through the mid-1960s and then declined through the mid-1970s. In 2015 dollars, the budget increased during most years through 2008 and then generally declined.

A NEW AGENCY IS BORN

NASA was founded on October 1, 1958, following enactment of the National Aeronautics and Space Act of 1958 (http://history.nasa.gov/spaceact.html). The stated purpose of the act was "to provide for research into problems of flight within and outside the earth's atmosphere, and for other purposes."

The act specifically mandated that NASA would be a civilian agency with control over all nonmilitary aeronautical and space activities by the United States. The research and development of weapons and national defense systems remained under the control of the U.S. Department of Defense (DOD). However, the act called for the sharing of information between the two agencies. Cooperation by NASA in space ventures with other countries was allowed if the purpose was "peaceful application of the results."

In "NACA—90 Years Later" (*X-Press*, vol. 47, no. 2, March 25, 2005), Peter Merlin notes that NASA consolidated the resources of several government organizations,

FIGURE 2.1

NASA funding in current and 2015 dollars, fiscal years 1958–2015

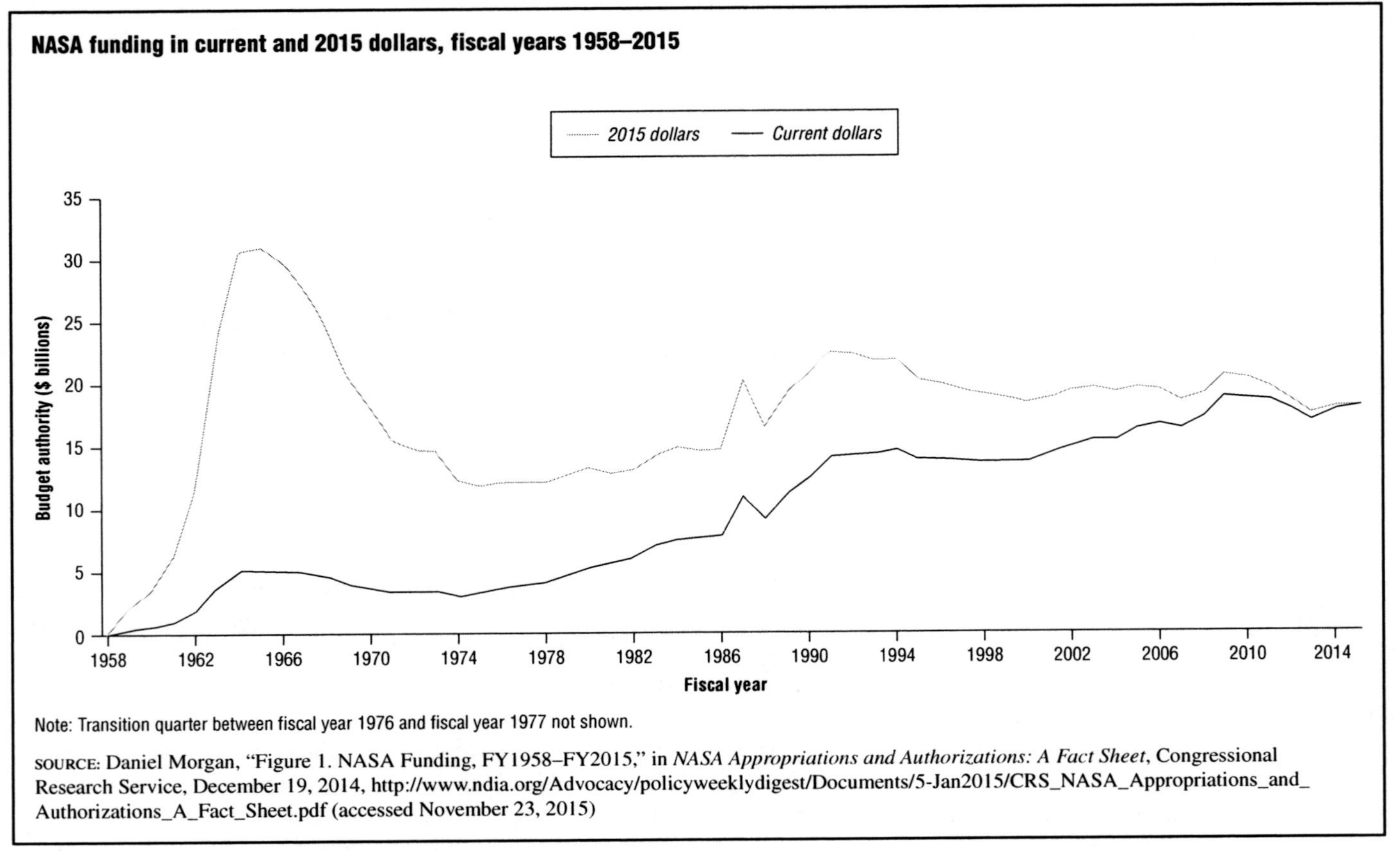

Note: Transition quarter between fiscal year 1976 and fiscal year 1977 not shown.

SOURCE: Daniel Morgan, "Figure 1. NASA Funding, FY1958–FY2015," in *NASA Appropriations and Authorizations: A Fact Sheet*, Congressional Research Service, December 19, 2014, http://www.ndia.org/Advocacy/policyweeklydigest/Documents/5-Jan2015/CRS_NASA_Appropriations_and_Authorizations_A_Fact_Sheet.pdf (accessed November 23, 2015)

chiefly the National Advisory Committee for Aeronautics (NACA). The NACA was perhaps best known for its aeronautical research and development efforts, which benefited both military and civilian aviation. For example, the agency played an integral role in developing and testing the X-series of experimental aircraft, which is described in Chapter 1.

In October 1957 the Soviet Union launched *Sputnik 1*, the world's first artificial satellite. The event helped prompt President Dwight D. Eisenhower (1890–1969) to establish a new agency to assume responsibility for all U.S. nonmilitary space projects. NASA absorbed the personnel and facilities of the NACA, which ceased to exist.

According to Steven J. Dick and Steve Garber of NASA, in *Celebrating NASA's Fortieth Anniversary 1958–1998: Pioneering the Future* (February 8, 2005, http://www.history.nasa.gov/40thann/40home.htm), NASA began with approximately 8,000 employees and an annual budget of $100 million. About half of the agency's first employees were civilian personnel working on space projects at the U.S. Army's Redstone Arsenal in Huntsville, Alabama. They included a rocketry team that was headed by Wernher von Braun (1912–1977). Von Braun was a German rocket scientist who moved to the United States following World War II and played a major role in building the U.S. rocket program at NASA.

John A. Pitts of NASA explains in *The Human Factor: Biomedicine in the Manned Space Program to 1980* (1985, http://history.nasa.gov/SP-4213/contents.htm) that President Eisenhower believed the civilian space program should be "small in scale and limited in its objectives."

NASA SHOOTS FOR THE MOON

NASA did not stay small for long, because it had grand plans. In February 1960 NASA presented to Congress a 10-year plan for the nation's space program. It included an array of scientific satellites; robotic probes to the moon, Mars, and Venus; development of new and powerful rockets; and manned spaceflights to orbit Earth and the moon. NASA estimated the program would cost around $12 billion.

Congress was politically motivated to support the program. The Soviet Union had already landed a probe (*Luna 2*) on the moon as part of its Luna Program. The United States was behind in the space race. After Kennedy was elected president in November 1960, he charged Vice President Lyndon B. Johnson (1908–1973) with finding a way for the United States to beat the Soviets to a major space goal. NASA pushed for a manned lunar landing, and Johnson agreed. On May 25, 1961, in the speech "Special Message to the Congress on

Urgent National Needs" (http://www.nasa.gov/pdf/59595 main_jfk.speech.pdf), President Kennedy asked Congress to provide financial support to NASA to put a man on the moon before the end of the decade.

NASA had a lot of work to do just to catch up with the Soviets in space. In April 1961 the Soviets had put the first human in Earth orbit. Cosmonaut Yuri Gagarin (1934–1968) circled Earth one time in a flight that lasted 1 hour and 49 minutes.

Project Mercury

The program that would enable the United States to draw even in the space race, Project Mercury, had actually began in 1958, only a week after NASA was created. The official announcement was made on December 17, 1958—the 55th anniversary of the Wright brothers' flight. The program was named after the mythical Roman god Mercury, the winged messenger.

The Mercury project had three specific objectives:

- Put a manned spacecraft into Earth orbit
- Investigate the effects of space travel on humans
- Recover the spacecraft and humans safely

On May 5, 1961, the astronaut Alan Shepard (1923–1998) became the first American in space when he made a suborbital flight that lasted 15 minutes and 28 seconds. Shepard's flight was far shorter than Gagarin's had been and included only five minutes of weightlessness. NASA desperately needed more data on the effects of weightlessness on the human body. This was considered a key element to manned flights to the moon. Between 1961 and 1963 six Mercury astronauts made six successful space flights and spent a total of 53.9 hours in space. (See Table 2.1.) NASA explains in "In the Beginning: Project Mercury" (July 31, 2015, http://www.nasa.gov/multimedia/imagegallery/image_feature_1203.html) that each astronaut picked the name of his space capsule and added a "7" to the name to honor the seven members of NASA's first astronaut corps.

Is It Worth It?

The United States paid a high price for NASA's moon program. It was conducted during one of the most turbulent times in U.S. history. The 1960s were characterized by social unrest, protest, and national tragedies.

On November 22, 1963, President Kennedy was assassinated in Dallas, Texas, and Vice President Johnson assumed the presidency. Johnson had always supported the space program and had been instrumental in passing the bill that created NASA. He assured NASA that the Apollo Program would continue as planned. On November 29, 1963, Johnson announced that portions of the U.S. Air Force missile testing range on Merritt Island, Florida, would be designated the John F. Kennedy Space Center (KSC).

In 1964 the social scientist Amitai Etzioni (1929–) published *The Moon-Doggle: Domestic and International Implications of the Space Race*, a book that was extremely critical of NASA. The title was a play on the word *boondoggle*, which means a wasteful and impractical project. Etzioni criticized the agency for spending too much money on manned space flights when unmanned satellites could achieve more for less money. He also questioned the scientific value (and costs) of sending astronauts to the moon. Etzioni was not alone in feeling this way. American society was increasingly concerned with pressing social and national issues, including the escalating war in Vietnam and civil rights.

Project Gemini

NASA scientists realized during the Mercury missions that they needed an intermediate step before the moon flights. They had to be sure that humans could survive and function in space for up to 14 days. This was the amount of time estimated for a round trip to the moon. The program that was designed to test human endurance in space was named Project Gemini, after the Gemini constellation, which is represented by the twin stars Castor and Pollux. The name was chosen because the Gemini space capsule was designed to hold two astronauts, rather than one.

TABLE 2.1

Project Mercury manned flights

Date of launch	Mercury flight number	Spacecraft name	Flight type	Highest altitude	Time in space	Astronaut
5/5/1961	3	Freedom 7	Suborbital	116 miles	15 min 28 sec	Alan Shepard
7/21/1961	4	Liberty Bell 7	Suborbital	118 miles	15 min 37 sec	Virgil Grissom
2/20/1962	6	Friendship 7	3 orbits	162 miles	4 hr 55 min	John Glenn
5/24/1962	7	Aurora 7	3 orbits	167 miles	4 hr 56 min	Scott Carpenter
10/3/1962	8	Sigma 7	6 orbits	176 miles	9 hr 13 min	Walter Schirra
5/15/1963	9	Faith 7	22.5 orbits	166 miles	1 day 10 hr 19 min	Gordon Cooper

Notes: hr = hour(s); min = minutes; sec = seconds.

SOURCE: Created by Kim Masters Evans for Gale, © 2016

A major goal of the Gemini project was to rendezvous orbiting vehicles into one unit and maneuver that unit with a propulsion system. This was a feat considered crucial to conduct the moon landings. A final Gemini goal was to perfect atmospheric reentry of the spacecraft and to perform a ground landing, rather than a landing at sea. All the goals except a ground landing were achieved.

Between 1965 and 1966 NASA completed 10 Gemini missions with 16 astronauts, who spent a total of more than 40 days in space. (See Table 2.2.) The Gemini IV mission featured the first extravehicular activity by an American. The astronaut Edward Higgins White (1930–1967) spent 22 minutes outside his spacecraft during a "space walk." The longest Gemini flight (Gemini VII) took place in December 1965, lasting 14 days.

Moon Resources

By 1965 NASA scientists and engineers had been studying the details of a moon landing for years. Numerous probes had been launched under the Pioneer and Ranger programs to collect data. On June 2, 1966, NASA achieved a milestone when the *Surveyor 1* spacecraft made a controlled "soft landing" on the moon in the Ocean of Storms. The ability to do a soft landing was considered crucial to putting a human safely on the moon. *Surveyor 1* returned a host of high-quality photographs. However, NASA was still running behind the Soviet space program. The Soviet spacecraft *Luna 9* had soft-landed in the Ocean of Storms four months before *Surveyor 1* got there. *Luna 9* also provided the first television transmission from the lunar surface.

TABLE 2.2

Gemini program manned flights

Dates	Gemini flight number	Astronauts	Achievements
March 23, 1965	III	Virgil I. Grissom John W. Young	3 orbits. Only Gemini spacecraft to be named (Molly Brown).
June 3–7, 1965	IV	James A. McDivitt Edward H. White	First American EVA—a 22 minute space walk by White
August 21–29, 1965	V	Gordon Cooper Charles Conrad Jr.	120 orbits. First use of fuel cells for electrical power.
December 4–18, 1965	VII	Frank Borman James A. Lovell Jr.	Longest mission at 14 days
December 15–16, 1965	VI-A	Walter M. Schirra Jr. Thomas Stafford	First space rendezvous (with Gemini VII)
March 16, 1966	VIII	Neil A. Armstrong David R. Scott	First space docking (with unmanned craft)
June 3–6, 1966	IX-A	Thomas Stafford Eugene A. Cernan	2 hours of EVA
July 18–21, 1966	X	John W. Young Michael Collins	Rendezvous with unmanned craft
September 12–15, 1966	XI	Charles Conrad Jr. Richard F. Gordon Jr.	Record altitude (739.2 miles)
November 11–15, 1966	XII	James A. Lovell Jr. Edwin E. (Buzz) Aldrin Jr.	Record EVA by Aldrin (5 hours 30 minutes)

Note: EVA is extravehicular activity.

SOURCE: Created by Kim Masters Evans for Gale, © 2016

In all, NASA sent seven Surveyor spacecraft to the moon between 1966 and 1968. Two lost control and crashed, whereas the remaining five achieved soft landings. In 1967 *Surveyor 6* was particularly successful. During its mission NASA controllers were able to lift the spacecraft about 10 feet (3 m) off the lunar surface and set it softly back down again. NASA was ready to put humans aboard a lunar lander.

The Apollo Program

The massive Saturn V rocket developed by von Braun was the launch vehicle selected for the Apollo missions. Each mission carried three astronauts. The Apollo spacecraft had three parts:

- Command module containing the crew quarters and flight control section
- Service module for the propulsion and spacecraft support systems
- Lunar module to take two of the crew to and from the lunar surface

Figure 2.2 shows the three modules stacked atop a rocket for launch. The astronauts rode in the command module during launch. The tubular structure sitting on top of the command module was called the Launch Escape Subsystem (LES). In "Launch Escape" (November 29, 2009, http://www.hq.nasa.gov/alsj/CSM15_Launch_Escape_Subsystem_pp137-146.pdf), NASA notes that the LES provided a means of escape for the astronauts in case the Saturn V rocket malfunctioned during or soon after launch. The LES included a rocket that could propel the command module away from the remainder of the spacecraft. Then the command module would parachute to Earth. The LES was never deployed during an Apollo flight; each time the assembly was jettisoned after the spacecraft reached the appropriate altitude.

Each Apollo flight began with the spacecraft going into Earth orbit. There, the lunar module was disconnected from the aft of the service module, turned, and docked to the front of the command module. A rocket burn called a trans-lunar injection sent the spacecraft out of Earth orbit and on its way to the moon. When the three modules reached lunar orbit, two of the astronauts entered the lunar module, which was then detached for the journey to the moon's surface. The third astronaut remained aboard the combined command module and service module. The command module and the service module were frequently referred to as a single entity known as the CSM.

FIGURE 2.2

Apollo Program launch configuration for lunar landing mission

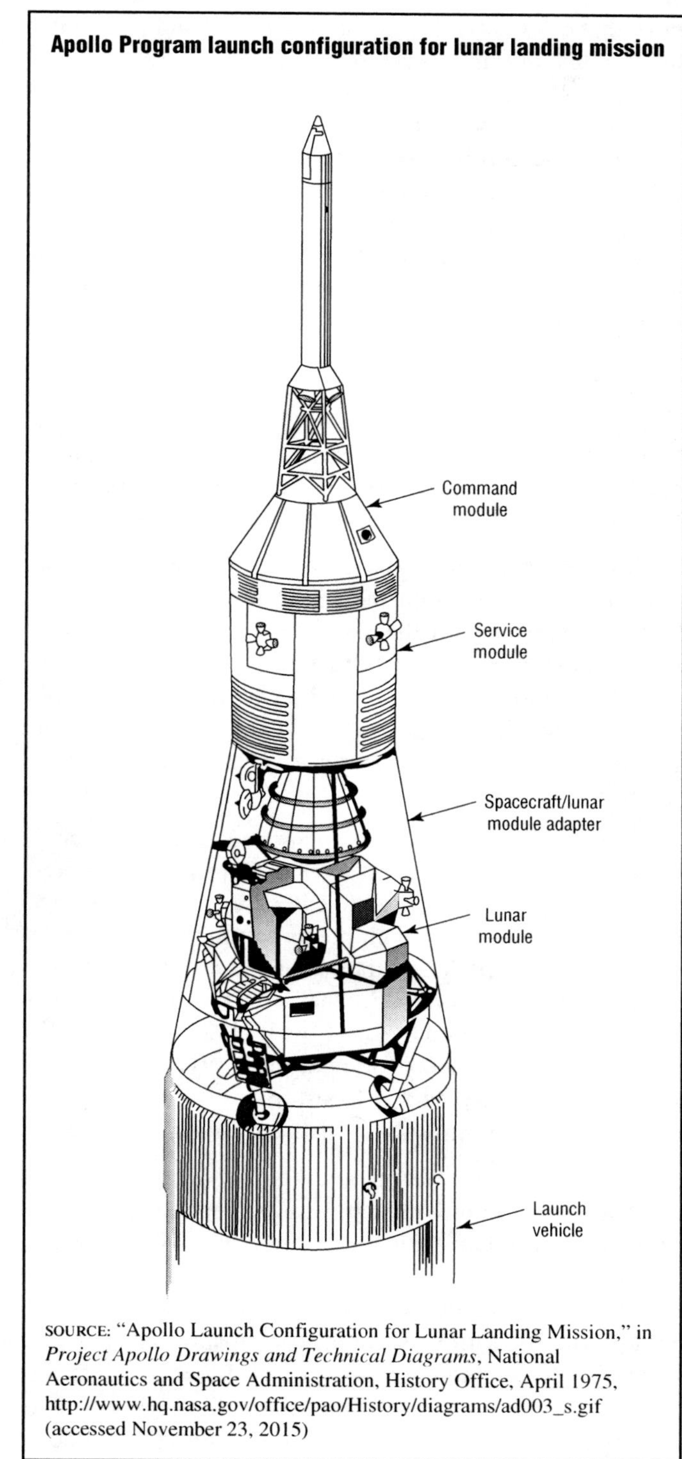

SOURCE: "Apollo Launch Configuration for Lunar Landing Mission," in *Project Apollo Drawings and Technical Diagrams*, National Aeronautics and Space Administration, History Office, April 1975, http://www.hq.nasa.gov/office/pao/History/diagrams/ad003_s.gif (accessed November 23, 2015)

After the two astronauts finished their work on the moon, they ascended in the lunar module to dock with the CSM. Once the two astronauts were safely aboard the CSM, the lunar module was jettisoned away from the spacecraft. Only the CSM made the journey back to Earth. The service module was jettisoned just before reentry into Earth's atmosphere. The command module with all three astronauts aboard then splashed down in the sea with the aid of parachutes.

A Tragic Setback

NASA lost its first astronauts during the Apollo Program. In 1966 three unmanned Apollo spacecraft were launched to test the structural integrity of the spacecraft and the flight systems. These were called the Apollo-Saturn missions and were numbered AS-201, AS-202, and AS-203.

On January 27, 1967, NASA was preparing a spacecraft for mission AS-204, the first manned test flight. The agency (1967, http://www.hq.nasa.gov/office/pao/History/SP-4009/v4p1h.htm) notes that the mission was also called Apollo 1. During a launchpad test of the spacecraft, a flash fire broke out and killed all three astronauts in the command module.

The tragedy temporarily devastated morale at NASA, and the agency was not treated kindly by the media. Many newspapers questioned whether a manned lunar mission was worth the risk. Rumors even circulated that the astronauts had been murdered by NASA for criticizing the agency or for other sinister reasons.

A review board convened by NASA investigated the accident. In its report (April 5, 1967, http://history.nasa.gov/Apollo204/find.html), the Apollo 204 Review Board notes that "no single ignition source of the fire was conclusively identified." However, the fire is generally thought to have started in a bundle of wires running through the forward portion of the left-hand equipment bay in a section to the left of and well below the command pilot's seat.

The board found that a variety of factors contributed to the astronauts' deaths. Some were operational problems: a hatch that was difficult to open, the presence of 100% oxygen in the module, and the use of flammable materials inside the module. The investigation also revealed a number of management and contractor problems. NASA decided to redesign the Apollo modules and reorganize the top management staff. The moon landing that was scheduled for late 1968 was delayed until 1969 because of the Apollo 1 tragedy.

One Giant Leap

Following the Apollo 1 tragedy NASA (1967, http://www.hq.nasa.gov/office/pao/History/SP-4009/v4p1h.htm) decided to name the next mission Apollo 4. Thus, there were officially no Apollo 2 or Apollo 3 missions. Apollo 4, Apollo 5, and Apollo 6 were unmanned missions that tested various spacecraft components. By late 1968 the Apollo Program was making tremendous strides. The first manned flight, *Apollo 7*, was launched on October 11, 1968. Apollo 7 included the first live television broadcast from a manned spacecraft. Watching the astronauts on television helped rekindle a feeling of excitement about the space program. The American public grew more excited as one Apollo mission after another was successful.

In May 1969 the Apollo 10 mission featured the first live color television pictures broadcast from outer space.

Two months later *Apollo 11* was launched into space with three astronauts aboard: Neil Armstrong (1930–2012), Michael Collins (1930–), and Edwin (Buzz) Aldrin (1930–). On July 20, 1969, at 4:18 p.m. eastern daylight time, the lunar module softly landed near the Sea of Tranquility. Armstrong reported, "Houston, Tranquility Base here. The Eagle has landed."

At 10:56 p.m. eastern daylight time Armstrong opened the door of the lunar module and climbed down a short ladder. As he put his left foot onto the surface of the moon, he said, "That's one small step for [a] man, one giant leap for mankind." It was the first time in history that a human being had set foot on another celestial body.

The event was televised live to a worldwide audience that was estimated at 528 million people. They watched as Armstrong and Aldrin explored the lunar surface for 2 hours and 31 minutes. The astronauts planted an American flag in the dusty soil and collected 48 pounds (22 kg) of moon rocks. They unveiled a plaque that was attached to the descent stage (the lower part) of the lunar module. The plaque read: "Here men from the planet Earth first set foot upon the Moon, July 1969 A.D. We came in peace for all mankind." The plaque bore the signatures of all three astronauts and President Richard M. Nixon (1913–1994).

The two astronauts climbed back into the lunar module. On July 21, 1969, the ascent portion of the module lifted off the lunar surface, leaving the descent stage behind. Armstrong and Aldrin had spent 21 hours and 36 minutes on the surface of the moon. They then docked with the CSM piloted by Collins. Once reunited, the three astronauts headed for Earth, leaving the lunar module ascent stage in orbit around the moon (it eventually crashed into the moon). On July 24, 1969, their command module safely splash landed in the Pacific Ocean.

The Can-Do Culture

NASA had achieved something that many people thought impossible. The agency was heaped with praise and congratulations. Putting a man on the moon was considered to be an enormous milestone in technological progress. In addition, it had been done before the end of the decade, just as President Kennedy had requested. The achievement fostered a tremendous sense of pride and confidence among NASA personnel. The agency was left with an optimistic conviction that it could do anything, an attitude known as NASA's "can-do culture."

NASA's critics believe the agency's can-do culture caused it to make many overly optimistic promises during the following decades. NASA continued to set bold goals for the nation's space program and promise Congress that it could achieve them. The problem was that these goals did not receive nearly as much financial support as the Apollo Program received. The moon landing was possible because NASA was given the necessary resources. According to NASA, in *Project Apollo: A Retrospective Analysis* (April 21, 2014, http://history.nasa.gov/Apollomon/Apollo.html), putting a man on the moon within a decade had taken the talents of hundreds of thousands of people and cost approximately $25.4 billion of taxpayers' money.

Apollo Fizzles Out

NASA launched six more Apollo missions following Apollo 11. In November 1969 the crew of *Apollo 12* landed near the Ocean of Storms and found the *Surveyor 3* lunar probe that had been sent several years before.

Five months later the *Apollo 13* spacecraft was launched. Two days into the flight an oxygen tank suddenly ruptured aboard the service module. The pressure in the cabin dropped quickly. Fearing the crew would otherwise be lost, NASA devised a way for the astronauts to rely on the limited resources in the lunar module to limp back to Earth. The spacecraft splashed down safely in April 1970.

Once again NASA had achieved a near-miracle. Although on the surface the Apollo 13 mission appeared like a failure, NASA classified it as a success, because the agency learned so much about handling emergencies during space flight. The experience was later captured in the 1995 movie *Apollo 13*. The movie made famous a phrase that was uttered by Jim Lovell (1928–), the Apollo 13 mission commander, following the oxygen tank rupture. Lovell calmly said, "Houston, we've had a problem," which was misquoted in the media as "Houston, we have a problem."

The next Apollo launch was postponed as NASA worked on problems that were brought to light by the *Apollo 13* incident. In February 1971 *Apollo 14* successfully reached the moon for a lunar exploration mission at Fra Mauro. The astronauts took along a new cart that was specially designed to hold moon rocks. Later that year the crew of *Apollo 15* took a lunar rover that resembled a dune buggy. The astronauts zoomed around the Hadley-Apennine region at a top speed of 8 miles per hour (13 km/h). They collected nearly 170 pounds (77 kg) of moon rocks. The lunar rover concept was so effective that it was used on all the remaining Apollo missions.

In April 1972 *Apollo 16* set down in the Descartes Highlands of the moon. It was the first mission to explore the highlands and the southern-most landing site of all the Apollo missions. In December 1972 the *Apollo 17* crew explored the highlands and a valley in the Taurus-Littrow area of the moon. For the first time the mission crew

TABLE 2.3

Apollo Program manned missions

Name	Dates	Spacecraft call signs	Crew	Mission time	Note
Apollo 1	January 27,1967	Not used	Virgil I. Grissom (commander), Edward H. White, Roger B. Chaffee		Spacecraft caught on fire on landing pad during practice drill. All astronauts killed.
Apollo 7	October 11–22 1968	Not used	Walter M. Schirra Jr. (commander), Donn F. Eisele (CM pilot), R. Walter Cunningham (LM pilot)	10 days, 20 hours	CSM piloted flight demonstration in Earth orbit. First live TV from manned spacecraft.
Apollo 8	December 21–27, 1968	Not used	Frank Borman (commander), James A. Lovell Jr. (CM pilot), William A. Anders (LM pilot)	6 days, 3 hours	First manned lunar orbital mission. Live television broadcasts.
Apollo 9	March 03–13, 1969	CM: Gumdrop LM: Spider	James A. McDivitt (commander), David R. Scott (CM pilot), Russell L. Schweickart (LM pilot)	10 days, 1 hour	First manned flight of all lunar hardware in Earth orbit. Schweickart performed 37 minutes EVA. First manned flight of lunar module.
Apollo 10	May 18–26, 1969	CM: Charlie Brown LM:Snoopy	Thomas P. Stafford (commander), John W. Young (CM pilot), Eugene A. Cernan (LM pilot)	8 days, 3 minutes	Practice for moon landing. First manned CSM/LM operations in cislunar and lunar environment; First live color TV from space.
Apollo 11	July 16–24, 1969	CM: Columbia LM: Eagle	Neil A. Armstrong (commander), Michael Collins (CM pilot), Edwin E. (Buzz) Aldrin Jr. (LM pilot)	8 days, 3 hours, 18 minutes	First manned lunar landing mission and lunar surface EVA.
Apollo 12	November 14–24, 1969	CM: Yankee Clipper LM: Intrepid	Charles Conrad Jr. (commander), Richard F. Gordon Jr. (CM pilot), Alan L. Bean (LM pilot)	10 days, 4 hours, 36 minutes	Lunar landing and lunar exploration.
Apollo 13	April 11–17, 1970	CM: Odyssey LM: Aquarius	James A. Lovell Jr. (commander), John L. Swigert Jr. (CM pilot), Fred W. Haise Jr. (LM pilot)	5 days, 22.9 hours	Mission aborted before spacecraft reached moon.
Apollo 14	January 31–February 09, 1971	CM: Kitty Hawk LM: Antares	Alan B. Shepard Jr. (commander), Stuart A. Roosa (CM pilot), Edgar D. Mitchell (LM pilot)	9 days	Lunar landing and lunar exploration.
Apollo 15	July 26–August 07, 1971	CM: Endeavor LM: Falcon	David R. Scott (commander), Alfred M. Worden (CM pilot), James B. Irwin (LM pilot)	12 days, 17 hours, 12 minutes	Lunar landing and lunar exploration.
Apollo 16	April 16–27, 1972	CM: Casper LM: Orion	John W. Young (commander), Thomas K. Mattingly II (CM pilot), Charles M. Duke Jr. (LM pilot)	11 days, 1 hour, 51 minutes	Lunar landing and lunar exploration.
Apollo 17	December 07–19, 1972	CM: America LM: Challenger	Eugene A. Cernan (commander), Ronald E. Evans (CM pilot), Harrison H. Schmitt (LM pilot)	12 days, 13 hours, 52 minutes	Last lunar landing mission.

Note: EVA is extravehicular activity. CM is command module. CSM is command and service module. LM is lunar module.

SOURCE: Created by Kim Masters Evans for Gale, © 2016

included a scientist—the geologist Harrison Schmitt (1935–). The astronauts collected 243 pounds (110 kg) of moon rocks, the most of any Apollo mission. On December 19, 1972, *Apollo 17* splashed down safely in the Pacific Ocean. With the successful completion of the Apollo 17 mission, the Apollo Program was over.

Table 2.3 summarizes information about all the manned Apollo missions. In all, NASA put 12 astronauts on the moon. According to the agency, in "NASA Gives Media, Public Look inside Apollo Moon Rock Vault" (June 25, 2009, http://www.nasa.gov/home/hqnews/2009/jun/HQ_M09-116_Moon_rocks.html), the astronauts collected 842 pounds (382 kg) of rocks, soil, and other geological samples from the moon.

The Apollo missions that followed Apollo 11 never captured the public's imagination the same way that the first moon landing did. The feeling was that the United States had already achieved its goal of beating the Soviet Union to the moon, and continued lunar exploration held little appeal for many people. NASA had to cancel its planned remaining Apollo missions: Apollo 18, Apollo 19, and Apollo 20.

SPACE SCIENCE SUFFERS

Putting a man on the moon was conducted mostly for political purposes. It bolstered national pride and prestige and was largely a symbolic endeavor. Many scientists thought the Apollo Program achieved far less in scientific terms than unmanned probes could have accomplished. One reason the program was so expensive was that so many resources had to be devoted to keeping fragile humans alive and safe in the harsh environment of space. Critics said this money could have been invested in robotics research and development to produce a fleet of unmanned probes and sample collectors to explore the moon and far beyond.

NASA's Ranger and Surveyor probes of the early 1960s were originally designed to collect data to support many research goals within astronomy and space science. Once the Apollo Program began, these probes were retooled to gather data important to the manned program.

This was called human factors research and was a small part of the discipline called space biology. NASA's focus on human factors at the expense of broader research in space biology, space science, and astronomy brought harsh criticism from scientists.

NASA'S FIRST SPACE STATION

According to Edward Clinton Ezell and Linda Neuman Ezell, in *On Mars: Exploration of the Red Planet, 1958–1978* (1984, http://history.nasa.gov/SP-4212/contents.html), as early as the 1950s NASA made plans to put a manned space station in orbit around Earth. These plans took center stage at the agency when the Apollo Program ended. For its next project NASA envisioned an orbiting space station that was devoted to scientific research and a fleet of reusable space planes to carry humans to and from the station. In 1969 President Nixon and Congress were not interested in extending the Apollo Program, let alone pursuing a new and costly endeavor. NASA found its budget cut year after year. Nevertheless, the agency devoted many of its resources to developing a new manned space program.

The Apollo Applications Program (AAP) began in 1963 with a plan to use leftover Apollo hardware in some kind of orbiting station including a laboratory, workshop, and space telescope. When the Apollo 20 mission was canceled in January 1970, the AAP inherited a Saturn V rocket. It used the rocket as the launch vehicle for a newly developed station called *Skylab*.

Skylab was a small scientific laboratory and solar observatory that could hold three crew members at a time. The station weighed nearly 100 tons (90.7 t) on Earth and was about the size of a small three-bedroom house.

The station was designed with two solar panels that were folded flat against the rocket during launch. Once in orbit, they were to open up like wings and harness the sun's energy to provide electricity for the station. On May 14, 1973, the unmanned *Skylab* station was launched into orbit. It was damaged during liftoff when a protective shield came loose and smashed against the solar panels, ripping one of them off and damaging the other. Astronauts later repaired some of the damaged components and deployed a temporary sail-like shield.

In total, the Skylab program included four different missions as shown in Table 2.4. Three separate crews visited and lived in *Skylab* between May 1973 and February 1974. The first mission lasted 28 days, the second lasted 59 days, and the third lasted 84 days.

According to Lee B. Summerlin of NASA, in *Skylab, Classroom in Space* (1977), *Skylab* crews conducted several experiments that were suggested by high school students from around the country. For example, the Skylab 3 mission included two spiders named Anita and Arabella. The spiders were part of an experiment that was suggested by Judith Miles, a high school student from Lexington, Massachusetts. She wondered if spiders would be able to spin their webs in microgravity. NASA scientists seized on the idea and sent the spiders into space in cages that were equipped with still cameras and television cameras. The public became enthralled in hearing about the two spiders.

Neither spider adjusted well to the new environment. Arabella's initial webs were sloppy and lopsided. However, after a few days the spider began spinning web patterns as it would on Earth. Both spiders died during the mission, apparently from dehydration. Their bodies were turned over to the Smithsonian Institution and as of 2016 were still kept there or lent temporarily to other museums.

The *Skylab* was not designed for long-term use. It had no method of independent reboost to keep it from falling out of orbit. As a result, on July 11, 1979, the

TABLE 2.4

Skylab statistics

	Skylab 1	Skylab 2	Skylab 3	Skylab 4
Launch date	5/14/1973	5/25/1973	7/28/1973	11/16/1973
Launch vehicles	Saturn V	Saturn 1B	Saturn 1B	Saturn 1B
Orbital parameters	268.1 × 269.5 miles	268.1 × 269.5 miles	268.1 × 269.5 miles	268.1 × 69.5 miles
Orbital inclination	50 degrees	50 degrees	50 degrees	50 degrees
Orbital period (approximate)	93 minutes	93 minutes	93 minutes	93 minutes
Distance orbit	26,575 miles	26,575 miles	26,575 miles	26,575 miles
Crew's mission distance		11.5 million miles	24.5 million miles	34.5 million miles
Crew's number of revolutions		404	585	1,214
Crew's mission duration		28 days 49 min	59 days 11 hrs 9 min	84 days 1 hr 16 min
Crew's experiment time		392 hr	1,081 hr	1,563 hr
Crew's EVA time		6 hr 20 min	13 hr 43 min	22 hr 13 min

Note: EVA is extravehicular activity. hr = hours; min = minutes.

SOURCE: Adapted from "Skylab Statistics," in *Skylab*, National Aeronautics and Space Administration, Kennedy Space Center, December 12, 2000, http://www-pao.ksc.nasa.gov/history/skylab/skylab-stats.htm (accessed November 23, 2015)

station reentered Earth's atmosphere and broke apart over the Pacific Ocean.

Despite its early mechanical problems, *Skylab* was considered to be a great success. The total number of hours spent in space by *Skylab* astronauts was greater than the combined totals of all space flights made up to that time. NASA gained valuable knowledge about human performance under microgravity conditions. However, the United States lagged behind the Soviet Union in this area. *Salyut 1*, the first Soviet space station, was put into orbit two years before *Skylab*, as described in Chapter 3.

THE SPACE SHUTTLE PROGRAM

During the early 1960s NASA planners envisioned a space station program as the next step after Apollo. It was assumed that the United States would establish large space stations in orbit around Earth and develop a new type of reusable space plane or "shuttle" to carry cargo and personnel back and forth.

However, these grand plans did not mesh with the realities of the times. By the late 1960s the nation was heavily engaged in the Vietnam War (1954–1975). Domestic unrest and social issues dominated the political agenda into the early 1970s. Furthermore, historians report that Nixon, who was president of the United States from 1969 to 1974, was not interested in pursuing any large or expensive vision for space exploration.

In *CAIB Report: Volume 1* (August 2003, http://s3.amazonaws.com/akamai.netstorage/anon.nasa-global/CAIB/CAIB_lowres_full.pdf), the *Columbia* Accident Investigation Board (CAIB) relates the early history of the space shuttle program (SSP). NASA promoted the project as a transport business, rather than as an exploratory adventure. By the early 1970s satellites were increasingly being used by the government for defense, intelligence, and scientific purposes. They were also being used by commercial companies, particularly in the communications industry. These satellites were launched into orbit using nonreusable rockets. The agency convinced Nixon that reusable shuttles would save the country money, because they could haul all the nation's government and commercial satellites into space. According to the CAIB, the shuttles were anticipated to haul up to 50 satellites per year into space.

This economic argument was successful. The CAIB notes that in 1971 NASA was given a $5 billion budget over a five-year period for the development of a shuttle program. This was later increased to $5.5 billion. It was expected that the shuttle program would be operational by the end of the 1970s.

This projection proved to be overly optimistic; the first shuttle did not launch until 1981. By that point the Soviet space station *Mir* had been in orbit for several years. However, the Soviet space program had pursued—but failed to develop—a reusable space plane. As a result, transportation to and from *Mir* was accomplished using expendable rocket boosters.

Space Shuttle Design

At first, U.S. engineers hoped to develop a fully reusable vehicle. Budget constraints soon made it obvious that this was not going to be possible. Instead, NASA designed a three-part vehicle for the shuttle:

- A reusable space plane called an orbiter
- An expendable external liquid fuel tank for the orbiter's three main engines
- A reusable pair of external rocket boosters containing a powdered fuel

Figure 2.3 shows the major components of the space shuttle design. The orbiter held the crew compartment and payload bay. During the ascent phase the solid rocket boosters and the external fuel tank were jettisoned away from the orbiter. The rocket boosters were designed to be recovered, refurbished, and refilled with fuel for the next launch. The external fuel tank was jettisoned above Earth's atmosphere and burned up during reentry. It should be noted that the word *shuttle* was commonly used to refer both to the entire vehicle and to the orbiter alone.

FIGURE 2.3

Space shuttle components

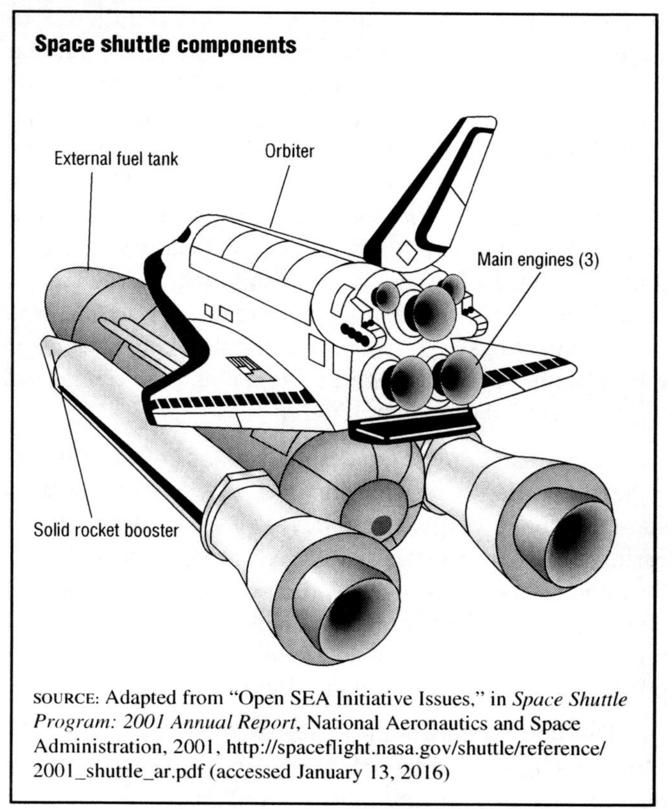

SOURCE: Adapted from "Open SEA Initiative Issues," in *Space Shuttle Program: 2001 Annual Report*, National Aeronautics and Space Administration, 2001, http://spaceflight.nasa.gov/shuttle/reference/2001_shuttle_ar.pdf (accessed January 13, 2016)

The orbiter was designed to travel into low Earth orbit (LEO). NASA explains in "What Is Orbit" (September 10, 2003, http://www.nasa.gov/audience/forstudents/5-8/features/orbit_feature_5-8.html) that LEO begins 100 miles (160 km) above Earth's surface, which is the lowest altitude that an object can remain in orbit around the planet. Although there is no set definition for the upper altitude of LEO, it is commonly said to extend between 600 and 1,200 miles (965 and 1,900 km) above Earth's surface.

The orbiter had to be capable of maneuvering while in space and during landing. Unlike the Apollo capsules, the orbiter was intended to be reusable. It had to land on the ground, rather than splash down in the ocean. Ultimately, the orbiter was designed to glide through the air to its landing site.

One of the most difficult design problems for the orbiter was a thermal protection system that could be reused. Previous spacecraft had been well protected from the intense heat of reentry, but their thermal protection materials were rendered unusable after one reentry. The orbiter's underside was covered with high-tech thermal blankets and more than 20,000 heat-resistant tiles.

Too Many Design Demands

The CAIB explains in *CAIB Report* that the U.S. military's demands for certain shuttle capabilities made it extremely difficult for the SSP to be economical. The DOD wanted the shuttle to carry larger and heavier payloads than those that were typically flown into space. The primary launch and landing facility for the shuttle was to be the KSC in Florida. However, the DOD also wanted a West Coast launch and landing facility. Vandenberg Air Force Base (AFB) in California was the chosen site. The DOD insisted that the shuttle be designed to return to the Vandenberg AFB after only one polar orbit. This was a technological challenge because it meant that the shuttle had to fly more than 1,000 miles (1,600 km) to the east during reentry. Engineers call this the "cross-range requirement." To meet this requirement, the shuttle was given delta wings (symmetrical triangular wings designed for subsonic and supersonic flight) and an enhanced thermal protection system.

NASA had to meet the design demands of the military to keep the project moving forward. However, this added substantially to the spacecraft development costs. As noted earlier, NASA was under an approximate budget ceiling of $5 billion. The space agency reportedly promised that it would develop shuttles each capable of flying 100 total missions at a cost of less than $8 million per launch. These projections proved to be wildly optimistic.

The first space shuttle launch was supposed to be in March 1978, but the launch was postponed several times due to budget and equipment problems. The shuttle's main engines and thermal protection tiles proved to be particularly troublesome. In 1979 President Jimmy Carter (1924–) reassessed the need for the SSP and considered canceling it. According to the CAIB, he decided to continue shuttle development because the United States wanted to launch intelligence satellites to monitor the Soviet Union's nuclear missile program. As a result, the White House and Congress put their support behind the space shuttle. In early 1981 NASA declared that development was complete, and the shuttle was "finished."

The Trade-Offs

The CAIB notes in *CAIB Report* that NASA produced a spacecraft that was revolutionary in several ways. The shuttle was:

- The first reusable spacecraft
- The first spacecraft with wings
- The first spacecraft with a reusable thermal protection system
- The first spacecraft with reusable engines that were powered by hydrogen and oxygen
- The first winged vehicle that was able to travel from orbital speed to a hypersonic glide during reentry

However, the space agency made many trade-offs during development to keep costs down. The spacecraft was originally supposed to be completely reusable. NASA soon abandoned this goal and settled for a partially reusable spacecraft. The boosters were originally supposed to burn liquid propellants, because they were much cheaper to use than solid propellants, but it was easier (and hence less expensive) to develop boosters that would fly using solid propellants. NASA's most fateful cost-cutting decision was to accept a spacecraft without a crew escape system. The CAIB explains that all these trade-offs resulted in a vehicle that "proved difficult and costly to operate, riskier than expected, and, on two occasions, deadly."

Space Shuttle Overview

The shuttle was designed to carry payloads into space and to serve as a short-term laboratory for science experiments. Payloads including satellites and heavy equipment needed for space station construction were transported in its large payload bay. Shuttle crews were trained to deploy, retrieve, and service LEO satellites from their spacecraft. The shuttle could also deploy satellites that required higher orbits. Built-in propulsion systems boosted these satellites into their orbits once they were a safe distance away from the shuttle.

The shuttle could be equipped with specialized laboratories for conducting experiments related to astronomy, earth science, medicine, and other fields. A space shuttle crew normally consisted of five members: a commander, a pilot, and three mission specialists. However, some missions included one or two "guest" crew members called payload specialists. These non-NASA crew members were from private companies or universities or were astronauts from foreign space agencies. The crew module on the shuttle was pressurized and maintained at a comfortable temperature to provide what is called a "shirt-sleeve environment."

The SSP was administered and operated by NASA, with the help of thousands of contract employees. The Johnson Space Center (JSC) in Houston, Texas, was home to the operational office of the program.

Space Shuttle Missions

On April 12, 1981, *Columbia* became the first shuttle to fly into space. The flight's purpose was to test the shuttle's systems, and the mission lasted only two days. It was considered to be a major success. Three more test flights were conducted in 1981 and 1982, all with the orbiter *Columbia*. On July 4, 1982, President Ronald Reagan (1911–2004) announced that shuttle testing was completed. The next flight of the shuttle was to begin its operational phase.

The orbiters *Columbia*, *Challenger*, *Discovery*, and *Atlantis* carried out 24 missions before disaster struck.

On January 28, 1986, *Challenger* broke apart only 73 seconds after liftoff. The seven crew members who were killed in the disaster were Gregory B. Jarvis (1944–1986), Christa McAuliffe (1948–1986), Ronald McNair (1950–1986), Ellison S. Onizuka (1946–1986), Judith A. Resnik (1949–1986), Francis R. Scobee (1939–1986), and Michael J. Smith (1945–1986). McAuliffe was a teacher who had been selected for the mission by NASA to capture the imagination of U.S. schoolchildren.

According to John A. Logsdon, in "Return to Flight: Richard H. Truly and the Recovery from the *Challenger* Accident" (Pamela E. Mack, ed., *From Engineering Science to Big Science: The NACA and NASA Collier Trophy Research Project Winners*, 1998), NASA investigators quickly pinpointed the direct cause of the disaster—a ruptured joint in a solid rocket motor (or solid rocket booster). The rupture allowed hot gases to escape from the booster and ignite the hydrogen fuel within the external tank. The resulting explosion tore the shuttle apart.

On February 3, 1986, President Reagan issued Executive Order 12546 (http://www.archives.gov/federal-register/executive-orders/1986.html), which established the Presidential Commission on the Space Shuttle Challenger Accident to investigate the accident. Four months later, in June 1986, the commission published its findings in *Report of the Presidential Commission on the Space Shuttle Challenger Accident* (http://history.nasa.gov/rogersrep/genindex.htm). Logsdon notes that the commission quickly concurred with NASA's finding that the explosion had been caused by the ruptured joint. However, the commission was shocked to learn that engineers from Morton Thiokol Inc., the company that manufactured the solid rocket motor, had warned NASA the night before the launch that cold weather could compromise the seal in the joint. The engineers had recommended that the launch be canceled, but NASA officials overruled this recommendation. Logsdon notes that "this was a 'turning point' in the investigation." The commission "decided to broaden the scope of its investigation to include NASA's management practices, Center-Headquarters relationships, and the chain of command for launch decisions—in effect, shifting the focus of the inquiry from a technical failure to NASA itself." In its report, the commission blamed NASA for fostering an overall culture that put schedule ahead of safety concerns.

To reduce the scheduling pressure, it was decided that the shuttle would cease carrying commercial satellites and phase out military missions as soon as possible. The U.S. Air Force had hoped to stage the first shuttle launch ever from the Vandenberg AFB in 1986. However, the *Challenger* disaster and the resulting decision to cease carrying military payloads ended these plans. The U.S. military and private companies returned to using nonreusable rockets to launch their payloads into space. Thus, one purpose of the shuttle—to fly commercial and military satellites into space—was completely lost. By not having these so-called customers, NASA's SSP became even less economical to operate than originally anticipated.

A number of organizational changes were made within NASA in response to the *Challenger* disaster. Shuttle management was moved from the JSC to NASA headquarters in Washington, D.C. In addition, NASA created a new office to oversee safety, reliability, and quality assurance. The entire orbiter fleet was grounded and upgraded with new equipment and systems. A new orbiter named *Endeavour* was built to replace *Challenger*. A White House committee later estimated that the shuttle disaster cost the nation approximately $12 billion. This included the cost of building a new orbiter.

Table 2.5 provides general information about each orbiter in the shuttle fleet.

The space shuttle flew again on September 29, 1988, with the successful launch of *Discovery* 32 months after the *Challenger* disaster. Space shuttles flew 87 successful missions between 1988 and 2002.

In October 1998 NASA achieved a public relations boost when Senator John Glenn (1921–; D-OH) flew into space aboard the space shuttle *Discovery*. The 77-year-old

TABLE 2.5

Orbiter vehicles

Orbiter name	Date completed	Date of first launch	Date of last launch	Named after	Note
Enterprise	September 1976	Not applicable	Not applicable	The starship Enterprise in the television series "Star Trek"	Used for testing only during the 1970s, never launched into space
Columbia	March 1979	April 12, 1981	January 16, 2003	A ship captained by American explorer Robert Gray during the 1790s	Exploded during reentry, February 1, 2003
Challenger	July 1982	April 4, 1983	January 28, 1986	A British Naval research vessel that sailed during the 1870s	Exploded shortly after launch, January 28, 1986
Discovery	November 1983	August 30, 1984	February 24, 2011	A ship captained by British Explorer James Cook during the 1770s	First shuttle to dock with the International Space Station (1999)
Atlantis	April 1985	October 3, 1985	July 8, 2011	A research vessel used by the Woods Hole Oceanographic Institute in Massachusetts from 1930 to 1966	First shuttle to dock with the Russian spacecraft Mir (1995)
Endeavour	May 1991	September 12, 1992	May 16, 2011	A ship captained by British Explorer James Cook during the 1760s	Built to replace Challenger. Endeavour was the first shuttle to deliver an American module to the International Space Station (1998)

SOURCE: Adapted from Jim Dumoulin, *Orbiter Vehicles*, National Aeronautics and Space Administration, February 1, 2003, http://science.ksc.nasa.gov/shuttle/resources/orbiters/orbiters.html (accessed November 23, 2015), and "Shuttle Missions," in *Missions*, National Aeronautics and Space Administration, August 29, 2011, http://www.nasa.gov/mission_pages/shuttle/shuttlemissions/index.html (accessed November 23, 2015)

Glenn was already a hero for his participation in Project Mercury during the early 1960s. In 1998 he became the oldest person ever to travel into space. NASA scientists conducted extensive medical tests before, during, and after his flight to monitor his well-being. They were particularly eager to learn about the effects of weightlessness on an older person. Prolonged weightlessness in space is known to weaken human bones, a condition also seen on Earth in older people who suffer from osteoporosis.

Then, tragedy struck the SSP again. On February 1, 2003, *Columbia* broke apart during reentry over the western United States. Seven crew members were killed: Michael P. Anderson (1959–2003), David M. Brown (1956–2003), Kalpana Chawla (1962–2003), Laurel B. Clark (1961–2003), Rick D. Husband (1957–2003), William C. McCool (1961–2003), and Ilan Ramon (1954–2003). Ramon was a colonel from the Israeli air force who traveled on the shuttle as a guest payload specialist. Following the disaster, the shuttle fleet was grounded for more than two years.

The *Columbia* Disaster

Immediately after the *Columbia* disaster, President George W. Bush (1946–) appointed the CAIB to investigate what happened. In August 2003 the CAIB released its report, *CAIB Report: Volume 1*, which concluded that the most likely cause of the disaster was a damaged thermal protection tile on the orbiter's left wing. Video clips of the launch showed a large piece of foam falling off the external tank and striking the left wing 82 seconds after liftoff. This piece of foam fell a distance of only 58 feet (18 m), but the space shuttle was traveling very fast at the time, so the foam struck with extreme force.

NASA engineers knew about the foam strike, but were unsure whether it had caused any damage. While *Columbia* was in orbit, some engineers suggested that high-resolution photographs be taken of the orbiter using DOD satellites or NASA's ground-based telescopes. This suggestion was overruled by NASA officials, who believed that the foam strike did not endanger mission safety.

During reentry to Earth's atmosphere, one or more damaged thermal tiles along the left wing likely allowed hot gases to breach the shuttle structure. Aerodynamic stresses then tore it apart. Debris from the shuttle was spread along a corridor across southeastern Texas and into Louisiana.

The CAIB was extremely critical of the entire SSP and complained that NASA shuttle managers had once again become preoccupied with schedule, rather than safety. Beginning in 1998 the SSP was under tremendous pressure to meet construction deadlines for the *International Space Station* (*ISS*).

Although the CAIB acknowledged that the shuttle was an "engineering marvel," the board concluded that "the Shuttle has few of the mission capabilities that NASA originally promised." Besides the fact that the shuttle could not be as easily launched as had been promised, the discontinuation of its role as a carrier of military and commercial payloads contributed to the program's inability to be cost effective. Furthermore, concerns about the ongoing dangers associated with the shuttle led the board to conclude that "despite efforts to improve its safety, the Shuttle remains a complex and risky system."

The CAIB recommended major changes within the SSP and within NASA management. One recommendation was that NASA develop a means for the shuttle crew to inspect the orbiter while docked at the *ISS* and repair any damage. Such a procedure might have saved the *Columbia* crew.

The Return to Flight

NASA developed new safety strategies. The most critical technical issue was debris shedding from the external fuel tank during ascent and subsequent damage to the orbiter's thermal protection system. The procedures were changed for applying foam insulation to the external fuel tank, and quality control and inspection programs were expanded. Equipment changes were implemented to provide a smoother surface for foam application and to impede ice formation.

Discovery was selected for the first return to flight (RTF) mission. More than 100 cameras were installed on exterior spacecraft surfaces and at ground locations to provide an array of observation angles during ascent. The 50-foot-long (15.2-m-long) orbital boom sensor system (OBSS) was installed on the end of the shuttle remote manipulator system to allow visual inspection of the wing tips and most of the orbiter underbelly once the shuttle obtained orbit or docked with the *ISS*. A team of analysts was selected to inspect the images for any signs of damage. In addition, dozens of sensors were installed on the wing edges of *Discovery* to take temperature readings and record the time and location of any debris impacts.

On July 26, 2005, *Discovery* launched for a 14-day mission. The orbiter, with seven crew members onboard, docked with the *ISS*. The shuttle astronauts used the OBSS and conducted spacewalks to inspect the orbiter's underbelly. The shuttle landed safely on August 9, 2005. NASA proclaimed the first RTF a success. However, camera footage showed that foam debris from the external tank had fallen but did not hit the orbiter during its ascent. NASA and the public realized that the hazard that had doomed *Columbia* had not been eliminated, but merely avoided by chance this time.

On July 4, 2006, the second RTF mission began with the launch of *Discovery* on a 13-day mission. Over 100 high-definition cameras recorded the launch and ascent phases so the images could be scoured for signs of damage to the orbiter. In addition, the shuttle crew used the OBSS to carefully inspect the craft while it was docked at the *ISS*. No significant damage was detected. *Discovery* landed safely on July 17, 2006.

Missions after the Return to Flight

The space shuttle conducted 20 successful missions following the second RTF flight. All the missions, but one, were dedicated to *ISS* assembly. Extensive imaging and visual inspections were conducted during each shuttle flight to identify any damage to the thermal protection system because of foam debris impacts during launch. In all cases the orbiters were deemed structurally sound for reentry.

Accomplishments of the SSP

A historical summary of all the space shuttle missions is presented in Table 2.6.

NASA referred to each shuttle flight using a Space Transportation System (STS) number. Thus, STS-1 was the first shuttle flight into space. NASA assigned numbers to space shuttle flights in the order in which they were planned (or manifested). There was typically a period of several years between the time a mission was planned and the time of its scheduled launch. During this period priorities could change, and missions were often reshuffled or canceled. This explains why the STS numbers shown in Table 2.6 do not always match the flight order number. For example, the 2003 flight of *Columbia* was called STS-107, yet it was actually the 113th flight of a space shuttle. Missions STS-108 through STS-113 launched before STS-107 because they moved up in priority as launch time approached.

Shuttle flights deployed more than 50 satellites for military, governmental, and commercial clients before such practices were halted. In addition, between 1989 and 1990 three interplanetary craft were launched from shuttles: the *Magellan* spacecraft that traveled to Venus, the *Galileo* spacecraft that traveled to Jupiter, and the *Ulysses* spacecraft that traveled to the sun. Shuttles also deployed space observatories, including the *Hubble Space Telescope* in 1990, the *Compton Gamma Ray Observatory* in 1991, the *Diffuse X-Ray Spectrometer* in 1993, and the *Chandra X-Ray Observatory* in 1999.

The shuttle carried more than 1,500 tons (1,361 t) of cargo and over 600 crew members into space. Hundreds of scientific experiments were conducted in orbit. Shuttle crews also serviced and repaired satellites as needed, particularly the *Hubble Space Telescope*. Between 1995 and 1998 shuttles docked nine times with the Russian space station *Mir*. Flights to construct the *ISS* began in 1998. Shuttles carried major pieces of the *ISS* into space and traveled to the station 37 times.

Despite these accomplishments, as previously mentioned, the shuttle failed to meet several of the original goals that NASA set for the SSP. NASA planners had promised that the shuttle would fly dozens of times per year. As shown in Figure 2.4, the most shuttle flights ever accomplished in one year was nine flights in 1985. During

TABLE 2.6

Space shuttle missions, 1981–2011

Flight order	STS number	Orbiter name	Mission and/or primary payload	Launch date	Landing date
1	STS-1	Columbia	Shuttle systems test	4/12/1981	4/14/1981
2	STS-2	Columbia	OSTA-1	11/12/1981	11/14/1981
3	STS-3	Columbia	Office of Space Science-1 (OSS-1)	3/22/1982	3/30/1982
4	STS-4	Columbia	Department of Defense and Continuous Flow Electrophoresis System (CFES)	6/27/1982	7/4/1982
5	STS-5	Columbia	Canadian Satellite ANIK C-3; SBS-C	11/11/1982	11/16/1982
6	STS-6	Challenger	TDRS-1	4/4/1983	4/9/1983
7	STS-7	Challenger	Canadian Satellite ANIK C-2; PALAPA B1	6/18/1983	6/24/1983
8	STS-8	Challenger	India Satellite INSAT-1B	8/30/1983	9/5/1983
9	STS-9	Columbia	Spacelab-1	11/28/1983	12/8/1983
10	STS-41-B	Challenger	WESTAR-VI; PALAPA-B2	2/3/1984	2/11/1984
11	STS-41-C	Challenger	LDEF deploy	4/6/1984	4/13/1984
12	STS-41-D	Discovery	SBS-D; SYNCOM IV-2; TELSTAR	8/30/1984	9/5/1984
13	STS-41-G	Challenger	Earth Radiation Budget Satellite (ERBS); OSTA-3	10/5/1984	10/13/1984
14	STS-51-A	Discovery	Canadian Communications Satellite TELESAT-H; SYNCOM IV-1	11/8/1984	11/16/1984
15	STS-51-C	Discovery	DOD	1/24/1985	1/27/1985
16	STS-51-D	Discovery	Canadian Satellite TELESAT-I; SYNCOM IV-3	4/12/1985	4/19/1985
17	STS-51-B	Challenger	Spacelab-3	4/29/1985	5/6/1985
18	STS-51-G	Discovery	MORELOS-A; Arab Satellite ARABSAT-A; AT&T Satellite TELSTAR-3D	6/17/1985	6/24/1985
19	STS-51-F	Challenger	Spacelab-2	7/29/1985	8/6/1985
20	STS-51-I	Discovery	American Satellite ASC-1; AUSSAT-1; SYNCOM IV-4	8/27/1985	9/3/1985
21	STS-51-J	Atlantis	DOD	10/3/1985	10/7/1985
22	STS-61-A	Challenger	D-1 Spacelab mission (first German-dedicated Spacelab)	10/30/1985	11/6/1985
23	STS-61-B	Atlantis	MORELOS-B; AUSSAT-2; RCA Americom Satellite SATCOM KU-2	11/26/1985	12/3/1985
24	STS-61-C	Columbia	RCA Americom Satellite SATCOM KU-1	1/12/1986	1/18/1986
25	STS-51-L	Challenger	TDRS-B; SPARTAN-203	1/28/1986	Vehicle broke apart 73 seconds after liftoff
26	STS-26	Discovery	TDRS-C	9/29/1988	10/3/1988
27	STS-27	Atlantis	DOD	12/2/1988	12/6/1988
28	STS-29	Discovery	TDRS-D	3/13/1989	3/18/1989
29	STS-30	Atlantis	Magellan	5/4/1989	5/8/1989
30	STS-28	Columbia	DOD	8/8/1989	8/13/1989
31	STS-34	Atlantis	Galileo; SSBUV	10/18/1989	10/23/1989
32	STS-33	Discovery	DOD	11/22/1989	11/27/1989
33	STS-32	Columbia	SYNCOM IV-F5; LDEF Retrieval	1/9/1990	1/20/1990
34	STS-36	Atlantis	DOD	2/28/1990	3/4/1990
35	STS-31	Discovery	HST deploy	4/24/1990	4/29/1990
36	STS-41	Discovery	Ulysses; SSBUV; INTELSAT Solar Array Coupon (ISAC)	10/6/1990	10/10/1990
37	STS-38	Atlantis	DOD	11/15/1990	11/20/1990
38	STS-35	Columbia	ASTRO-1	12/2/1990	12/10/1990
39	STS-37	Atlantis	Gamma Ray Observatory (GRO)	4/5/1991	4/11/1991
40	STS-39	Discovery	DOD; Air Force Program-675 (AFP675); Infrared Background Signature Survey (IBSS); Shuttle Pallet Satellite-II (SPAS-II)	4/28/1991	5/6/1991
41	STS-40	Columbia	Spacelab Life Sciences-1 (SLS-1)	6/5/1991	6/14/1991
42	STS-43	Atlantis	TDRS-E; SSBUV	8/2/1991	8/11/1991
43	STS-48	Discovery	Upper Atmosphere Research Satellite (UARS)	9/12/1991	9/18/1991
44	STS-44	Atlantis	DOD; Defense Support Program (DSP)	11/24/1991	12/1/1991
45	STS-42	Discovery	IML-1	1/22/1992	1/30/1992
46	STS-45	Atlantis	ATLAS-1	3/24/1992	4/2/1992
47	STS-49	Endeavour	Intelsat VI repair	5/7/1992	5/16/1992
48	STS-50	Columbia	USML-1	6/25/1992	7/9/1992
49	STS-46	Atlantis	TSS-1; EURECA deploy	7/31/1992	8/8/1992
50	STS-47	Endeavour	Spacelab-J	9/12/1992	9/20/1992
51	STS-52	Columbia	USMP-1; Laser Geodynamic Satellite-II (LAGEOS-II)	10/22/1992	11/1/1992
52	STS-53	Discovery	DOD; Orbital Debris Radar Calibration Spheres (ODERACS)	12/2/1992	12/9/1992
53	STS-54	Endeavour	TDRS-F; Diffuse X-ray Spectrometer (DXS)	1/13/1993	1/19/1993
54	STS-56	Discovery	ATLAS-2; SPARTAN-201	4/8/1993	4/17/1993
55	STS-55	Columbia	D-2 Spacelab mission (second German-dedicated Spacelab)	4/26/1993	5/6/1993
56	STS-57	Endeavour	SPACEHAB-1; EURECA retrieval	6/21/1993	7/1/1993
57	STS-51	Discovery	Advanced Communications Technology Satellite (ACTS)/Transfer Orbit Stage (TOS)	9/12/1993	9/22/1993

TABLE 2.6

Space shuttle missions, 1981–2011 [CONTINUED]

Flight order	STS number	Orbiter name	Mission and/or primary payload	Launch date	Landing date
58	STS-58	Columbia	Spacelab SLS-2	10/18/1993	11/1/1993
59	STS-61	Endeavour	1st HST servicing	12/2/1993	12/13/1993
60	STS-60	Discovery	WSF; SPACEHAB-2	2/3/1994	2/11/1994
61	STS-62	Columbia	USMP-2; Office of Aeronautics and Space Technology-2 (OAST-2)	3/4/1994	3/18/1994
62	STS-59	Endeavour	SRL-1	4/9/1994	4/20/1994
63	STS-65	Columbia	IML-2	7/8/1994	7/23/1994
64	STS-64	Discovery	LIDAR In-Space Technology Experiment (LITE); SPARTAN-201	9/9/1994	9/20/1994
65	STS-68	Endeavour	SRL-2	9/30/1994	10/11/1994
66	STS-66	Atlantis	ATLAS-03	11/3/1994	11/14/1994
67	STS-63	Discovery	SPACEHAB-3; Mir rendezvous	2/3/1995	2/11/1995
68	STS-67	Endeavour	ASTRO-2	3/2/1995	3/18/1995
69	STS-71	Atlantis	First Shuttle-Mir docking	6/27/1995	7/7/1995
70	STS-70	Discovery	TDRS-G	7/13/1995	7/22/1995
71	STS-69	Endeavour	SPARTAN 201-03; WSF-2	9/7/1995	9/18/1995
72	STS-73	Columbia	USML-2	10/20/1995	11/5/1995
73	STS-74	Atlantis	Second Shuttle-Mir docking	11/12/1995	11/20/1995
74	STS-72	Endeavour	Space Flyer Unit (SFU); Office of Aeronautics and Space Technology Flyer (OAST-Flyer)	1/11/1996	1/20/1996
75	STS-75	Columbia	TSS-1 Reflight; USMP-3	2/22/1996	3/9/1996
76	STS-76	Atlantis	Third Shuttle-Mir docking; SPACEHAB	3/22/1996	3/31/1996
77	STS-77	Endeavour	SPACEHAB; SPARTAN (Inflatable Antenna Experiment)	5/19/1996	5/29/1996
78	STS-78	Columbia	Life and Microgravity Spacelab (LMS)	6/20/1996	7/7/1996
79	STS-79	Atlantis	Fourth Shuttle-Mir docking	9/16/1996	9/26/1996
80	STS-80	Columbia	Orbiting and Retrievable Far and Extreme Ultraviolet Spectrograph-Shuttle Pallet Satellite II (ORFEUS-SPAS II)	11/19/1996	12/7/1996
81	STS-81	Atlantis	Fifth Shuttle-Mir docking	1/12/1997	1/22/1997
82	STS-82	Discovery	Second HST servicing	2/11/1997	2/21/1997
83	STS-83	Columbia	MSL-1	4/4/1997	4/8/1997
84	STS-84	Atlantis	Sixth Shuttle-Mir docking	5/15/1997	5/24/1997
85	STS-94	Columbia	MSL-1 Reflight	7/1/1997	7/17/1997
86	STS-85	Discovery	Cryogenic Infrared Spectrometers and Telescopes for the Atmosphere-Shuttle Pallet Satellite-2 (CRISTA-SPAS-2)	8/7/1997	8/19/1997
87	STS-86	Atlantis	Seventh Shuttle-Mir docking	9/25/1997	10/6/1997
88	STS-87	Columbia	USMP-4, Spartan-201 rescue	11/19/1997	12/5/1997
89	STS-89	Endeavour	Eighth Shuttle-Mir docking	1/22/1998	1/31/1998
90	STS-90	Columbia	Final Spacelab mission	4/17/1998	5/3/1998
91	STS-91	Discovery	Ninth and final Shuttle-Mir docking	6/2/1998	6/12/1998
92	STS-95	Discovery	John Glenn's Flight; SPACEHAB	10/29/1998	11/7/1998
93	STS-88	Endeavour	First ISS Flight	12/4/1998	12/15/1998
94	STS-96	Discovery	1st ISS docking	5/27/1999	6/6/1999
95	STS-93	Columbia	Chandra X-Ray Observatory	7/22/1999	7/27/1999
96	STS-103	Discovery	HST repair - 3A	12/19/1999	12/27/1999
97	STS-99	Endeavour	Shuttle Radar Topography Mission (SRTM)	2/11/2000	2/22/2000
98	STS-101	Atlantis	ISS Assembly Flight 2A.2a	5/19/2000	5/29/2000
99	STS-106	Atlantis	ISS Assembly Flight 2A.2b	9/8/2000	9/20/2000
100	STS-92	Discovery	ISS Assembly Flight 3A, Z1 Truss and PMA 3	10/11/2000	10/24/2000
101	STS-97	Endeavour	ISS Assembly Flight 4A, P6 Truss	11/30/2000	12/11/2000
102	STS-98	Atlantis	ISS Assembly Flight 5A, U.S. Destiny Laboratory	2/7/2001	2/20/2001
103	STS-102	Discovery	ISS Assembly Flight 5A.1, Crew Exchange, Leonardo Multi-Purpose Logistics Module	3/8/2001	3/21/2001
104	STS-100	Endeavour	ISS Assembly Flight 6A, Canadarm2, Raffaello Multi-Purpose Logistics Module	4/19/2001	5/1/2001
105	STS-104	Atlantis	ISS Assembly Flight 7A, Quest Airlock, High Pressure Gas Assembly	7/12/2001	7/24/2001
106	STS-105	Discovery	ISS Assembly Flight 7A.1, Crew Exchange, Leonardo Multi-Purpose Logistics Module	8/10/2001	8/22/2001
107	STS-108	Endeavour	ISS Flight UF-1, Crew Exchange, Raffaello Multi-Purpose Logistics Module, STARSHINE 2	12/5/2001	12/17/2001
108	STS-109	Columbia	HST Servicing Mission 3B	3/1/2002	3/12/2002
109	STS-110	Atlantis	ISS Flight 8A, S0 (S-Zero) Truss, Mobile Transporter	4/8/2002	4/19/2002
110	STS-111	Endeavour	ISS Flight UF-2, Crew Exchange, Mobile Base System	6/5/2002	6/19/2002
111	STS-112	Atlantis	ISS Flight 9A, S1 (S-One) Truss	10/7/2002	10/16/2002
112	STS-113	Endeavour	ISS Flight 11A, P1 (P-One) Truss	11/23/2002	12/7/2002
113	STS-107	Columbia	SpaceHab-DM Research Mission, Freestar module	1/16/2003	Vehicle broke up during reentry 2/1/2003

TABLE 2.6

Space shuttle missions, 1981–2011 [CONTINUED]

Flight order	STS number	Orbiter name	Mission and/or primary payload	Launch date	Landing date
114	STS-114	Discovery	ISS Assembly Flight LF1, External Stowage Platform-2, Raffaello Multi-Purpose Logistics Module	7/26/2005	8/9/2005
115	STS-121	Discovery	ISS Flight ULF-1, Leonardo Multi-Purpose Logistics Module; One ISS Crew member	7/4/2006	7/17/2006
116	STS-115	Atlantis	ISS Flight 12A, P3/P4 trusses w/solar arrays, Photovoltaic radiator	9/9/2006	9/21/2006
117	STS-116	Discovery	ISS Flight 12A.1, P5 spacer truss, SpaceHab cargo module, resupply	12/9/2006	12/22/2006
118	STS-117	Atlantis	ISS Flight 13A, S3/S4 trusses w/solar arrays, Photovoltaic radiator, Crew Exchange	6/8/2007	6/22/2007
119	STS-118	Endeavour	ISS Flight 13A.1, S5 truss, resupply, SpaceHab cargo module	8/8/2007	8/21/2007
120	STS-120	Discovery	ISS Flight 10A, Harmony node, Crew Exchange	10/23/2007	11/7/2007
121	STS-122	Atlantis	ISS Assembly Flight 1E, Columbus Lab, Crew Exchange	2/7/2008	2/20/2008
122	STS-123	Endeavour	ISS Assembly Flight 1J/A, Kibo Logistics Module, Dextre Robotics System, Crew Exchange	3/11/2008	3/26/2008
123	STS-124	Discovery	ISS Assembly Flight 1J, Kibo's Japanese Pressurized Module, Crew Exchange	5/31/2008	6/14/2008
124	STS-126	Endeavour	ISS Flight ULF2, Leonardo Multi Purpose Logistics Module, Crew Exchange	11/14/2008	11/30/2008
125	STS-119	Discovery	ISS Flight 15A, S6 truss, Crew Exchange	3/15/2009	3/28/2009
126	STS-125	Atlantis	HST Repair-4	5/11/2009	5/24/2009
127	STS-127	Endeavour	ISS Assembly Flight 2J/A, Kibo Exposed Facility and Experiment Logistics Module-Exposed, Crew Exchange	7/15/2009	7/31/2009
128	STS-128	Discovery	ISS Flight 17A, Leonardo Multi Purpose Logistics Module, Experiment Support Structure Center, Crew Exchange	8/28/2009	9/11/2009
129	STS-129	Atlantis	ISS Flight ULF3, EXPRESS Logistics Centers 1 & 2, Crew Exchange	11/16/2009	11/27/2009
130	STS-130	Endeavour	ISS Flight 20A, Tranquility node 3, cupola	2/8/2010	2/21/2010
131	STS-131	Discovery	ISS Flight 19A, Multi-Purpose Logistics Module	4/5/2010	4/20/2010
132	STS-132	Atlantis	ISS Flight ULF4, Integrated Cargo Carrier, Rassvet ("Dawn") Mini Research Module	5/14/2010	5/26/2010
133	STS-133	Discovery	ISS Flight ULF5, EXPRESS Logistics Carrier 4, Permanent Multi-Purpose Module	2/24/2011	3/9/2011
134	STS-134	Endeavour	ISS Flight ULF6, EXPRESS Logistics Carrier 3, Alpha Magnetic Spectrometer	5/16/2011	6/1/2011
135	STS-135	Atlantis	ISS Flight ULF7, Raffaello Multi Purpose Logistics Module, Lightweight Multi-Purpose Experiment Support Structure Carrier	7/8/2011	7/21/2011

ATLAS = Atmospheric laboratory for applications and science.
AUSSAT = Australian satellite.
DOD = Department of defense.
EURECA = European retrievable carrier.
HST = Hubble space telescope.
IML = International microgravity laboratory.
ISS = International space station.
LDEF = Long duration exposure facility.
MORELOS = Mexican satellite.
MSL = Microgravity science laboratory.
OSTA = Office of space and terrestrial applications.
PALAPA = Indonesian satellite.
SBS = Satellite business systems.
SRL = Space radar laboratory.
SSBUV = Shuttle solar backscatter ultraviolet.
SYNCOM = Synchronous communication satellite.
TDRS = Tracking and data relay satellite.
TSS = Tethered satellite system.
USML = United States microgravity laboratory.
USMP = U.S. microgravity payload.
WSF = Wake shield facility.

SOURCE: Adapted from "Shuttle Missions," in *Missions*, National Aeronautics and Space Administration, August 29, 2011, http://www.nasa.gov/mission_pages/shuttle/shuttlemissions/index.html (accessed November 23, 2015)

the 30-year period between 1981 and 2011, the shuttle averaged 4.5 flights per year.

NASA also originally promised that each shuttle orbiter would be good for 100 flights. However, far fewer flights were achieved by each orbiter in the shuttle fleet:

- *Discovery* made 39 flights
- *Atlantis* made 33 flights

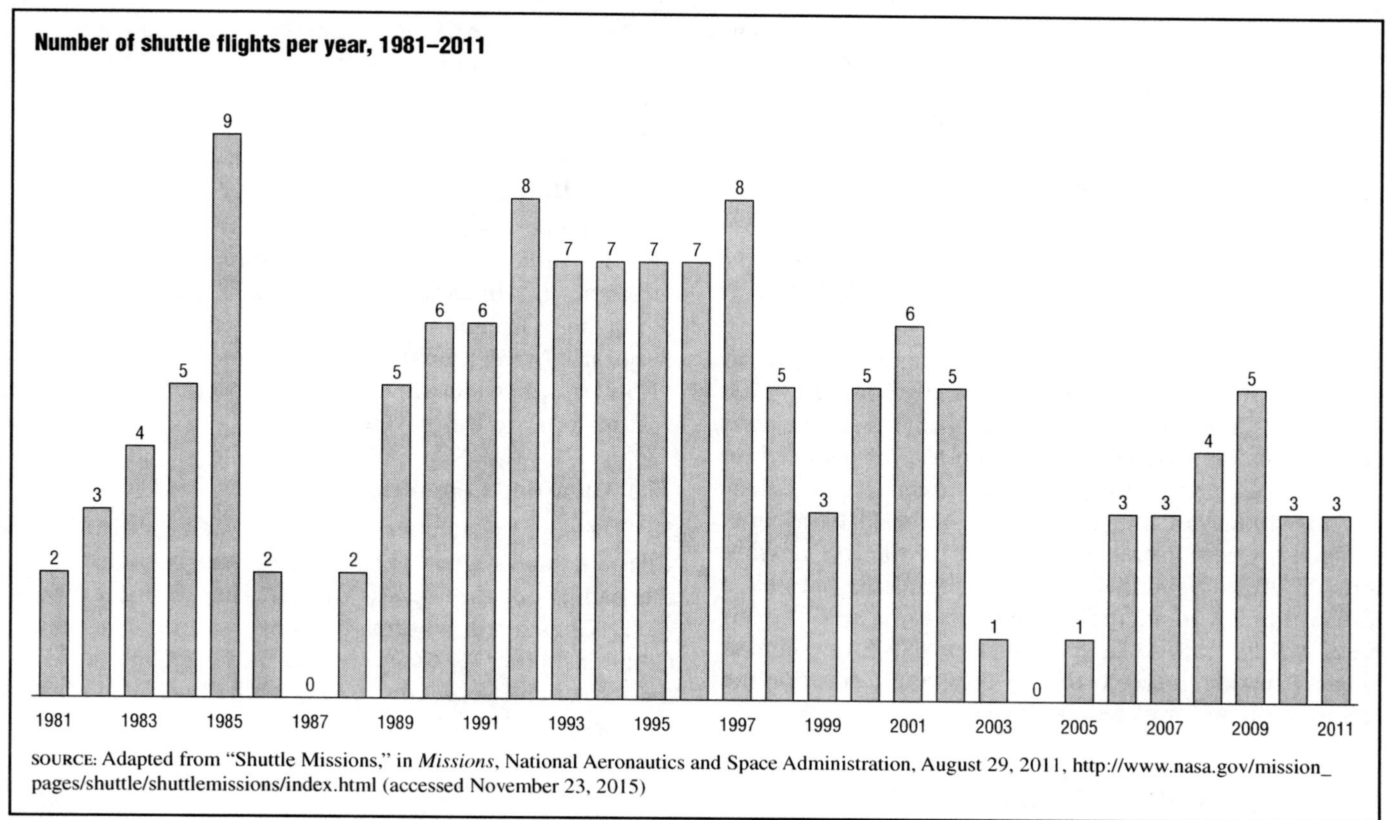

FIGURE 2.4

Number of shuttle flights per year, 1981–2011

SOURCE: Adapted from "Shuttle Missions," in *Missions*, National Aeronautics and Space Administration, August 29, 2011, http://www.nasa.gov/mission_pages/shuttle/shuttlemissions/index.html (accessed November 23, 2015)

- *Columbia* made 28 flights before it was lost in 2003
- *Endeavour* made 25 flights
- *Challenger* made 10 flights before it was lost in 1986

The Space Shuttle Program Ends

In January 2004 President Bush announced a new goal for the nation's space program: to return to the moon and travel to Mars and beyond. The so-called Vision for Space Exploration required the development of a new fleet of spacecraft that were capable of carrying astronauts to the moon, Mars, and elsewhere in the solar system. The Bush administration decided to end the SSP by 2010, assuming that existing U.S. commitments to build the *ISS* were completed by then.

NASA embarked on the Constellation Program, which is described later in this chapter, as the successor to the SSP. In 2010, however, President Barack Obama (1961–) declined to fund the Constellation Program in his proposed federal budget for fiscal year (FY) 2011 even as he maintained the 2010 deadline for the last shuttle flight. This deadline slipped to 2011 as NASA struggled to complete the flights needed to transport the last major components to the *ISS*. On July 8, 2011, *Atlantis* was launched on the last space shuttle flight. More than 1 million people gathered in or around the KSC to witness the historic launch. The shuttle crew docked at the *ISS* and delivered some final components to the station. On July 21, 2011, *Atlantis* returned to Earth safely. The space shuttle era was over.

NASA dispatched the retired orbiters to exhibitions around the country. *Enterprise* was donated to the Intrepid Sea, Air, and Space Museum (http://www.intrepidmuseum.org/shuttle/) in New York City. *Discovery* went to the Steven F. Udvar-Hazy Center (http://airandspace.si.edu/visit/udvar-hazy-center/) in Chantilly, Virginia. The center serves as a satellite location for the Smithsonian National Air and Space Museum in Washington, D.C. *Endeavour* was donated to the California Science Center (http://www.californiasciencecenter.org/Exhibits/AirAndSpace/endeavour/endeavour.php) in Los Angeles, California. The final surviving orbiter, *Atlantis*, went into a permanent exhibit at the KSC (http://www.kennedyspacecenter.com/the-experience/atlantis-shuttle-experience.aspx).

THE *INTERNATIONAL SPACE STATION*

In 1988 the United States and 15 other nations embarked on a new space venture called the *International Space Station* (*ISS*). The United States and Russia collaborated throughout the 1990s to lead construction of an orbiting space station that was designed for prolonged inhabitation by scientists engaged in space research. They invited other countries to participate by contributing parts, components, and scientific facilities or sending researchers to the station.

The Russian Space Agency was determined to play a major role in the *ISS*, but it had even less funding than NASA. Both agencies struggled to put U.S. and Russian modules into place and keep them operational. The station was repeatedly scaled back in size and capability due to budget restrictions in both countries. The space shuttles performed the bulk of the heavy lifting required to put *ISS* modules into space. The 2003 grounding of the shuttle fleet halted *ISS* construction, and NASA astronauts had to travel aboard Russian spacecraft until the space shuttles resumed operating.

As the end of the SSP loomed, plans were made to finish construction on the station to a level dubbed "core complete," a point less than what NASA and scientists had originally anticipated. This point was reached in 2011. NASA and its international partners began debating the future of the station. In early 2014 Obama pledged his administration's support for the ISS mission through 2024. Nevertheless, future presidents may have different plans. In addition, it remains to be seen if Congress will allocate sufficient funds to NASA to continue its participation in the costly ISS program. Details on the ISS mission are provided in Chapter 5.

THE VISION FOR SPACE EXPLORATION

As noted earlier, in 2004 President Bush set a new course for NASA with his Vision for Space Exploration. It called for the agency to send astronauts to the moon and then to Mars, but a space shuttle could not serve this purpose, because it was designed only for LEO. Furthermore, NASA no longer had any of the massive Saturn V rockets that lifted Apollo spacecraft into space. Under the Constellation Program, NASA began development of new rockets called the Ares I and Ares V to propel crew and cargo, respectively, into space. Ares was the god of war in Greek mythology and was called Mars by the ancient Romans. NASA began development of a new crew exploration vehicle named Orion after the heroic hunter in Greek mythology (and a constellation of stars).

Another component of the new vision involved greater engagement by NASA with the private sector. In *The Vision for Space Exploration* (February 2004, http://www.nasa.gov/pdf/55583main_vision_space_exploration2.pdf), the agency lists one of its new goals: "Pursue commercial opportunities for providing transportation and other services supporting the International Space Station and exploration missions beyond low Earth orbit." In 2005 NASA established its Commercial Crew and Cargo Program (http://www.nasa.gov/offices/c3po/home/index.html) to facilitate private development of spacecraft capable of flying to and from LEO. Subsequently, two private companies—the Space Exploration Technologies Corp. (SpaceX) and Orbital ATK (formerly the Orbital Sciences Corporation)—developed spacecraft that have made cargo deliveries to the *ISS*. These projects are described in detail in Chapter 4.

A NEW AGENDA FOR HUMAN SPACEFLIGHT

In January 2009 President Obama began his first term in office. During his first year he gave few clues about his administration's plans for the U.S. space program. At the time, NASA continued development under the Constellation Program of the *Orion* space capsule and the Ares lift vehicles. However, when Obama's FY 2011 budget request was released in February 2010, it did not include further funding for the Constellation Program, indicating his intention to terminate it. Joel Achenbach reports in "Obama Overhauls NASA's Agenda in Budget Request" (WashingtonPost.com, February 11, 2010) that 11,500 people were working on the program in 2010, the vast majority in the private sector.

The Augustine Committee

The Obama administration's decision to cancel the Constellation Program is believed to have been driven by the findings of the Review of U.S. Human Space Flight Plans Committee, informally known as the Augustine Committee after its chair, Norman R. Augustine (1935–), the former chief executive officer of the Lockheed Martin Corporation. The committee consisted of former astronauts, scientists, representatives of the aerospace industry, and scientific advisers to the Obama administration.

In October 2009 the Augustine Committee published its findings in *Seeking a Human Spaceflight Program Worthy of a Great Nation* (http://www.nasa.gov/pdf/396093main_HSF_Cmte_FinalReport.pdf). The committee concluded that the United States' existing plan for human spaceflight—the Constellation Program—was "on an unsustainable trajectory" and noted that "it is perpetuating the perilous practice of pursuing goals that do not match allocated resources." In other words, the committee believed that NASA was once again pursuing a lofty space goal without the financial funding to actually make it happen.

The committee also noted problems with schedule delays in the Constellation Program. The program originally planned to transport astronauts to the *ISS* by 2012, but that date had slid to at least 2017. Overall, the committee was strongly in favor of turning over some traditional NASA duties to the commercial sector, primarily the transporting of cargo and crew into orbit. The committee examined three destination alternatives for NASA's human spaceflight program:

- Mars First, with a Mars landing, perhaps after a brief test of equipment and procedures on the moon.
- Moon First, with lunar surface exploration focused on developing the capability to explore Mars.
- A Flexible Path to inner solar system locations, such as lunar orbit, Lagrange points, near-Earth objects, and the moons of Mars, followed by exploration of the lunar surface and/or Martian surface.

The Flexible Path option offers the greatest opportunity for scientific exploration and observation. It not only includes previously visited destinations such as the moon and Earth orbits and the lunar surface but also provides the potential opportunity to visit Mars and its orbit, asteroids, and other objects in the inner solar system.

National Space Policy

In June 2010 President Obama announced his National Space Policy, which was published in *National Space Policy of the United States of America* (http://www.whitehouse.gov/sites/default/files/national_space_policy_6-28-10.pdf). His policy laid out six broad goals for U.S. space endeavors:

- Energize competitive domestic industries
- Expand international cooperation
- Strengthen stability in space
- Increase assurance and resilience of mission-essential functions
- Pursue human and robotic initiatives
- Improve space-based Earth and solar observation

The first goal—energize competitive domestic industries—reiterated the president's intention to integrate greater commercial participation into NASA's activities. The role of private-sector entities in space exploration is described in detail in Chapter 4.

President Obama also listed eight specific directives for NASA to follow, including the instruction to "set far-reaching exploration milestones. By 2025, begin crewed missions beyond the moon, including sending humans to an asteroid. By the mid-2030s, send humans to orbit Mars and return them safely to Earth."

National Aeronautics and Space Administration Authorization Act

In October 2010 Congress passed the National Aeronautics and Space Administration Authorization Act. Since NASA's founding in 1958, several authorization acts for the agency have been passed. Amy Klamper explains in "NASA Authorization Bill Still Leaves Questions Unanswered" (SpaceNews.com, October 4, 2010) that "authorization bills provide policy guidance and offer recommended funding levels for congressional appropriators to consider in crafting their funding bills." Klamper notes that passage of the 2010 law followed much "contentious bipartisan wrangling" in Congress over NASA's future.

Overall, the act approved the main elements of Obama's National Space Policy and authorized appropriations of $58.5 billion over a three-year period for NASA: $19 billion in FY 2011, $19.5 billion in FY 2012, and $20 billion in FY 2013. However, NASA's actual budgets for these years were set much lower as the appropriations process unfolded over subsequent years. The authorization act did not authorize the termination of the Constellation Program. This authorization came in April 2011 via passage by Congress of an FY 2011 appropriations act for NASA; at that point the Constellation Program was officially ended.

NASA's New Human Exploration Goals

In *Voyages: Charting the Course for Sustainable Human Space Exploration* (June 7, 2012, http://www.nasa.gov/sites/default/files/files/ExplorationReport_508_6-4-12.pdf), NASA discusses its general strategies for reaching its new exploration destinations:

- Cis-lunar space (the space between Earth and the moon)
- Near-Earth asteroids (defined by the International Astronomical Union [October 7, 2013, http://www.iau.org/public/themes/neo/nea2/] as asteroids that come within 0.3 of an astronomical unit [27.9 million miles (44.9 million km)] of Earth's orbit)
- Moon
- Mars

Figure 2.5 shows some of the steps that NASA believes will be involved in these endeavors. The agency plans to develop new spacecraft and accessories to facilitate future human exploration missions.

The earliest voyages will likely be in cis-lunar space. NASA is particularly interested in investigating Earth-moon Lagrange points. Lagrange points are points in space where the gravitational pulls of nearby large bodies are balanced. Thus, a spacecraft located at a Lagrange point would require little fuel to stay "parked" there. Figure 2.6 shows the five Lagrange points in Earth-moon space. These points could provide future locations for space stations from which missions could be launched into even deeper space.

NASA's New Human Exploration Spacecraft

As noted earlier, NASA began development of new launch vehicles (then called Ares I and Ares V) and the *Orion* crewed spacecraft during the Constellation Program. In accordance with President Obama's revised agenda for the agency, these development activities have continued, but with some design and name changes. The launch vehicles have been redubbed Space Launch System (SLS). As shown in Figure 2.7, there were two SLS versions under development as of 2015. The smaller vehicle can carry payloads weighing up to 77 tons (70 t). The larger vehicle is designed for payloads as large as 143 tons (130 t). Even the smaller vehicle will provide more thrust than the powerful Saturn V rocket that was used during the Apollo era.

FIGURE 2.5

NASA's human space exploration plan

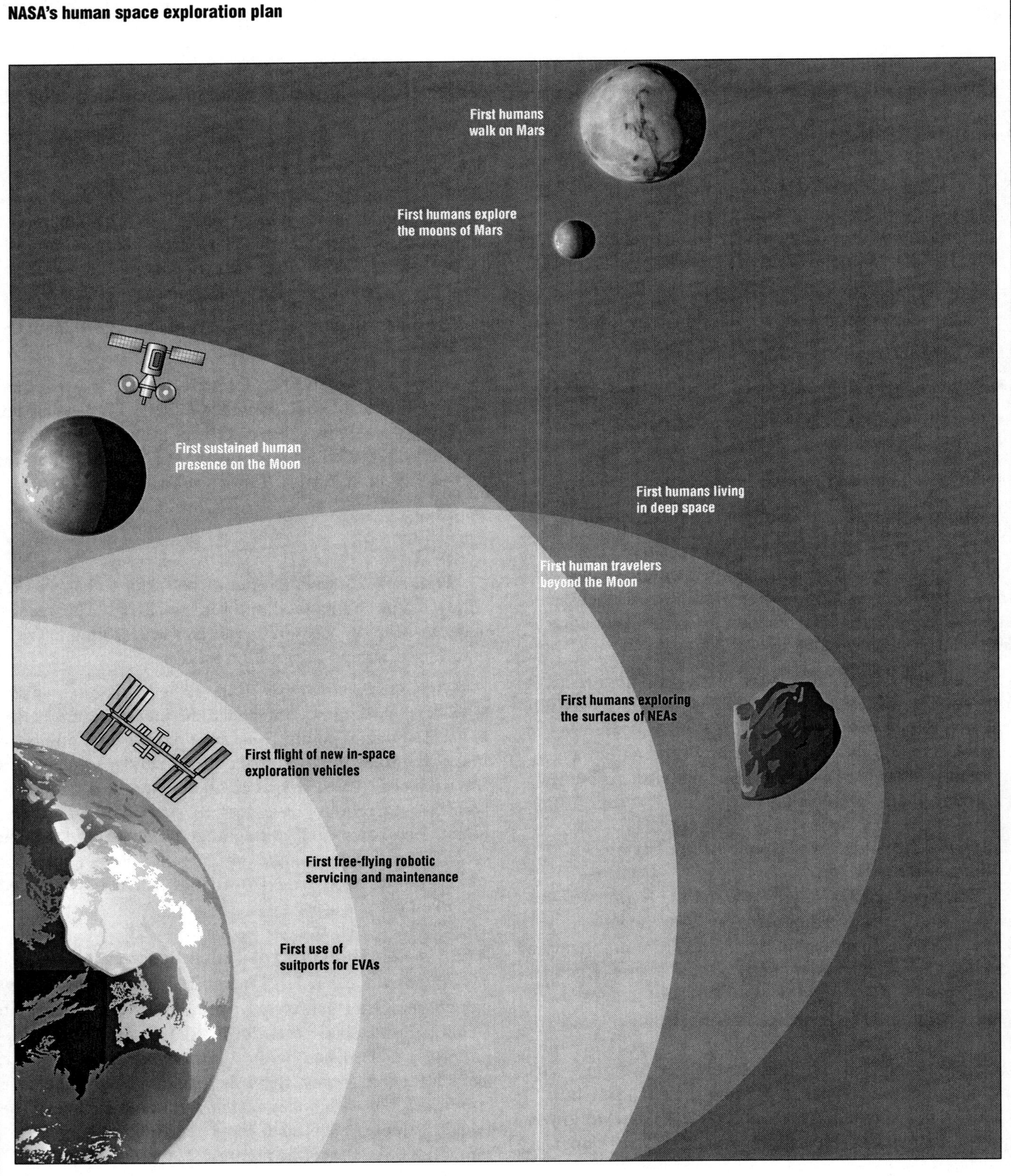

Notes: NEA = Near-earth asteroid. EVA = Extra-vehicular activity.

SOURCE: "Small Steps on the Path from LEO to Mars," in *Voyages: Charting the Course for Sustainable Human Space Exploration*, National Aeronautics and Space Administration, June 4, 2012, http://www.nasa.gov/sites/default/files/files/ExplorationReport_508_6-4-12.pdf (accessed November 23, 2015)

FIGURE 2.6

Lagrange points in Earth-moon space

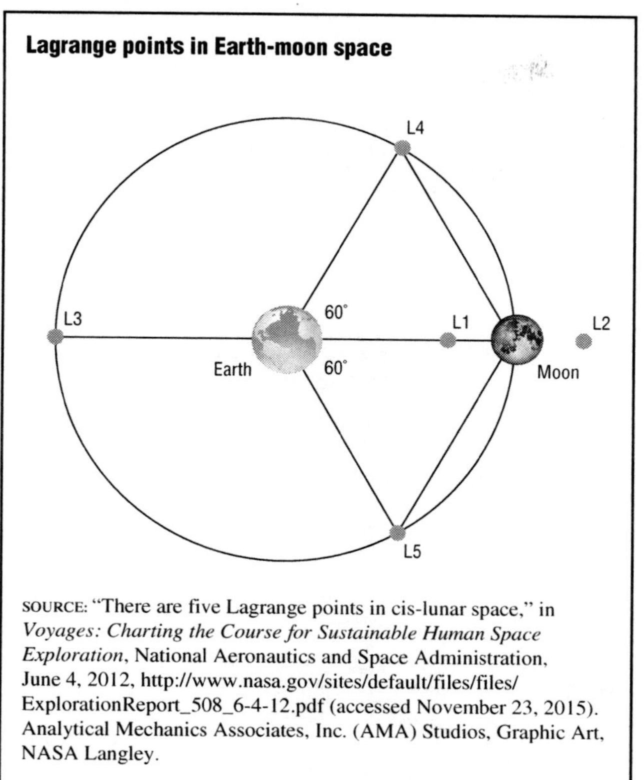

SOURCE: "There are five Lagrange points in cis-lunar space," in *Voyages: Charting the Course for Sustainable Human Space Exploration*, National Aeronautics and Space Administration, June 4, 2012, http://www.nasa.gov/sites/default/files/files/ExplorationReport_508_6-4-12.pdf (accessed November 23, 2015). Analytical Mechanics Associates, Inc. (AMA) Studios, Graphic Art, NASA Langley.

NASA provides some design information about the launch vehicles in *NASA Facts: Space Launch System* (2015, http://www.nasa.gov/sites/default/files/atoms/files/sls_october_2015_fact_sheet.pdf). Figure 2.8 shows the components of the 77-ton SLS, which NASA calls a Block 1 configuration. It features a stacked module configuration similar to that used in the Apollo spacecraft. Like the space shuttles, it will have a propulsion system based on liquid hydrogen and liquid oxygen and will include RS-25 engines and solid rocket boosters. Figure 2.9 shows three evolved configurations for the 70-ton SLS and one configuration for the 143-ton SLS, which NASA calls a Block 2 configuration. This latter launch system was still in the early stages of development in 2015.

Figure 2.10 provides a detailed view of the *Orion* spacecraft, which will have room for four crew members. Like the Apollo command module, the *Orion* will have a launch abort system to provide the crew a means of escape in case of a rocket malfunction. In addition, the command module will splash down in the ocean on its return to Earth, as did the Apollo command module.

The first test flight—Exploration Flight Test (EFT) 1—of the *Orion* took place in December 2014. It was an uncrewed flight. Because the SLS was not completed at that time, the spacecraft was launched by a Delta IV heavy rocket from the Cape Canaveral Air Force Station in Florida. As shown in Figure 2.11, the *Orion* made nearly two Earth orbits and traveled a peak distance of approximately 3,600 miles (5,800 km) from Earth's surface. By contrast, the *ISS* orbits approximately 220 to 250 miles (350 to 400 km) from Earth. The EFT 1 mission ended successfully when *Orion* splash landed in the Pacific Ocean.

In "NASA Completes Key Milestone for Orion Spacecraft in Support of Journey to Mars" (September 16, 2015, http://www.nasa.gov/press-release/nasa-completes-key-milestone-for-orion-spacecraft-in-support-of-journey-to-mars), NASA indicates that the next expected flight of *Orion* will be during the Exploration Mission 1. It is an uncrewed mission to be launched by the SLS. Significant upgrades are under way at the KSC to modernize its infrastructure, software, and ground support equipment for the launch. NASA notes that these upgrades are expected to be completed by the fall of 2018. This is also the target date for completion of the 77-ton SLS. The first crewed *Orion*/SLS flight is expected to occur no later than 2023.

NASA'S ROBOTIC SPACE PROGRAMS

NASA has sent dozens of robotic spacecraft into outer space. These machines have achieved some incredible milestones in space exploration. Satellites have been put into Earth orbit since the earliest days of NASA's space program to collect weather data or serve military purposes. During the 1960s and 1970s lunar probes were sent to the moon to support the Apollo Program. At the same time, NASA began launching robotic explorers that traveled to other planets. These were followed by sophisticated observatories and other robotic spacecraft that were placed in orbit around Earth or the sun or sent to intercept asteroids. All these missions are examined in detail in subsequent chapters.

NASA'S ORGANIZATION AND FACILITIES

Agency-level management takes place at NASA headquarters in Washington, D.C. People at this level interact with national leaders and NASA customers regarding overall agency concerns, such as budget, strategy, policies, and long-term investments. The headquarters are considered to be the centralized point of accountability and communication between NASA and people outside the agency.

As of February 2016 (http://www.nasa.gov/about/org_index.html#.UvOsQp5dWrY), NASA's organizational structure included four major divisions called mission directorates:

- Aeronautics Research
- Human Exploration and Operations
- Science
- Space Technology

FIGURE 2.7

Space Launch System facts

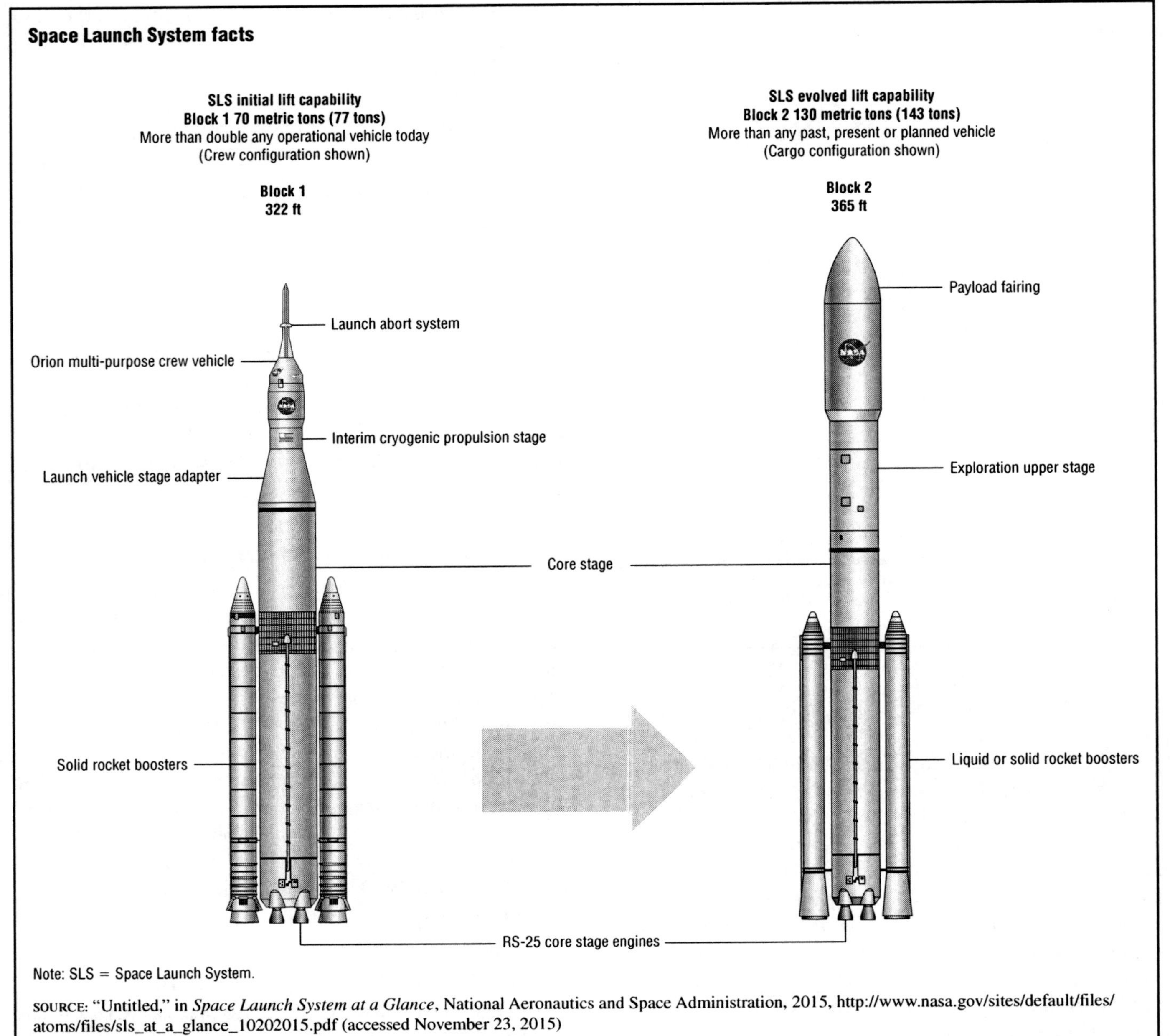

Note: SLS = Space Launch System.

SOURCE: "Untitled," in *Space Launch System at a Glance*, National Aeronautics and Space Administration, 2015, http://www.nasa.gov/sites/default/files/atoms/files/sls_at_a_glance_10202015.pdf (accessed November 23, 2015)

Major NASA Facilities

Some of the major NASA facilities are:

- Ames Research Center—located in Moffett Field, California, the center conducts research in science and technologies important to space exploration. It also performs wind tunnel testing and flight simulations.
- Neil A. Armstrong Flight Research Center (named for the first human to walk on the moon; previously called the Dryden Flight Research Center)—it is located at Edwards AFB in Edwards, California, and is NASA's primary installation for flight research.
- Glenn Research Center (GRC)—it is located in Cleveland, Ohio, at Lewis Field next to the Cleveland Hopkins International Airport. The GRC researches and develops technologies related to aeronautics and space applications. Its facilities include the nearby Plum Brook Station at which large-scale testing is conducted.
- Goddard Space Flight Center (GSFC)—it is located in Greenbelt, Maryland, a suburb of Washington, D.C., and is a major laboratory for developing robotic scientific spacecraft. The center also operates the Wallops Flight Facility near Chincoteague, Virginia, which is NASA's principal installation for managing and implementing suborbital research programs. GSFC also operates the Independent Verification

FIGURE 2.8

Initial configuration of the 70-ton Space Launch System

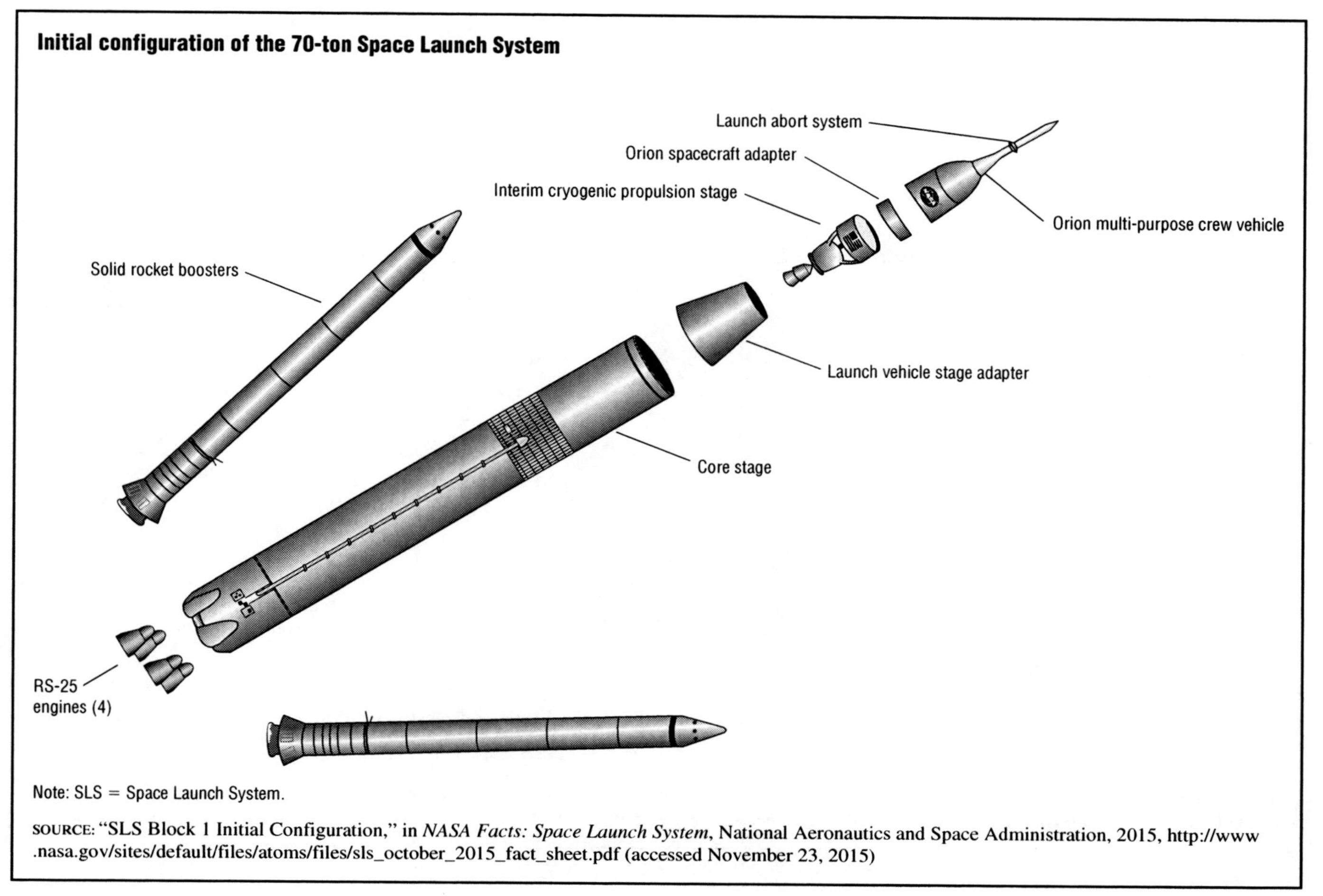

Note: SLS = Space Launch System.

SOURCE: "SLS Block 1 Initial Configuration," in *NASA Facts: Space Launch System*, National Aeronautics and Space Administration, 2015, http://www.nasa.gov/sites/default/files/atoms/files/sls_october_2015_fact_sheet.pdf (accessed November 23, 2015)

and Validation Facility in Fairmont, West Virginia, which ensures that mission-critical software is safe and cost effective. The GSFC houses the National Space Science Data Center, which is the archive center for data from NASA's space science missions. The data are made available to researchers and, in some cases, to the general public.

- Johnson Space Center (JSC)—located in Houston, Texas, it houses the program offices and mission control centers for the *ISS*. Until 2011 it performed similar services for the SSP. JSC facilities are also used for astronaut training and spaceflight simulations.
- Kennedy Space Center (KSC)—it is located on Merritt Island, Florida, next to the Cape Canaveral Air Force Station, which was the site of the Mercury and Gemini launches during the early 1960s. The KSC was created specifically for the Apollo missions to the moon. The center also provided facilities and services for the SSP. As of 2016, its infrastructure was being upgraded to support launches of the SLS and *Orion* spacecraft and to support other users, including private entities.
- Langley Research Center—located in Hampton, Virginia, it designs and develops military and civilian aircraft, conducts atmospheric flight research, and tests structures and materials in wind tunnels and other testing facilities.
- Marshall Space Flight Center—it is located near Huntsville, Alabama, at the Redstone Arsenal. The center conducts research in microgravity (an environment in which there is a perceived minimal gravitational force) and space optics.
- Stennis Space Center—located in Bay St. Louis, Mississippi, it is home to the largest rocket propulsion test complex in the United States. It is NASA's primary installation for testing and flight-certifying rocket propulsion systems. The center also works with government and commercial partners to develop remote sensing technology.

Other NASA Facilities

There are many facilities and installations that provide support to the field centers and are either operated by NASA or under contract to NASA. Administrative functions, such as payroll, human resources, procurement, and information technology coordination, are performed at the NASA Shared Services Center, which is

FIGURE 2.9

Evolved configurations of the Space Launch System

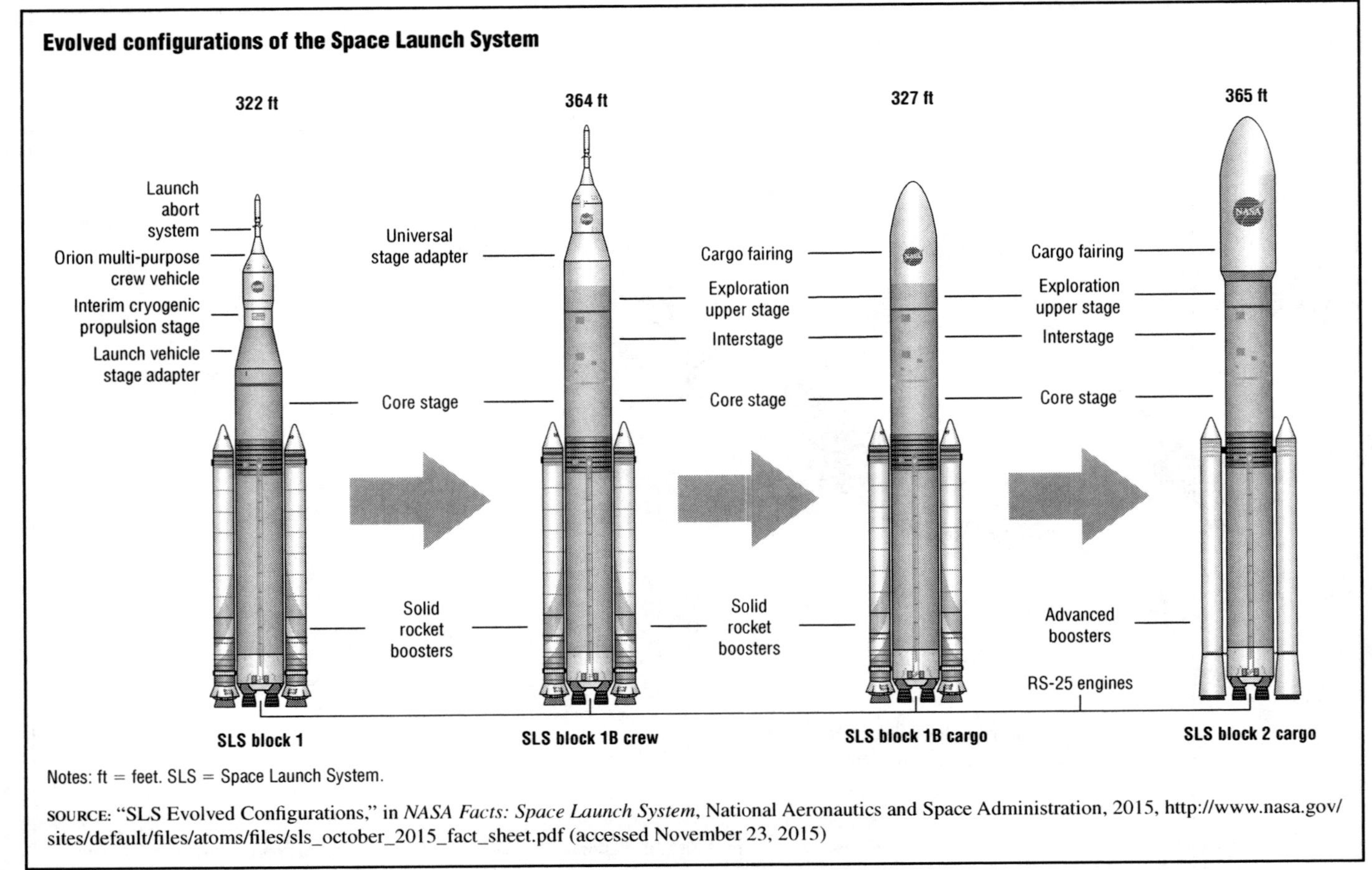

Notes: ft = feet. SLS = Space Launch System.

SOURCE: "SLS Evolved Configurations," in *NASA Facts: Space Launch System*, National Aeronautics and Space Administration, 2015, http://www.nasa.gov/sites/default/files/atoms/files/sls_october_2015_fact_sheet.pdf (accessed November 23, 2015)

located at the Stennis Space Center. NASA's Office of Inspector General (OIG) is actually a collection of offices at various field facilities. The OIG conducts audits and investigations that are designed to prevent fraud, crime, waste, and mismanagement and to promote efficient use of resources within the agency.

JET PROPULSION LABORATORY. The Jet Propulsion Laboratory (JPL) is located in Pasadena, California. This facility is owned by NASA but operated under a contractual agreement by the California Institute of Technology. It serves as NASA's primary operator of robotic exploration missions. It also manages and operates NASA's Deep Space Network, an international network of antennas that enables NASA mission teams to communicate with distant spacecraft. The Deep Space Network communications complexes are situated at three locations around the world (roughly 120 degrees apart): Goldstone, California; Robledo near Madrid, Spain; and Tidbinbilla near Canberra, Australia. This placement allows the JPL operations control center to maintain constant contact with spacecraft as Earth rotates.

WHITE SANDS TEST FACILITY. The White Sands Test Facility is located in Las Cruces, New Mexico, a remote desert location. The facility provides services to military and government clients. It is NASA's primary facility for testing and evaluating rocket propulsion systems, spacecraft components, and hazardous materials that are used in space travel. The facility also supports the *ISS* program.

NASA'S WORKFORCE

People employed by federal agencies (excluding the military) are called civil servants. As of February 2016, NASA (https://wicn.nssc.nasa.gov/wicn_cubes.html) employed 17,120 civil servants. The agency divides its civil service workforce into categories. The largest component (10,949 employees, or 64% of the total) of the workforce consists of scientists and engineers. These are highly educated professionals who conduct aerospace research and development or perform biological, life science, or medical research or services. This category includes space scientists, biologists, aerospace engineers, physicians, nurses, and psychologists. The second-largest component of NASA's workforce (5,028 employees, or 29% of the total) is devoted to professional administration. These employees work in nontechnical functions such as management, legal affairs, public relations, and human resources. The remaining employee categories are technicians, clerical workers, and wage earners.

FIGURE 2.10

Orion spacecraft

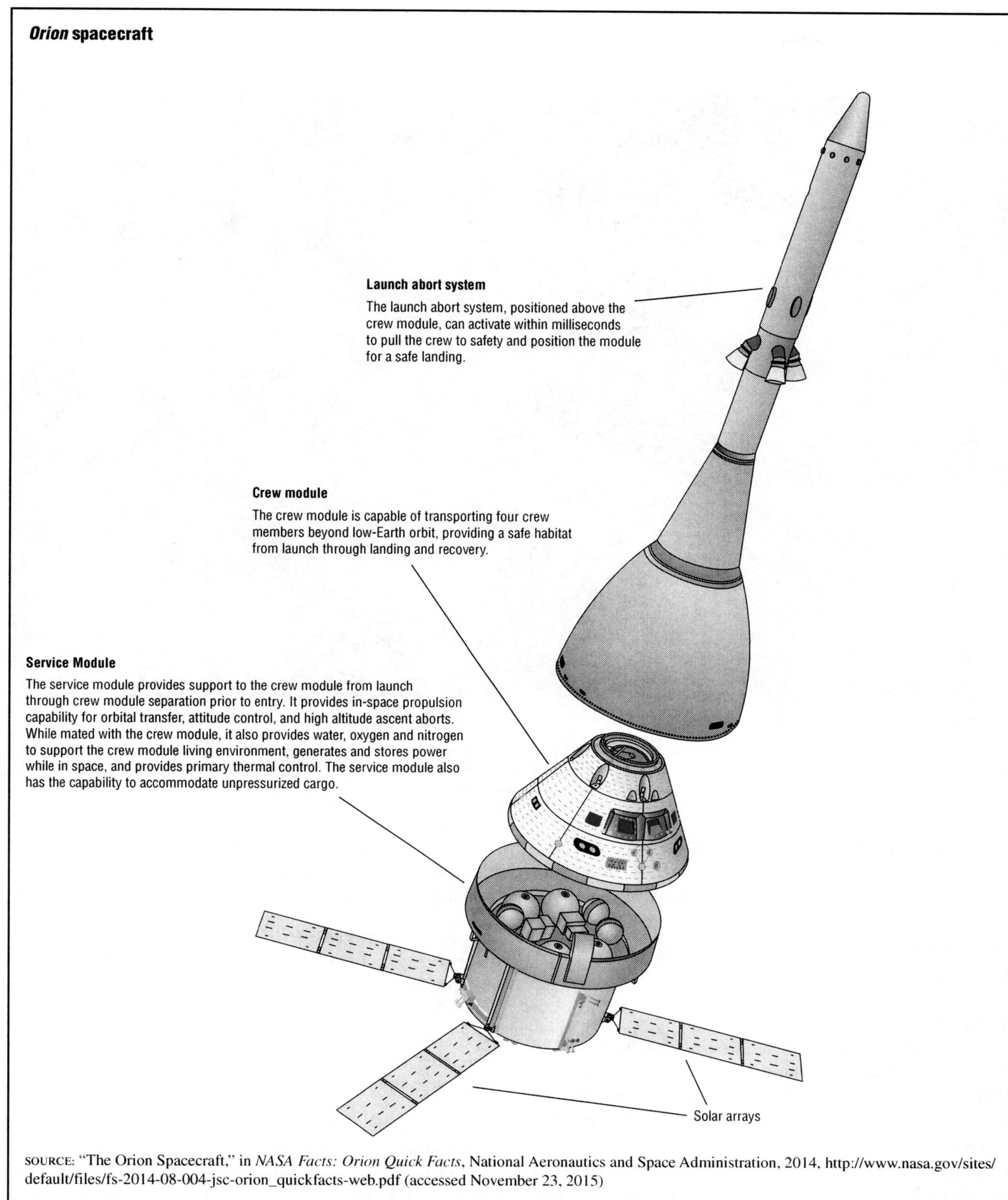

SOURCE: "The Orion Spacecraft," in *NASA Facts: Orion Quick Facts*, National Aeronautics and Space Administration, 2014, http://www.nasa.gov/sites/default/files/fs-2014-08-004-jsc-orion_quickfacts-web.pdf (accessed November 23, 2015)

The U.S. General Accounting Office (now the U.S. Government Accountability Office) explains in *NASA Personnel: Challenges to Achieving Workforce Reductions* (August 1996, http://www.gao.gov/archive/1996/ns96176.pdf) that in 1967, the height of Apollo development, the agency employed 35,900 civil servants. By the early 1990s this number had dropped to 25,000 and continued to decrease over the next several years. NASA

FIGURE 2.11

First flight path of *Orion*, December 2014

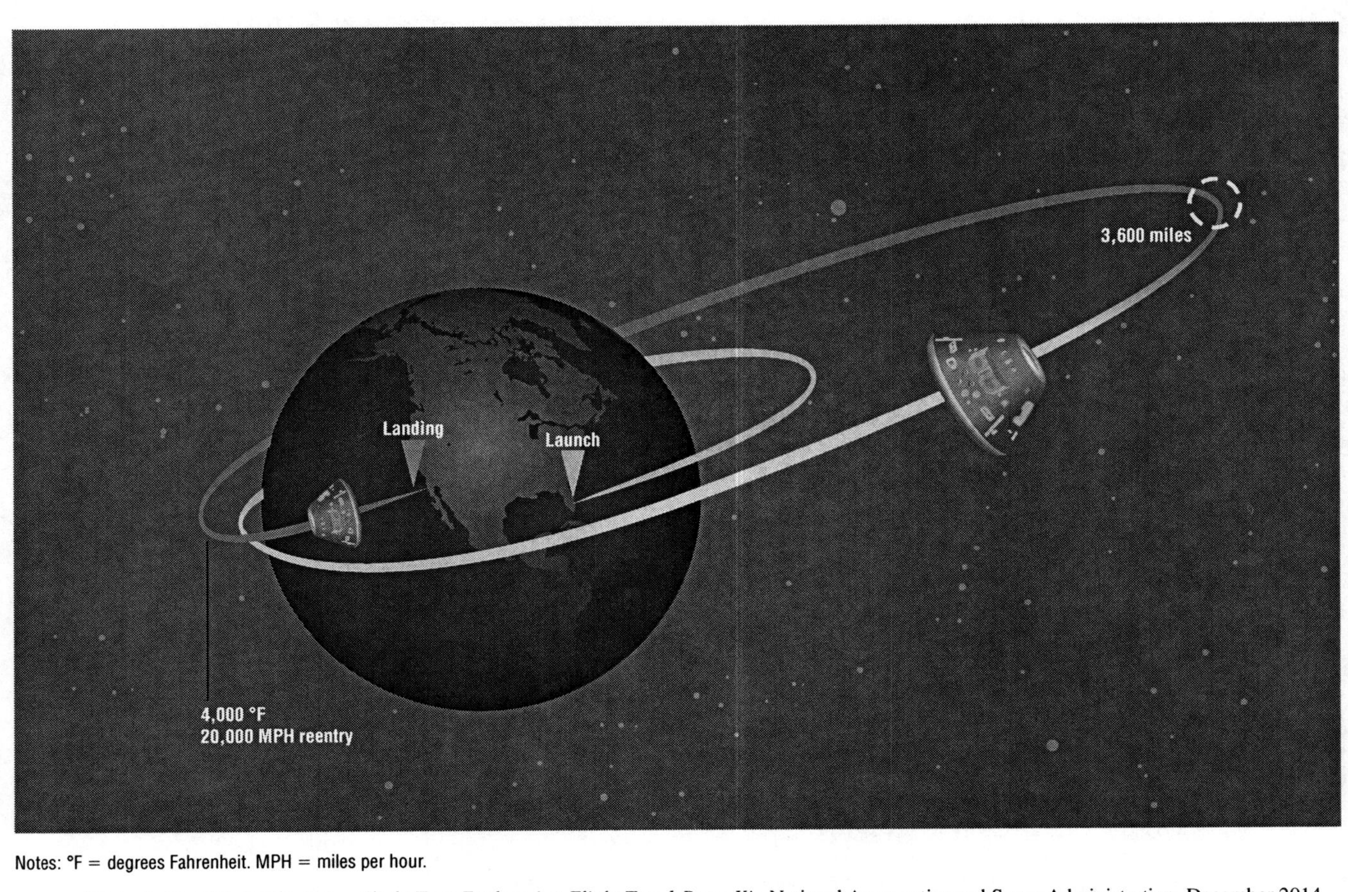

Notes: °F = degrees Fahrenheit. MPH = miles per hour.

SOURCE: "Orion's First Flight," in *Orion Flight Test: Exploration Flight Test-1 Press Kit*, National Aeronautics and Space Administration, December 2014, http://www.nasa.gov/sites/default/files/files/JSC_OrionEFT-1_PressKit_accessible.pdf (accessed November 23, 2015)

reduced its workforce by offering employees cash bonuses to retire early and through normal attrition (not replacing workers who leave). During most of the 1990s the agency operated under a hiring freeze. One consequence of this was that few young people entered the NASA workforce during this period.

Astronauts

Astronauts are the most famous NASA workers. In 1959 the first group of seven astronauts was chosen from 500 candidates. All were military men with experience flying jets. At the time, spacecraft restrictions required that astronauts be less than 5 feet 11 inches (180.3 cm) tall. During the early days of the Apollo Program, all astronauts were chosen from the military services. This soon changed, and NASA began including civilian pilots with extensive flight experience. During the mid-1960s NASA expanded the astronaut corps to include nonpilots with academic qualifications in science, engineering, or medicine.

In 1978 the first group of space shuttle astronauts was selected. For the first time the trainees included women and minorities. The unique environment aboard the space shuttle permitted even more opportunities for nonpilots to fly into space.

The 1980s witnessed several firsts in NASA's astronaut corps. In June 1983 Sally Ride (1951–2012) became the first U.S. woman in space when she served as a mission specialist aboard the space shuttle *Challenger*. It was the shuttle's seventh mission. Two months later the mission specialist Guion Bluford (1942–) became the first African American in space as part of the shuttle's next mission.

During the early to mid-1980s NASA was enthusiastic about including private citizens on space shuttle flights. This was viewed as a way to better interest the public, particularly children, in space travel. One of the most famous participants was Christa McAuliffe, the first schoolteacher selected to go into space. On January 28, 1986, she died along with her crewmates when the space shuttle *Challenger* exploded shortly after launch. This disaster ended NASA's policy of inviting private citizens on shuttle flights.

TABLE 2.7

Summary of NASA appropriations, requests, and estimates for fiscal years 2014–20

[Budget authority ($ in millions)]

	Fiscal year						
	Actual	Enacted	Request	Notional			
	2014	2015	2016	2017	2018	2019	2020
NASA total	**17,646.5**	**18,010.2**	**18,529.1**	**18,807.0**	**19,089.2**	**19,375.5**	**19,666.1**
Science	**5,148.2**	**5,244.7**	**5,288.6**	**5,367.9**	**5,488.4**	**5,530.2**	**5,613.1**
Earth science	1,824.9	—	1,947.3	1,966.7	1,988.0	2,009.3	2,027.4
Planetary science	1,345.7	—	1,361.2	1,420.2	1,458.1	1,502.4	1,527.8
Astrophysics	678.3	—	709.1	726.5	769.5	1,005.5	1,138.3
James Webb Space Telescope	658.2	645.4	620.0	569.4	534.9	305.0	197.5
Heliophysics	641.0	—	651.0	685.2	697.9	708.1	722.1
Aeronautics	**566.0**	**651.0**	**571.4**	**580.0**	**588.7**	**597.5**	**606.4**
Space technology	**576.0**	**596.0**	**724.8**	**735.7**	**746.7**	**757.9**	**769.3**
Exploration	**4,113.2**	**4,356.7**	**4,505.9**	**4,482.2**	**4,298.7**	**4,264.7**	**4,205.4**
Exploration systems development	3,115.2	3,245.3	2,862.9	2,895.7	2,971.7	3,096.2	3,127.1
Commercial spaceflight	696.0	805.0	1,243.8	1,184.8	731.9	173.1	1.1
Exploration research and development	302.0	306.4	399.2	401.7	595.1	995.4	1,077.2
Space operations	**3,774.0**	**3,827.8**	**4,003.7**	**4,191.2**	**4,504.9**	**4,670.8**	**4,864.3**
International Space Station	2,964.1	—	3,105.6	3,273.9	3,641.0	3,826.0	4,038.3
Space and flight support	809.9	—	898.1	917.3	863.8	844.8	826.1
Education	**116.6**	**119.0**	**88.9**	**90.2**	**91.6**	**93.0**	**94.4**
Safety, security, and mission services	**2,793.0**	**2,758.9**	**2,843.1**	**2,885.7**	**2,929.1**	**2,973.0**	**3,017.5**
Center management and operations	2,041.5	—	2,075.2	2,105.0	2,136.6	2,168.6	2,201.0
Agency management and operations	751.5	—	767.9	780.7	792.5	804.4	816.5
Construction and environmental compliance and restoration	**522.0**	**419.1**	**465.3**	**436.1**	**442.6**	**449.3**	456.0
Construction of facilities	455.9	—	374.8	344.3	349.3	354.6	359.9
Environmental compliance and restoration	66.1	—	90.5	91.8	93.3	94.7	96.1
Inspector general	**37.5**	**37.0**	**37.4**	**38.0**	**38.5**	**39.1**	**39.7**
NASA total	**17,646.5**	**18,010.2**	**18,529.1**	**18,807.0**	**19,089.2**	**19,375.5**	**19,666.1**

NASA = National Aeronautics and Space Administration.
Notes: Fiscal year 2014 reflects funding amounts specified in the June 2014 Operating Plan per P.L. 113-76.
Fiscal year 2015 reflects only funding amounts specified in P.L. 113-235, the Consolidated and Further Continuing Appropriations Act, 2015. For projects in development, NASA's tentatively planned fiscal year 2015 funding level is shown. Fiscal year 2015 funding levels are subject to change pending finalization of the fiscal year 2015 operating plan.
The totals for the Exploration and Space Operations accounts in this document supersede the figures in the draft appropriations language.

SOURCE: "FY 2016 President's Budget Request Summary," in *FY 2016 President's Budget Request Summary*, National Aeronautics and Space Administration, February 2015, http://www.nasa.gov/sites/default/files/files/NASA_FY_2016_Budget_Estimates.pdf (accessed November 23, 2015)

NASA'S BUDGET

NASA is a federal government agency. For accounting purposes, the federal government operates on a fiscal year that begins in October and runs through the end of September. Thus, FY 2016 covers the period October 1, 2015, through September 30, 2016. Each year by the first Monday in February the president must present a proposed budget to the U.S. House of Representatives. This is the amount of money that the president estimates will be required to operate the federal government during the next fiscal year. The proposed budget also includes estimates of future FY budget requests. These estimates are based on the president's long-range plans and policies.

It can take many months for the House to debate, negotiate, and approve a final budget. Then, the U.S. Senate must also approve the budget. This entire process can take longer than a year, which means that NASA can be well into a fiscal year (or even beyond it) before knowing the exact amount of money appropriated for that year. This spending uncertainty has become more common during the early 21st century because of fierce disagreements in Congress about government spending.

Table 2.7 shows that NASA was appropriated $17.6 billion in FY 2014 and $18 billion in FY 2015. The agency's requested budget for FY 2016 was $18.5 billion. Overall, the five line items in the FY 2016 budget request with the largest spending amounts were:

- *International Space Station*—$3.1 billion
- Exploration systems development—$2.9 billion
- Center management and operations—$2.1 billion
- Earth science—$1.9 billion
- Planetary science—$1.4 billion

The exploration systems development category covers development of the *Orion* spacecraft and the SLS.

CHAPTER 3
SPACE ORGANIZATIONS PART 2: U.S. MILITARY AND FOREIGN

Outer space, including the moon and other celestial bodies, shall be free for exploration and use by all States.

—United Nations, Treaty on Principles Governing the Activities of States in the Exploration and Use of Outer Space, Including the Moon and Other Celestial Bodies (October 10, 1967)

The National Aeronautics and Space Administration (NASA) is perhaps the best-known space organization in the world. However, the U.S. military and many foreign governments also have active space programs. Most modern military space ventures center around ballistic missiles and data-gathering satellites. The United States officially holds the policy that it will not develop space weapons, only defensive systems. Some critics complain that the line between the two is becoming increasingly vague.

The Russian space program continues the program begun by the Soviet Union decades ago. During the latter half of the 20th century the Soviet Union engaged in a bitter cold war rivalry for space supremacy with the United States. The Soviet Union achieved many milestones in space ahead of the United States, including the first manned space flight in April 1961.

In 1991 the Soviet Union splintered into individual republics (including the Russian Federation, or Russia) that were friendlier with the United States. Civilian space agencies in the United States and Russia struggled to carry on ambitious space programs as their funding was cut. To compensate, they began working together on many space ventures. Eventually, space programs were developed by other countries around the world. This presented opportunities for new alliances in space.

U.S. MILITARY SPACE PROGRAMS

The U.S. military had space aspirations long before spaceflight was possible. The three main branches of the military (the army, air force, and navy) began their space programs around the time of World War II (1939–1945). Even before the war ended the U.S. military began collaborating with the National Advisory Committee for Aeronautics (a predecessor agency to NASA) to develop rocket-powered planes—part of the famous X-series. Following the war a group of German rocket scientists led by Wernher von Braun (1912–1977) surrendered to U.S. forces and moved to the United States. On February 24, 1949, the von Braun team launched *BUMPER WAC Corporal Round 5*, the first U.S. rocket to travel beyond Earth's atmosphere and penetrate outer space. During the 1940s and 1950s the U.S. Air Force (USAF) launched rockets into the upper atmosphere that carried fruit flies, fungus spores, and mammals, including mice and monkeys.

On January 31, 1958, the U.S. Army successfully launched into space *Explorer 1*, the first U.S. satellite. It was 80 inches (203 cm) long and 6.2 inches (15.7 cm) wide and weighed 31 pounds (14 kg) on Earth. The scientific payload included temperature gauges and instruments to detect cosmic rays and the impacts of micrometeorites. The payload was developed under the direction of James Van Allen (1914–2006), a physics professor at the University of Iowa. Data from *Explorer 1* and the later *Explorer 3* satellite led to Van Allen's discovery of radiation belts around Earth. The existence of the belts was confirmed in 1958 by the Soviet satellite *Sputnik 3*. (See Figure 3.1.)

On March 17, 1958, the U.S. Navy launched the *Vanguard 1* satellite, which was the size of a grapefruit and weighed 3 pounds (1.4 kg) on Earth. It was the first orbiting satellite to be powered by solar energy. Solar cells also powered its radio until the radio failed in 1964. As of February 2016 *Vanguard 1* continued to orbit Earth. It has remained in orbit longer than any other human-made object in space.

In 1958 President Dwight D. Eisenhower (1890–1969) limited the military's role in space when he created

FIGURE 3.1

Sputnik 3

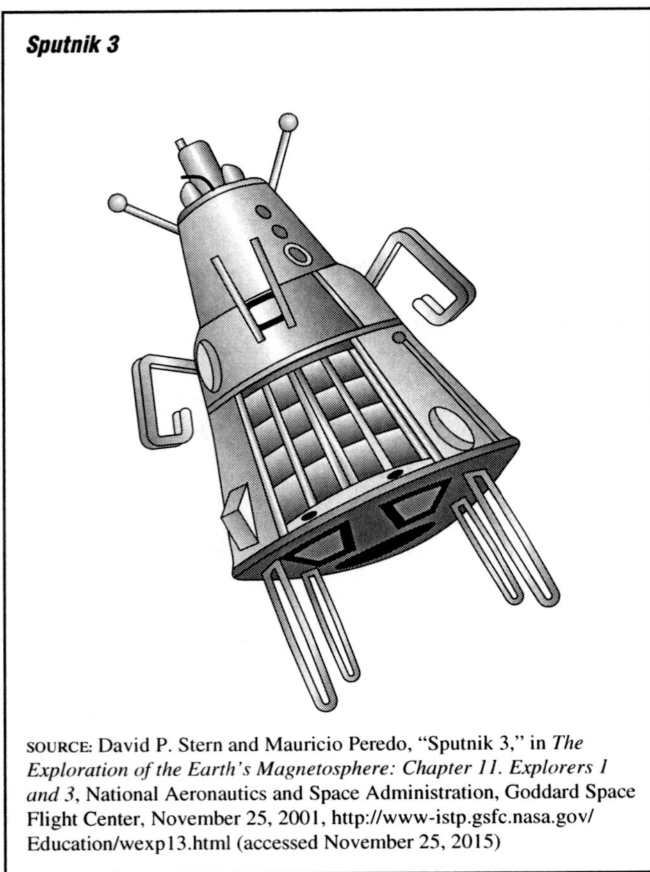

SOURCE: David P. Stern and Mauricio Peredo, "Sputnik 3," in *The Exploration of the Earth's Magnetosphere: Chapter 11. Explorers 1 and 3*, National Aeronautics and Space Administration, Goddard Space Flight Center, November 25, 2001, http://www-istp.gsfc.nasa.gov/Education/wexp13.html (accessed November 25, 2015)

NASA as a civilian agency. NASA was given responsibility for the nation's manned space program. The military was allowed to pursue space projects that benefited national defense. During the 1960s the USAF pursued its own version of a space station called the Manned Orbiting Laboratory (MOL). According to NASA, in "Proposed USAF Manned Orbiting Laboratory" (May 13, 2010, http://grin.hq.nasa.gov/ABSTRACTS/GPN-2003-00094.html), the U.S. government spent $1.3 billion researching and developing the MOL. The project suffered constant budget overruns and schedule delays and was finally canceled in 1969. By that time, unmanned reconnaissance satellites were available that could do much of what the MOL was supposed to accomplish. Military astronauts who had been training in the MOL program were transferred to NASA.

Since the formation of NASA, the U.S. military has focused most of its space resources on the development of ballistic missiles and satellites. Satellites are designed for a variety of purposes, including communications, navigation, weather surveillance, and reconnaissance.

During the late 1950s the USAF worked with the Central Intelligence Agency to develop a reconnaissance satellite that was capable of photographing Soviet installations on the ground. The project was code-named Corona. Publicly, the United States called the satellite *Discoverer* and claimed that it conducted scientific research. More than 100 Corona missions were flown during the 1960s and early 1970s. The Soviet Union orbited its own spy satellites and also claimed that they were for scientific purposes.

Launch Vehicles

Before the 1980s all national security satellites were launched aboard rockets called expendable launch vehicles (ELVs). Once above Earth's atmosphere, a satellite separated from its ELV, and the ELV burned up during reentry. During the 1970s the USAF used a number of ELVs including the Atlas, Delta, Scout, Thor, and Titan rockets.

The space shuttle introduced a new era in satellite deployment. The shuttle was reusable and included a crew of astronauts that could release, retrieve, and repair satellites as needed. As explained in Chapter 2, the U.S. military was heavily involved in the early design of the shuttle, which it hoped to use extensively to carry military payloads into space. However, only about a dozen military satellites were launched aboard shuttles because the shuttle program was plagued by problems. In 1984 USAF officials persuaded Congress to fund the development of a fleet of new ELVs for military missions. NASA protested strongly against this action, but was overruled.

During the 1990s the U.S. Department of Defense (DOD) invested heavily in a new generation of rockets called evolved ELVs (EELVs). According to the U.S. Government Accountability Office (GAO), in *Evolved Expendable Launch Vehicle: DOD Is Addressing Knowledge Gaps in Its New Acquisition Strategy* (July 2012, http://www.gao.gov/assets/600/593048.pdf), the EELV inventory includes two families of launch vehicles: the Delta IV family from the Boeing Company and the Atlas V family from the Lockheed Martin Corporation. Each family includes several types of rockets that are capable of carrying payloads of various weights into space. For example, Figure 3.2 shows a Delta IV medium rocket, which carries lighter loads than the powerhouse Delta IV heavy rocket shown in Figure 3.3. Originally, the DOD projected that the commercial satellite industry would help offset the costs of the EELV program, but demand from the commercial sector was weaker than anticipated.

FIGHTING FOR LAUNCH CONTRACTS. In 2006 Boeing and Lockheed Martin formed the United Launch Alliance (ULA), a joint venture that became the sole provider of EELV launches. According to the GAO, in *Evolved Expendable Launch Vehicle*, the Federal Trade Commission (FTC) was opposed to the venture because it allowed the two companies to monopolize the EELV industry. However, the GAO notes that the FTC relented when the "DOD stated the benefits of the joint venture to national security outweighed the loss of competition."

FIGURE 3.2

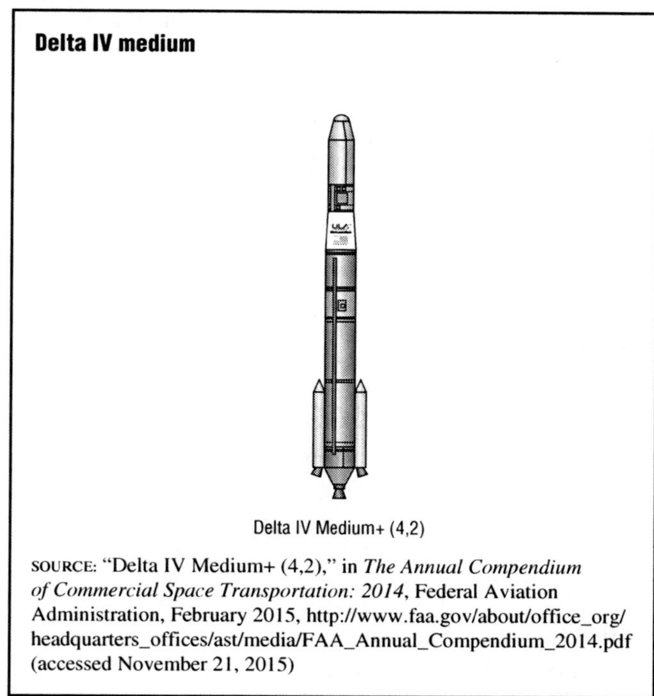

Delta IV medium

Delta IV Medium+ (4,2)

SOURCE: "Delta IV Medium+ (4,2)," in *The Annual Compendium of Commercial Space Transportation: 2014*, Federal Aviation Administration, February 2015, http://www.faa.gov/about/office_org/headquarters_offices/ast/media/FAA_Annual_Compendium_2014.pdf (accessed November 21, 2015)

FIGURE 3.3

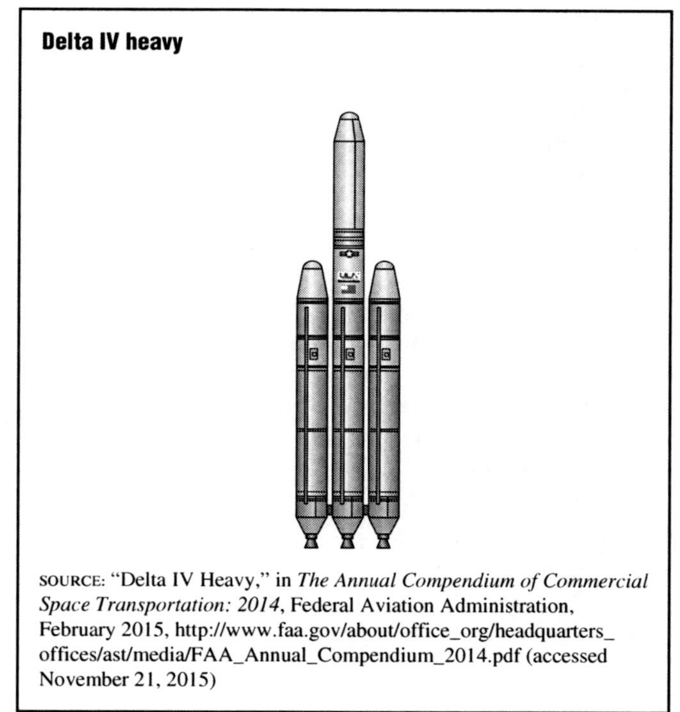

Delta IV heavy

SOURCE: "Delta IV Heavy," in *The Annual Compendium of Commercial Space Transportation: 2014*, Federal Aviation Administration, February 2015, http://www.faa.gov/about/office_org/headquarters_offices/ast/media/FAA_Annual_Compendium_2014.pdf (accessed November 21, 2015)

Even so, this monopoly came under fire as EELV launch prices increased and other private U.S. companies began developing launch capabilities.

One issue of contention has been the annual capability payments that the DOD pays the ULA. In *Space Launch Vehicle Competition: Briefing to the Senate Homeland Security and Governmental Affairs Committee Permanent Subcommittee on Investigations* (January 28, 2014, http://www.appropriations.senate.gov/imo/media/doc/hearings/FINAL_SPACE_LAUNCH_BRIEFING.PDF), the GAO explains that the payments guarantee that the ULA will keep a "standing army" of personnel (mostly engineers) who ensure that the DOD can achieve "access to space in a timely manner." The payments also cover overhead (ongoing business expenses not directly billable to clients) associated with the ULA's launchpads. Critics have decried the annual payments as a government subsidy that helps the ULA maintain its monopoly on EELV launches.

One of the most vocal critics has been Elon Musk (1971–), a cofounder of the PayPal online payment service. In 2002 he founded the Space Exploration Technologies Corp. (SpaceX). SpaceX has provided cargo transport to the *International Space Station* (*ISS*) for NASA. Chapter 4 describes these missions along with the company's commercial satellite launches. SpaceX complains in "EELV: The Right to Compete" (April 29, 2014, http://www.spacex.com/press/2014/04/29/eelv-right-compete) that "the United States pays ULA nearly $1 billion... per year just to maintain the ability to launch—regardless of whether or not they launch a single rocket." The GAO notes that the ULA is also paid for each national security launch that it conducts. From an accounting standpoint, this makes it difficult to ascertain the actual cost of an individual ULA launch. It also greatly complicates the task of comparing the affordability of the ULA's launch services to those of other companies.

The DOD indicates in *Selected Acquisition Report: Evolved Expendable Launch Vehicle (EELV) as of December 31, 2012* (May 21, 2013, http://www.globalsecurity.org/space/library/budget/fy2012/sar/eelv_december_2012_sar.pdf) that in late 2012 the USAF introduced a competitive procurement process for military launches. Beginning in 2014 commercial enterprises other than the ULA were allowed to compete for launches. Only SpaceX was a viable contender at the time. Its Falcon 9 launch vehicle was certified for military launches in May 2015.

THE RUSSIAN COMPLICATION. In 2014 the Russian military made excursions into neighboring Ukraine. The alarmed U.S. government instituted sanctions that were designed to hurt sectors of the Russian economy. In December 2014 Congress passed a military appropriations bill. Section 1608 of the law prohibits the DOD (after 2014) from awarding or renewing contracts to purchase Russian rocket engines under the EELV program. This prohibition is problematic because the ULA has long used Russian-made RD-180 engines on its Delta IV launch vehicles. However, the law includes a waiver provision under which the DOD can allow the ULA to continue purchasing the Russian engines if doing so is deemed in the best interests of U.S. national security.

According to Andrea Shalal, in "SpaceX Raps ULA Bid to Get U.S. Waiver for Russian Engines" (Reuters .com, October 8, 2015), during the fall of 2015 the ULA asked the DOD for a waiver. Shalal states that the company claimed it could not place a bid on an upcoming USAF launch contract unless it received the waiver. This elicited harsh criticism from SpaceX, which accused the ULA of being "deceptive." The DOD refused to grant the ULA the waiver. In November 2015 SpaceX was the only bidder on a contract to launch in 2018 a Global Positioning System (GPS) satellite for the USAF. As of February 2016, the contract had not been awarded; however, SpaceX was expected to win it by default.

In December 2015 Congress passed a massive spending bill to fund the federal government through the end of fiscal year (FY) 2016 (through September 2016). In "Lockheed-Boeing Venture Orders 20 More Russian Rocket Engines" (Reuters.com, December 23, 2015), Shalal notes that the bill "eased a ban on using Russian engines to launch U.S. military and intelligence satellites for fiscal 2016." The ULA promptly placed an order for 20 RD-180 rocket engines from Russia. However, Shalal indicates that the company "was moving forward" with plans to develop U.S.-made engines by the companies Blue Origin and Aerojet Rocketdyne Holdings, Inc. Blue Origin is headed by Jeff Bezos (1964–), the founder of the online shopping site Amazon.com. Since its founding in 2000, Blue Origin has been developing spacecraft technologies for NASA and the space tourism industry. Aerojet Rocketdyne dates back to 1915 and has long been a major aerospace contractor for NASA and the U.S. military.

WHAT LIES AHEAD? SpaceX's entry into the EELV business is projected to end the ULA monopoly on military launch services. The ULA's advantage in competitive bidding is its strong history of reliability and performance. However, it is widely expected that SpaceX can provide much cheaper EELV launches than the ULA. According to Shalal, in "SpaceX Raps ULA Bid to Get U.S. Waiver for Russian Engines," the "ULA has said it plans to discontinue production of all except the heaviest Delta 4 rockets because they are not and cannot ever be competitive with SpaceX's lower prices." As explained in Chapter 4, SpaceX has achieved substantial cost savings in its rocket development program compared with other companies. SpaceX's main liability is its relative newness and lack of an established track record. Falcon 9 rockets have undergone few flights compared with the ULA's long-lived Delta and Atlas lines. SpaceX has sold its launch services to commercial clients and to NASA. As of February 2016, there had been six successful launches of Falcon 9 rockets for cargo resupply missions to the *ISS*. A seventh mission ended in failure in June 2015, when a Falcon 9 exploded minutes after takeoff.

For its part, the ULA has continued development of a next-generation EELV called the Vulcan launch vehicle, which will have U.S.-made engines. The GAO explains in *Evolved Expendable Launch Vehicle: The Air Force Needs to Adopt an Incremental Approach to Future Acquisition Planning to Enable Incorporation of Lessons Learned* (August 2015, http://www.gao.gov/assets/680/671926.pdf) that the ULA does not expect to have the Vulcan line certified to military launch specifications prior to 2020.

The X37-B Space Plane

In 2010 the USAF conducted the first spaceflight of its unmanned X-37B space plane. Shaped like a miniature space shuttle, the reusable plane is launched vertically by a rocket, but lands horizontally on a runway. It is capable of orbiting Earth for months at a time. The X-37B is lofted into orbit by an Atlas V rocket, and its missions are classified.

In "US Air Force Launches X-37B Space Plane on 4th Mystery Mission" (Space.com, May 20, 2015), Mike Wall notes that as of May 2015 the USAF had built two of the space planes. Each plane is about 29 feet (8.8 m) long and has a payload bay "the size of a pickup-truck bed." The first X-37B mission in 2010 lasted for more than seven months. Subsequent missions were launched in March 2011, December 2012, and May 2015. As of May 2015, the plane's longest spaceflight, which began in December 2012, lasted 675 days. Wall indicates that there are suspicions that the X-37B missions may involve development and testing of space weapons. However, he states, "Air Force officials have long refuted that notion, saying the X-37B is simply testing out technologies for reusable vehicles and future spacecraft." It is known that the space plane that launched in May 2015 carried small payloads for NASA and the Planetary Society, a private space organization based in the United States. The NASA payload consisted of dozens of pieces of various materials undergoing testing for the effects of prolonged exposure to the space environment. The Planetary Society payload was a *LightSail* spacecraft, which is described in Chapter 4.

Star Wars

On March 23, 1983, President Ronald Reagan (1911–2004) announced a new military space venture called the Space Defense Initiative (SDI). Basically, the plan was to place a satellite shield in space to protect the United States from incoming Soviet nuclear missiles. The media nicknamed the SDI proposal the "Star Wars" program. (*Star Wars* was a hit 1977 movie that featured elaborate space weapons.)

In 1984 the army successfully tested an interceptor missile that was launched from the Kwajalein Missile

Range in the Marshall Islands. It flew above the atmosphere and then located and tracked a reentry missile that had been launched from the Vandenberg Air Force Base (AFB) in California. The interceptor missile homed in on the target using onboard sensors and computer targeting. It crashed into the target and destroyed it.

Over the decades the SDI program has evolved and grown under the administrations of various presidents. As of 2016, the program was overseen by the Missile Defense Agency (MDA) under the DOD. With the dissolution of the Soviet Union into numerous independent republics during the early 1990s, a more likely threat is considered to be missiles launched by unfriendly dictators and terrorists.

The MDA's (October 16, 2015, http://www.mda.mil/about/mission.html) mission is "to develop, test, and field an integrated, layered, ballistic missile defense system (BMDS) to defend the United States, its deployed forces, allies, and friends against all ranges of enemy ballistic missiles in all phases of flight." The MDA focuses on developing systems that can intercept and destroy ballistic missiles along their flight path, but preferably outside of Earth's atmosphere so that nuclear or biological warheads are destroyed during reentry.

The BMDS includes ground- and sea-based interceptor missiles and space-based tracking systems. Things that were once considered science fiction are slowly becoming viable components in the DOD arsenal. This is because of technological advances and a large influx of money to the program. The MDA (2015, http://www.mda.mil/global/documents/pdf/histfunds.pdf) reports that between FYs 1985 and 2015 the United States invested a total of $173.4 billion in missile defense.

Since the 1970s the United States has relied on a space-based early missile warning system called the Defense Support Program (DSP). According to the USAF, in the fact sheet "Defense Support Program" (April 2015, http://www.losangeles.af.mil/library/factsheets/factsheet_print.asp?fsID=5323&page=1), the DSP consists of a series of satellites in geosynchronous orbit 22,230 miles (35,780 km) above Earth's surface. Details about the DSP inventory and capabilities are classified.

In 1996 the DOD began development of a replacement for the DSP called the Space Based Infrared System. The GAO describes the history and status of this system in *Space Based Infrared System High Program and Its Alternative* (September 12, 2007, http://www.gao.gov/new.items/d071088r.pdf). According to the GAO, the new satellites are more sensitive and accurate than existing DSP satellites. The first new satellites were supposed to be deployed in 2004, but this schedule proved unfeasible. The ULA (2016, http://www.ulalaunch.com/site/pages/News.shtml#/70/) notes that the first Space Based Infrared System launch took place in May 2011.

Space Weapons?

Historically, the United States has focused on developing defensive, rather than offensive, space-based assets. It is a party to the 1967 Treaty on Principles Governing the Activities of States in the Exploration and Use of Outer Space, Including the Moon and Other Celestial Bodies (2002, http://www.unoosa.org/pdf/publications/STSPACE11E.pdf), which states that nations may not "place in orbit around the Earth any objects carrying nuclear weapons or any other kinds of weapons of mass destruction, install such weapons on celestial bodies, or station such weapons in outer space in any other manner." In addition, since the 1950s presidential space policy has focused on unarmed satellites to prevent a new arms race in space.

Some analysts, however, note that DOD defensive systems could be used offensively as antisatellite (ASAT) weapons. In *A History of Anti-satellite Programs* (January 2012, http://www.ucsusa.org/assets/documents/nwgs/a-history-of-ASAT-programs_lo-res.pdf), Laura Grego of the Union of Concerned Scientists reviews the development of ASAT systems around the world. Grego notes, "Because missile defense systems are intended to destroy ballistic missile warheads, which travel at speeds and altitudes comparable to those of satellites, such systems also have ASAT capabilities." Thus, it is conceivable that the BMDS could be used to destroy the satellites of other nations. Maneuverable satellites (as opposed to satellites in orbit) are also potential space weapons. In 2005 the USAF launched *Experimental Spacecraft System 11* into space. The satellite was about the size of a washing machine and weighed around 200 pounds (91 kg) on Earth. Various news reports indicate that its tests were highly successful. According to Grego, the United States and other countries continue to develop maneuverable satellites.

U.S. Strategic Command

In 1985 the Reagan administration established the U.S. Space Command to oversee military space operations. Its commander was also in charge of the North American Aerospace Defense Command, which protects the U.S. and Canadian air space. In 1992 President George H. W. Bush (1924–) established the U.S. Strategic Command (StratCom) to oversee the nation's nuclear arsenal.

Following the September 11, 2001, terrorist attacks against the United States, President George W. Bush (1946–) abolished the U.S. Space Command and assigned its responsibilities to StratCom, which is headquartered at the Offutt AFB in Nebraska. As the command and control center for U.S. strategic forces, it controls military space operations and is responsible for early warning and defense against missile attacks.

AIR FORCE SPACE COMMAND. Much of StratCom's space operations are carried out by the Air Force Space Command (AFSPC), which is headquartered at the Peterson AFB in Colorado. The AFSPC operates the GPS, which is described in Chapter 1, and launches and operates satellites that provide weather, communications, intelligence, navigation, and missile warning capabilities. The command also provides services, facilities, and aerospace control for NASA operations. In 2004 the AFSPC established the National Security Space Institute to provide education and training in space-based topics.

JOINT SPACE OPERATIONS CENTER. Another StratCom office is the Joint Space Operations Center (JSOC) at Vandenberg AFB. One of the main missions of the JSOC is to operate the U.S. Space Surveillance Network, which tracks and catalogs objects in Earth orbit. The objects include operative and inoperative satellites, fragments of rockets and other spacecraft, and various miscellaneous objects. The "dead" objects (i.e., those that are no longer maneuverable by their controllers on Earth) pose a collision hazard to operational spacecraft, including satellites, exploratory robotic spacecraft, and the *ISS*.

As shown in Table 3.1, the U.S. Space Surveillance Network had cataloged 17,063 objects in Earth orbit as of September 30, 2015. The objects originated from various countries and space-faring organizations. In 2007 the quantity of objects in orbit greatly increased after the Chinese government purposely rammed a missile into one of its old weather satellites. The action, which was purportedly done to show off the nation's military space capabilities, was highly criticized for creating thousands of fragments. In 2009 another increase in objects occurred when two satellites accidentally collided while in orbit.

TABLE 3.1

Objects in Earth orbit, by country or organization and type, as of September 30, 2015

[As of 30 September 2015, cataloged by the U.S. SPACE SURVEILLANCE NETWORK]

Country/organization	Payloads	Rocket bodies & debris	Total
China	202	3,571	3,773
CIS	1,461	4,827	6,288
ESA	53	54	107
France	60	457	517
India	60	109	169
Japan	137	72	209
USA	1,276	3,886	5,162
Other	727	111	838
Total	**3,976**	**13,087**	**17,063**

Notes: CIS = Commonwealth of Independent States; ESA = European Space Agency; USA = United States of America.

SOURCE: "Satellite Box Score," in *Orbital Debris Quarterly News*, vol. 19, no. 4, National Aeronautics and Space Administration, October 2015, http://orbitaldebris.jsc.nasa.gov/newsletter/pdfs/ODQNv19i4.pdf (accessed November 23, 2015)

SPACE AGENCIES AROUND THE WORLD

The United States and Russia operate the two most active space programs in the world. Other nations with the resources to do so have also ventured into space. Some have sent their astronauts aboard U.S., Soviet, or Russian spacecraft. Others have developed their own space vehicles and programs. This has created new opportunities for cooperation and competition among space-faring nations. A few of the major space programs are described in this chapter.

RUSSIA

The Russian Space Agency was officially created on February 25, 1992, by decree of the president of the Russian Federation. The agency inherited the technologies, programs, and facilities of the Soviet space program. In 1999 it was expanded to include the aviation industry, so its name was changed to the Russian Space and Aviation Agency (Rosaviakosmos). The aviation responsibilities were removed in 2004, and the agency was renamed the Russian Federal Space Agency (Roscosmos). In January 2016 the Russian president Vladimir Putin (1952–) dissolved the agency to form the Roscosmos State Corporation. It is a state-run company, rather than a federally financed agency. The move was prompted by a series of rocket and spacecraft mishaps and allegations of mismanagement and corruption at Roscosmos.

Sergei Korolev

Sergei Korolev (1906–1966) is considered to be the founder of the Soviet space program. He was born in Zhitomir, a town in what is now Ukraine. An engineer and aviator who began building rockets during the 1930s, Korolev founded the rocket organization Gruppa Isutcheniya Reaktivnovo Dvisheniya (Group for Investigation of Reactive Motion). Following World War II the government appointed him to develop Soviet missile systems.

In August 1957 his team successfully tested the R-7, the world's first intercontinental ballistic missile. The R-7 was powerful enough to carry a nuclear warhead to the United States or a satellite into outer space. In October 1957 an R-7 rocket carried *Sputnik 1* into orbit. The Soviet Union had beaten the United States into space.

VOSTOK. Korolev's next challenge was to beat the United States to the moon. In January 1959 the Soviet probe *Luna 1* flew past the moon. In September 1959 *Luna 2* was deliberately crashed into the lunar surface, making it the first human-made object to reach the moon. A month later *Luna 3* took the first photographs of the far side of the moon. Korolev was already working on a spacecraft for manned missions. It was a modified R-7 called Vostok.

Throughout 1960 and early 1961 the Vostok was tested unmanned, with dogs, small mammals, and a mannequin aboard. Vostok flying dogs included Belka, Chernushka, Mushka, Pchelka, Strelka, and Zvezdochka. Many of the dogs died during these tests. The mannequin was nicknamed Ivan Ivanovich, which is the Russian equivalent of "John Doe."

On April 12, 1961, the Soviets launched the first man into space aboard *Vostok 1*. His name was Yuri Gagarin (1934–1968). He was one of the 20 original cosmonauts who were selected by the Soviet Union in 1959 for manned spaceflights. In 1960 the cosmonauts began training at a sprawling new complex called Zvezdny Gorodok (Star City) in the Russian countryside. His flight made one orbit before he reentered the atmosphere and parachuted safely from the module, landing in a field. In total, his mission lasted 1 hour and 49 minutes. The following month, on May 5, 1961, the astronaut Alan Shepard (1923–1998) became the first American in space.

There were five more Vostok flights between 1961 and 1963, as shown in Table 3.2. *Vostok 6* carried the first woman into space, Valentina Vladimirovna Tereshkova (1937–).

The Vostok program and Project Mercury both took place between 1961 and 1963. The Soviet cosmonauts beat the U.S. astronauts into space and spent more time there. The longest Mercury flight lasted only 34 hours, whereas the longest Vostok flight lasted nearly five days.

VOSKHOD. In 1964 the Soviets began testing a multipassenger spacecraft called Voskhod. (See Table 3.2.) On October 12, 1964, *Voskhod 1* carried three men into space: Vladimir Mikhailovich Komarov (1927–1967), the pilot; Boris Yegorov (1937–1994), a physician; and Konstantin Petrovich Feoktistov (1926–2009), a scientist. It was the first flight to carry more than one person into space. The flight lasted just over 24 hours as the spacecraft orbited Earth 16 times. A few months later *Voskhod 2* was put into orbit with two cosmonauts aboard: the pilot Pavel Belyayev (1925–1970) and the copilot Alexei Arkhipovich Leonov (1934–). On March 18, 1965, Leonov conducted the first extravehicular activity (space walk) in history. It lasted 20 minutes.

Despite these successes, the *Voskhod 2* mission was plagued by life-threatening problems. Leonov's spacesuit and the vehicle's airlock and reentry rockets malfunctioned. The crew module spun out of control during reentry and landed in heavy woods far from its intended landing point. A number of crewed Voskhod missions were planned for the mid- to late 1960s, including one with an all-female crew. However, the problems of *Voskhod 2* and the death of Korolev in January 1966 shook the Soviet space agency. As a result, these planned missions were canceled.

N-1 ROCKET. During his lifetime, Korolev was relatively unknown outside the Soviet Union. The Soviets were very secretive about national affairs and provided scant information to the foreign media. This was particularly true for the inner workings of the Soviet space program. It was only following Korolev's death that the rest of the world learned about his many contributions to space travel. These included many rockets and launch

TABLE 3.2

Crewed space missions of the former Soviet Union, 1961–65

Date of launch	Spacecraft name	Flight type	Time in space	Cosmonauts	Notable events
04/12/61	Vostok 1	1 orbit	1 hr 49 min	Yuri Gagarin	First person to orbit Earth
08/06/61	Vostok 2	17 orbits	25 hr	German S. Titov	First person to spend a full day in orbit
08/11/62	Vostok 3	64 orbits	94 hr	Andriyan G. Nikolayev	First spaceflight including two spacecraft in orbit at once (with Vostok 4)
08/12/62	Vostok 4	48 orbits	70 hr 42 min	Pavel R. Popovich	First spaceflight including two spacecraft in orbit at once (with Vostok 3)
06/15/63	Vostok 5	81 orbits	119 hr 6 min	Valeriy F. Bykovskiy	
06/16/63	Vostok 6	48 orbits	70 hr 42 min	Valentina V. Tereshkova	First woman in space
10/12/64	Voskhod 1	16 orbits	24 hr 17 min	Vladimir M. Komarov (command pilot), Boris B. Yegorov (physician), and Konstatin P. Feoktisov (scientist)	First space flight to carry more than one person
03/18/65	Voskhod 2	17 orbits	25 hr 55 min	Pavel I. Belyayev (pilot) and Aleksey A. Leonov (co-pilot)	First extravehicular activity (space walk)

Notes: hr = hour(s); min = minutes.

SOURCE: Adapted from "Spacecraft," in *National Space Science Data Center*, National Aeronautics and Space Administration, 2015, http://nssdc.gsfc.nasa.gov/nmc/SpacecraftQuery.jsp (accessed November 25, 2015)

vehicles, satellites and probes of different types, and manned spacecraft. His most famous spacecraft was the Soyuz. Modified versions of Soyuz rockets are still being used in the 21st century.

Korolev is also remembered for his one great failure: the N-1 rocket. This was supposed to be the superbooster that would launch a Soviet spacecraft called the L1 to the moon. Korolev's design team created the L1 from a modified Soyuz spacecraft. The N-1 superbooster was similar in scope to von Braun's Saturn V rocket. Korolev worked on the N-1 project from 1962 until his death in 1966, but he never achieved an operational rocket. His successors continued the work after his death, but they, too, were unsuccessful.

On July 3, 1969, an unmanned N-1 rocket exploded only seconds before liftoff. The resulting fireball was so huge that it destroyed the launch facilities. Thirteen days later a Saturn V rocket launched *Apollo 11* on its way to the moon.

The Soviet space program was shrouded in secrecy. Successes were publicized, whereas failures and plans were not. Although the Soviets had ambitions to land a man on the moon, this goal was never announced publicly. It was only years later that the public learned about the failed Soviet moon program in Sergei Leskov's article "How We Didn't Get to the Moon" (*Izvestiya*, August 18, 1989). Most observers in the United States assumed the Soviet Union was aiming for the moon, but this was not certain. In fact, as early as 1963 NASA critics in the United States asserted that the "Moon race" was a hoax advanced by the U.S. government to further its own aims. When the United States reached the moon first, the Soviets insisted that they had never intended to go there. In "Yes, There Was a Space Race" (Forbes.com, July 16, 2009), Dwayne A. Day describes the photographs and other evidence the Central Intelligence Agency collected during the 1960s that convinced the agency that the Soviets tried to beat the United States to the moon. The U.S. government kept this information classified for decades.

Soyuz Flights

The Soviet space program continued after Korolev's death using spacecraft launched by Soyuz rockets. Table 3.3 lists the spacecraft (which were also named Soyuz) that traveled into space between 1967 and 1971. On April 23, 1967, the Soviet space agency launched *Soyuz 1* with Vladimir Mikhailovich Komarov (1927–1967) aboard. A day later the flight ended in tragedy when the module's parachute failed during descent. Komarov died during the descent. He was the first human to die during a spaceflight. Eight crewed Soyuz flights followed through 1971. As shown in Table 3.3, all these flights represented "first-in-space" achievements for the Soviet Union.

A New Focus

The Soviet's moon program continued well into the 1970s. However, neither the N-1 nor a competing rocket called the Proton ever became dependable enough for manned launches. During the early 1970s the Soviets concentrated on perfecting their Soyuz rockets and building a space station. Like NASA, the Soviet space agency had always envisioned an orbiting space station as the next step after a lunar visit.

On April 19, 1971, the Soviet space station *Salyut 1* was launched from the Baikonur Cosmodrome in what is now Kazakhstan. The station was cylindrically shaped

TABLE 3.3

Crewed space missions of the former Soviet Union, 1968–71

Dates	Spacecraft name	Flight type	Cosmonauts	Notable events
April 23–24, 1967	Soyuz 1	14 orbits	Vladimir M. Komarov	First person to die during a space flight
October 25–28, 1968	Soyuz 2	48 orbits	None	First attempted spacecraft docking (with Soyuz 3)
October 26–30, 1968	Soyuz 3	64 orbits	Georgiy Beregovoy	First attempted spacecraft docking (with Soyuz 2)
January 14–17, 1969	Soyuz 4	48 orbits	Vladimir A. Shatalov	First spacecraft docking (with Soyuz 5)
January 15–18, 1969	Soyuz 5	49 orbits	Boris Volynov (commander), Aleksey Yeliseyev (flight engineer), and Yevgeniy Khrunov (research engineer)	First spacecraft docking (with Soyuz 4)
October 11–16, 1969	Soyuz 6	80 orbits	Georgi Shonin (commander) and Valeri Kubasov (flight engineer)	First spacecraft to orbit with 2 others (Soyuz 7 and 8)
October 12–17, 1969	Soyuz 7	80 orbits	Anatoli Filipchenko (commander), Vladislav Volkov (flight engineer), and Viktor Gorbatko (research engineer)	First spacecraft to orbit with 2 others (Soyuz 6 and 8)
October 13–18, 1969	Soyuz 8	80 orbits	Vladimir Shatalov (commander) and Aleksey Yeliseyev (flight engineer)	First spacecraft to orbit with 2 others (Soyuz 6 and 7)
June 1–19, 1970	Soyuz 9	288 orbits	Andriyan Nikolayev (commander) and Vitali Sevastyanov (flight engineer)	Longest spaceflight to date
April 23–24, 1971	Soyuz 10	32 orbits	Vladimir Shatalov (commander), Aleksey Yeliseyev (flight engineer), and Nikolai Rukavishnikov (systems engineer)	First spacecraft to dock with a space station (Salyut 1)

SOURCE: Adapted from "Spacecraft," in *National Space Science Data Center*, National Aeronautics and Space Administration, 2015, http://nssdc.gsfc.nasa.gov/nmc/SpacecraftQuery.jsp (accessed November 25, 2015)

and approximately 39 feet (12 m) long and 13 feet (4 m) wide at its widest point. It was placed into orbit approximately 124 miles (200 km) above Earth.

The station was built so that Soviet scientists could study the long-term effects on humans living in space. A crew of three cosmonauts flew aboard *Soyuz 10* to the station a few days after the station was placed in orbit. However, they were unable to dock with it, so they were forced to end their mission early and return to Earth. In June 1971 the Soviet spacecraft *Soyuz 11* successfully docked with the station, and three cosmonauts—Georgi Timofeyevich Dobrovolsky (1928–1971), Vladislav Nikolayevich Volkov (1935–1971), and Viktor Patsayev (1933–1971)—inhabited it for 24 days. During their decent back to Earth a valve opened on their spacecraft, depressurizing it and killing the three crew members. At that time, cosmonauts did not wear pressurized space suits during launch or reentry. This was later changed to provide them greater safety.

The Soviet space agency canceled future flights to the station and began an extensive redesign of the Soyuz spacecraft. In October 1971 *Salyut 1* fell into Earth's atmosphere and was destroyed. In total, the Soviets put seven Salyuts into orbit. (See Table 3.4.) These stations were visited by cosmonauts and scientists from a number of countries, including Cuba, France, and India. In 1984 three Soviet cosmonauts spent 237 days aboard *Salyut 7*. This was a new record for human duration in space. *Salyut 7* was deorbited in February 1991. It had stayed in orbit for nearly nine years and hosted 10 crews of cosmonauts that spent a total of 861 days in space. The Soviet space program gained invaluable experience in long-duration exposure to weightlessness.

Mir

On February 19, 1986, the Soviet Union launched the new space station *Mir* into orbit. *Mir* was to be Russia's first continuously occupied space station. In "What's in a Name?" (October 3, 1996, http://spaceflight.nasa.gov/history/shuttle-mir/references/r-documents-mirmeanings.htm), Frank L. Culbertson Jr. notes that the word *mir* has a few different translations, including "peace," "world," and "village." Unlike the temporary Salyut stations, *Mir* was designed to last for years and to be continuously inhabited. Dozens of cosmonauts, astronauts, and space tourists traveled aboard Soyuz flights to the station during its 15-year lifetime in space. It finally tumbled to Earth in 2001.

Russian cosmonauts repeatedly set and broke space duration records aboard the *Mir* station. Vladimir Georgiyevich Titov (1947–) and Musa Manarov (1951–) reached the one-year milestone when they completed 366 days in space in 1988. By 1995 the record for a continuous stay was 437 days and 18 hours, set by the cosmonaut Valery Vladimirovich Polyakov (1942–). As of February 2016, this was still the record.

The *Mir* is also famous for its nongovernmental inhabitants. Beginning in the 1980s the Soviet space program suffered financial difficulties. To raise funds, the space agency sold seats on *Mir* to a variety of foreign astronauts and adventurers. In 1990 the Japanese journalist Toyohiro Akiyama (1942–) became the first citizen of Japan to fly in space and the first private citizen to pay for a space flight. Akiyama's television network paid $28 million to send him on a seven-day mission to *Mir*. In 1991 the British chemist Helen Patricia Sharman (1963–) spent eight days in space after winning a contest sponsored by a London bank.

Cooperative Missions

In 1972 the Soviet Union and the United States agreed to work together to achieve a common docking system for their respective spacecraft. This would permit docking in space of U.S. and Soviet spacecraft during future missions. On July 17, 1975, the Soviet *Soyuz 19*, which carried two cosmonauts, docked with *Apollo 18*, which carried three astronauts. (See Figure 3.4.)

The crew members conducted a variety of scientific experiments during the two-day docking period. Both spacecraft returned to Earth safely. The Apollo-Soyuz Test Rendezvous and Docking Test Project was the first union of spacecraft from two different countries.

TABLE 3.4

The Salyut series of Soviet space stations

Name	Launch date	Deorbit date	Total crew occupancy time	Note
Salyut 1	April 1971	October 1971	24 days	Three cosmonauts died on their return to Earth.
Salyut 2	April 1973	April 1973	0 days	Unmanned. Station fell apart soon after reaching orbit.
Salyut 3	June 1974	January 1975	15 days	Hosted 1 crew. One unsuccessful docking.
Salyut 4	December 1974	February 1977	92 days	Hosted 2 crews and 1 unmanned craft. One abort.
Salyut 5	June 1976	August 1977	67 days	Hosted 2 crews. One unsuccessful docking.
Salyut 6	September 1977	July 1982	676 days	Hosted 16 crews and 1 unmanned craft.
Salyut 7	April 1982	February 1991	861 days	Hosted 10 crews.

SOURCE: Created by Kim Masters Evans for Gale, © 2016

FIGURE 3.4

Apollo-Soyuz rendezvous and docking test project

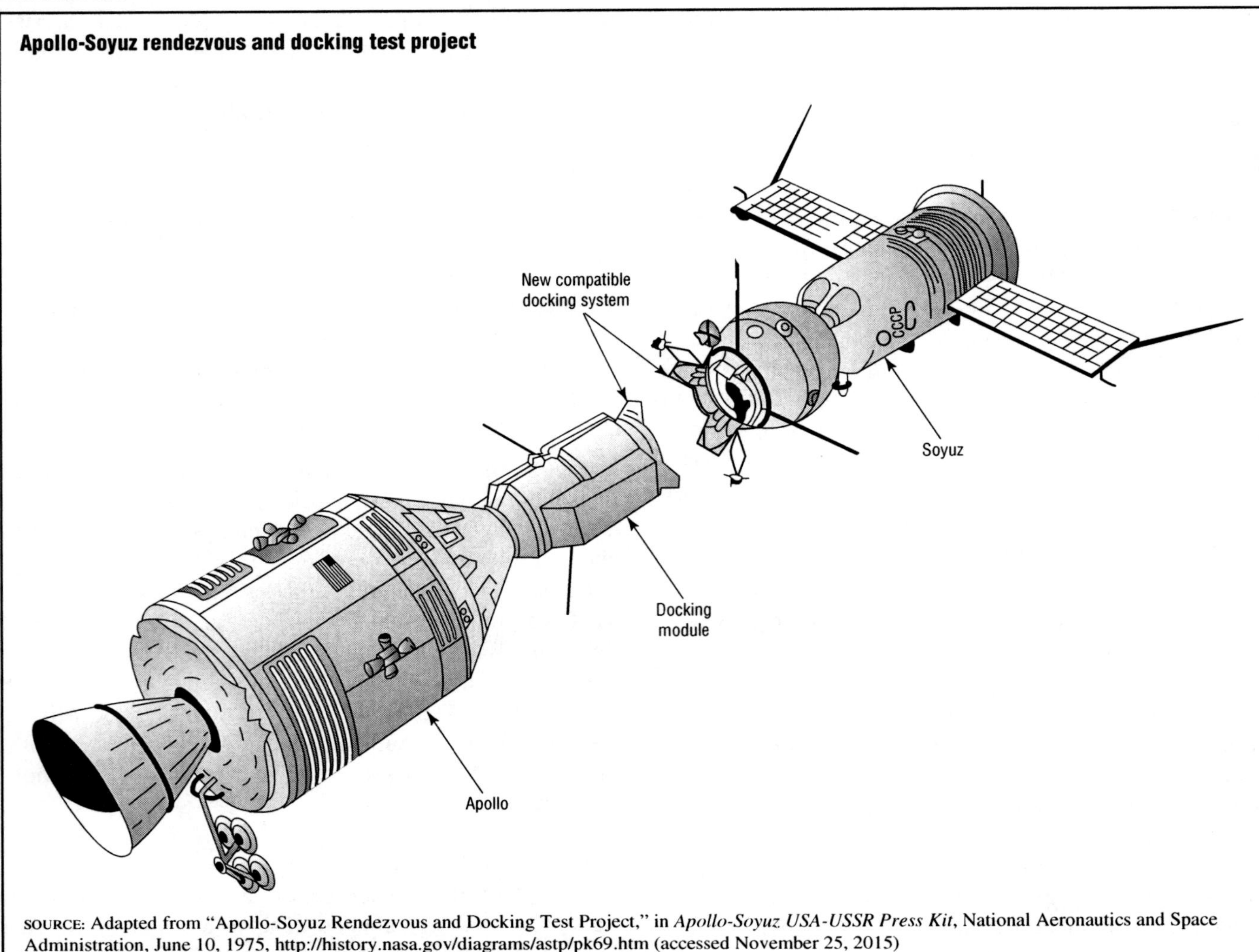

SOURCE: Adapted from "Apollo-Soyuz Rendezvous and Docking Test Project," in *Apollo-Soyuz USA-USSR Press Kit*, National Aeronautics and Space Administration, June 10, 1975, http://history.nasa.gov/diagrams/astp/pk69.htm (accessed November 25, 2015)

During the 1970s NASA proposed a joint U.S.-Soviet mission to a Salyut station. These scientific hopes were overshadowed by international politics. In 1979 the Soviet Union began a war in Afghanistan. Two years later the Soviet government imposed martial law in Poland to suppress dissenters. The U.S. response to both incidents was a sharp reduction in cooperative efforts between the two countries. Meanwhile, the Soviet space program was engrossed in a new project called Buran, a reusable space plane that was modeled after the U.S. space shuttle. The Buran program (like NASA's space shuttle program) was plagued by development, cost, and scheduling problems. Although an unmanned Buran orbited Earth twice and landed successfully in November 1988, the program was halted soon afterward because of funding cuts.

The 1980s were a tense time in U.S.-Soviet relations. The Soviets were at war with Afghanistan and continued to oppress dissidents in Poland. In 1983 the Soviet military shot down a South Korean jetliner that allegedly veered into Soviet air space. Sixty-two Americans were among the 269 passengers killed. The Soviet Union felt threatened by President Reagan's so-called Star Wars proposal to build a satellite shield. Furthermore, throughout the decade U.S.-Soviet arms talks repeatedly failed.

The Soviet empire began to dissolve during the late 1980s and was officially ended in 1991, when it separated into several independent republics. The largest of these was Russia, which inherited most of the Soviet space program.

SHUTTLE-*MIR* MISSIONS. In June 1992 the U.S. president George H. W. Bush and the Russian president Boris Yeltsin (1931–2007) signed the Agreement between the United States of America and the Russian Federation Concerning Cooperation in the Exploration and Use of Outer Space for Peaceful Purposes. NASA and the Russian Space Agency (which had been recently created) worked out a plan for joint shuttle-*Mir* missions. Both agencies considered this a prelude to a joint U.S.-Russian space station.

In 1993 President Bill Clinton (1946–) met with Yeltsin and agreed to continue cooperative efforts in space exploration. In June 1995 the space shuttle *Atlantis* docked with *Mir* for the first time. Over the next three years eight additional dockings took place.

Figure 3.5 depicts a shuttle docked with the *Mir* space station. In "Timeline of Shuttle-Mir" (August 23, 2013, http://spaceflight.nasa.gov/history/shuttle-mir/references/

FIGURE 3.5

Shuttle-*Mir* mated configuration

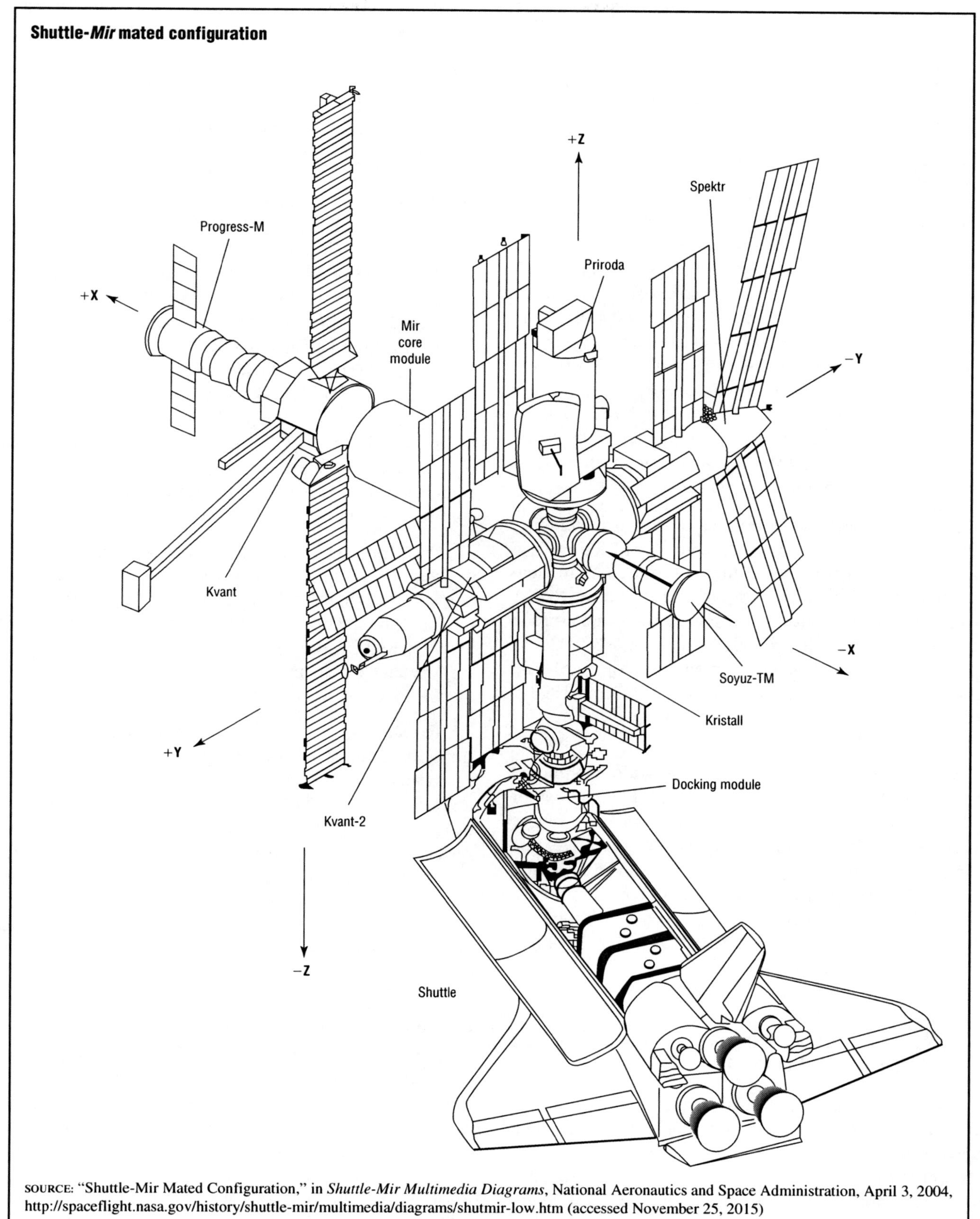

SOURCE: "Shuttle-Mir Mated Configuration," in *Shuttle-Mir Multimedia Diagrams*, National Aeronautics and Space Administration, April 3, 2004, http://spaceflight.nasa.gov/history/shuttle-mir/multimedia/diagrams/shutmir-low.htm (accessed November 25, 2015)

r-timeline.htm), NASA lists the major accomplishments of the missions:

- On February 3, 1994, Sergei Krikalev (1958?–) became the first cosmonaut to ride on a U.S. shuttle.
- On March 14, 1995, Norman E. Thagard (1943–) became the first U.S. astronaut to launch in a Soyuz rocket.
- In 1996 astronaut Shannon Lucid (1943–) set the women's record for space flight duration: 188 days.
- American astronauts logged nearly 1,000 days of orbit time.

***MIR* MISHAPS.** When the first Americans arrived at *Mir* in 1995, the station had already been in orbit for nine years. They found a cramped and crowded spacecraft bulging with hoses, cables, and scientific equipment. Every closet and storage space was crammed full. Some gear and tools floated around because there was no space left to stow or fasten them. Over the years, water droplets had escaped from environmental control systems and eventually clung to delicate electronics. *Mir*'s systems were plagued by computer crashes and battery problems. The cosmonauts spent the vast majority of their time doing repair and maintenance tasks.

In 1997 *Mir* suffered several major mishaps, including a fire (which was extinguished) and a collision with a visiting freighter that breached the hull of the *Mir* module called Spektr. U.S. politicians and media called for NASA to stop sending American astronauts to the trouble-prone station. Despite the pressure, NASA and the White House believed it was important to complete the project. Shuttle flights continued to *Mir* throughout 1997 and into 1998.

Russian Space Agency Takes Over

As noted earlier, in 1991 the Soviet Union splintered into a number of individual republics. The largest of these is Russia. In 1992 the new Russian government established the Russian Space Agency to take over the space programs of the former Soviet Union. Russia and the United States began a new era of cooperation in space. In 1993 the two countries agreed to work together to build an international space station. Between 1994 and 1998 U.S. space shuttles transported astronauts and cosmonauts to the *Mir* station. In 1998 *ISS* construction began when the Russians placed the first module (Zarya) into orbit. Construction and operation of the *ISS* are described in detail in Chapter 5.

The *ISS* receives regular visits from Russian Progress spacecraft. These are automated resupply vessels that bring consumables (food and water), spare parts, propellants, and other supplies to the station. The *ISS* has also relied heavily on the Russian Soyuz to ferry crew members to and from the station. Since the retirement of the U.S. space shuttles in 2011, the Soyuz has been the only spacecraft capable of conducting *ISS* crew transports and will continue to be so until the next generation of U.S. spacecraft begin operating.

Roscosmos controls all of Russia's nonmilitary space flights. Military space ventures are controlled by Russia's Military Space Forces (VKS). The two agencies share control of the Gagarin Cosmonaut Training Center in Star City, near Moscow. In addition, both agencies launch from the Baikonur Cosmodrome in Kazakhstan. The Russian government pays the Kazakhstani government for use of the facility. The Plesetsk Cosmodrome launch facility in northern Russia is under the control of the VKS. As of February 2016, a new space center called the Vostochny Cosmodrome was under construction in the Amur Region in southeastern Russia near the Pacific Ocean. It is expected to be operational before 2020. It is believed that the Russian government is building the new facility for both political and technical reasons. A spaceport on Russian soil allows the government to be less dependent on the Kazakhstani government. Also, the Vostochny site (like the Baikonur site) is much closer to the equator than the Plesetsk site. Launch sites near the equator are advantageous for launching satellites into equatorial orbits, which are particularly popular with commercial clients.

Past, present, and future Russian space missions devoted to scientific purposes (other than advancing human spaceflight) are described in Chapters 6 through 8. The Soviet Union developed the Global Navigation Satellite System (GLONASS), a space-based navigation system, that Roscosmos has continued to update to rival the GPS that is operated by the United States.

Russia, like the United States, conducts space missions for national security purposes. The Russian space-based missile warning system is called Oko, which translates as "Eye." The program began under the Soviet Union and is part of a comprehensive missile defense system. By early 2016 it was estimated to include dozens of satellites. Russia is also believed to have ASAT capabilities through its missile interceptors and other systems. According to Grego, Russia has under development maneuverable satellites capable of approaching satellites in orbit.

Russian Space Industry Woes

Between 2010 and 2015 the Russian space program suffered a series of problems including rocket and satellite failures. In addition, *Geo-IK 2*, a Russian military satellite, was put into the wrong Earth orbit by a rocket booster. On August 24, 2011, a Soyuz rocket carrying a Progress cargo ship bound for the *ISS* crashed soon after liftoff. Later that year a Chinese-Russian collaboration called *Yinghuo-1/Phobos-Grunt* was launched by a Zenit rocket on its way toward Mars's moon Phobos. However, the combined orbiter/lander failed to leave Earth orbit and was destroyed when it reentered Earth's atmosphere in January 2012. That same month a pressurization problem with a Soyuz spacecraft delayed a crewed mission to the *ISS*. In 2013 a Proton M rocket exploded soon after liftoff; likewise, three GLONASS satellites were lost.

As shown in Table 1.1 and Figure 1.8 in Chapter 1, Russia conducted 32 launches in 2014 to put spacecraft into Earth orbit. Nearly all (28) of the missions were noncommercial launches. The Federal Aviation Administration (FAA) provides in *The Annual Compendium of Commercial Space Transportation: 2014* (February 2015, http://www.faa.gov/about/office_org/headquarters_offices/ast/media/FAA_Annual_Compendium_2014.pdf) details about each of Russia's 32 launches. Only one was a failure. In May 2014 a Proton M rocket carrying a Russian communications satellite (*Express AM4R*) crashed back to Earth. In May 2015 another Proton M rocket crashed while attempting to transport a Mexican satellite (*Mexsat 1*) into orbit. This failure was followed by an incident in which an unmanned Soyuz rocket carrying cargo to the *ISS* veered off course before reaching orbit. It was destroyed when it reentered Earth's atmosphere.

Throughout 2015 news reports emerged about financial problems in Russia's aerospace industry. According to Ivana Kottasova, in "$1.8 Billion Disappears in Russian Space Program" (CNN.com, May 25, 2015), a Russian government audit of Roscosmos in 2014 revealed $1.8 billion in misspent funds and numerous corruption allegations. In "Russia's Space Program in Crisis after Decades of Brain Drain, Neglect" (NBCNews.com, August 23, 2015), Alexey Eremenko notes that Roscosmos receives far less funding per year than NASA. In addition, the Russian space agency has suffered key personnel losses as scientists and engineers have been lured away by higher pay in other nations. Sanctions imposed by U.S. and European governments over Russia's military intervention in Ukraine have negatively affected the Russian economy. The oil-rich country has also been hurt by low oil prices on world markets. The financial problems combined with the string of Proton and Soyuz mishaps since 2010 have raised troubling questions about the viability of Roscosmos and the Russian space industry as a whole.

Angara Launch Vehicles

One bright spot for the Russian space program may be its newly developed line of launch vehicles called Angara. (The name comes from the rapids-filled Angara River in Siberia.) Regular and heavy versions of the rockets were successfully flight tested in 2014. Matthew Bodner notes in "Russia's New Angara Rocket Launches on Global Market" (MoscowTimes.com, July 23, 2015) that in 2015 the sea-launch company International Launch Services began marketing Angara launches to commercial satellite clients. The first launch is expected to take place in 2017 using a lightweight version of the rocket. A heavy-lift model is to be available by the 2020s as the troubled Proton rocket is phased out.

Russia hopes the Angara will boost its commercial satellite launching business. It faces fierce competition from the longtime rival Arianespace (a subsidiary of the European Space Agency), which uses French-made Ariane rockets, and from the American newcomer SpaceX and its Falcon 9 rockets. According to Bodner, the Angara series is similar to the Falcon 9 family in terms of scalability, meaning it "allows the size of the rocket to be tailored to the weight of its payload." However, the success of the Angara line is also dependent on its launch site. Many commercial satellites are put into equatorial orbits. Bodner notes that Russia's existing spaceport at Plesetsk accommodates launches into non-equatorial orbits. Completion of the Vostochny Cosmodrome in southeastern Russia will likely be needed for the Angara launch vehicles to be commercially viable.

EUROPE

European countries can engage in space exploration through the European Space Agency (ESA), by collaborating with foreign space agencies, and/or through their own national space programs. Numerous European countries have national space programs. The French program is one of the oldest and most active in Europe.

French Space Program

Following World War II, France engaged in rocket research and development with the help of German engineers and scientists who had developed the V series of rockets. This work largely took place in Vernon, France. In 1961 the French government founded the agency Centre National d'Études Spatiales (CNES; National Space Study Center) to lead the nation's space program. The CNES is headquartered in Paris, but it has additional facilities throughout the country, including the Launcher Directorate in Evry and the Toulouse Space Centre. Launches take place at the Centre Spatial Guyanais (Guiana Space Centre) in Kourou, French Guiana. (See Figure 3.6.) Because of its nearness to the equator, Kourou is an ideal location from which to launch satellites into a geostationary transfer orbit.

During the 1960s France developed the Diamant, Berenice, and Véronique launch systems. In 1965 a Diamant rocket was used to launch the first French satellite, *Astérix*, into orbit. By the end of the 1970s France had developed what would become the primary launch vehicle for European spacecraft: the Ariane rocket. (Ariane is the French name for the Greek goddess

FIGURE 3.6

French Guiana map

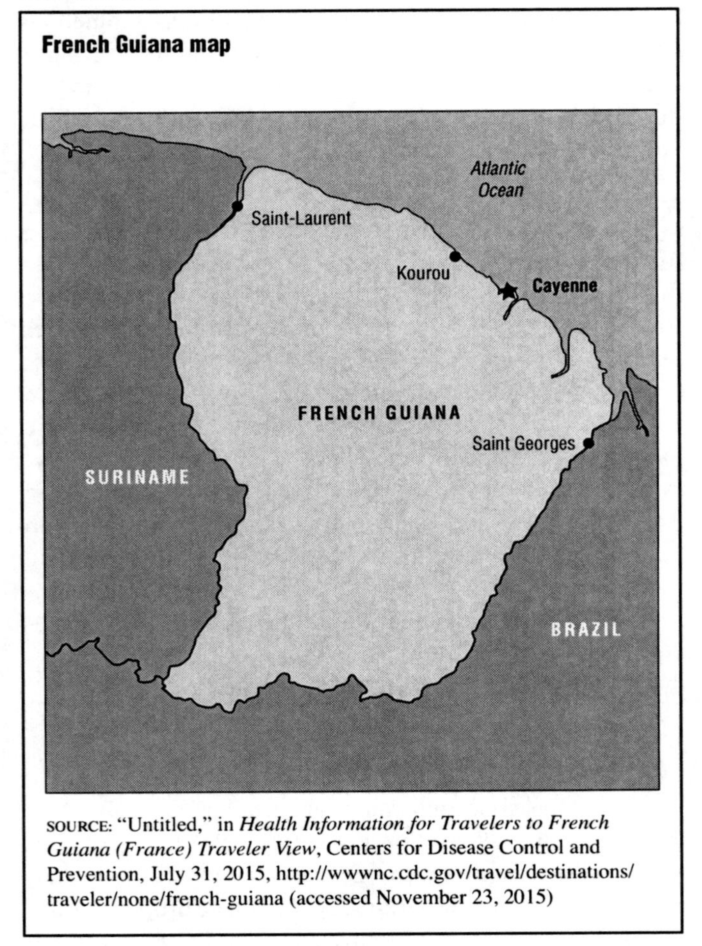

SOURCE: "Untitled," in *Health Information for Travelers to French Guiana (France) Traveler View*, Centers for Disease Control and Prevention, July 31, 2015, http://wwwnc.cdc.gov/travel/destinations/traveler/none/french-guiana (accessed November 23, 2015)

Ariadne.) In 1975 France joined with more than a dozen other European countries to form the ESA. The CNES continued to conduct national space ventures and collaborated with the ESA and other national space agencies.

On December 24, 1979, the first Ariane rocket was launched into space from the Kourou spaceport. Since that time the Ariane design has undergone several modifications. The latest generation is the Ariane 5. It is a robust launch system that is capable of carrying two satellites at once into orbit. According to Arianespace, in "Arianespace Starts the New Year with Record Operational Performance and Order Backlog" (http://www.arianespace.com/), as of January 2016, 227 Ariane rockets had been launched into space. The vast majority of the flights have transported commercial communications satellites into Earth orbit.

Ariane rockets have also been used to launch science payloads for various nations. The CNES's space science missions are discussed in Chapter 6. In 2015 the agency was embarrassed when a malfunction was discovered in an instrument it had provided for NASA's *InSight* spacecraft, a Mars lander. A planned March 2016 launch had to be postponed, and as of February 2016, had not been rescheduled. The mission is described in detail in Chapter 7.

TABLE 3.5

Member nations of the European Space Agency, December 2015

Austria	Italy
Belgium	Luxembourg
Czech Republic	Netherlands
Denmark	Norway
Estonia	Poland
Finland	Portugal
France	Romania
Germany	Spain
Greece	Sweden
Hungary	Switzerland
Ireland	United Kingdom

SOURCE: Created by Kim Masters Evans for Gale, © 2016

France is one of the partner nations in the ISS program, and Ariane rockets have launched the ESA's cargo carriers called automated transfer vehicles (ATVs) to the space station, as described in Chapter 5.

European Space Agency

The ESA was formed in 1975 from two existing organizations: the European Space Research Organisation and the European Launch Development Organisation. As of December 2015, the ESA had 22 member states. (See Table 3.5.) In addition, the ESA has agreements with Canada, Estonia, Hungary, Poland, and Slovenia to participate in some projects.

FACILITIES AND FUNDING. ESA headquarters are located in Paris, France. Other ESA facilities include the European Space Research and Technology Centre in Noordwijk, Netherlands; the European Space Operations Centre in Darmstadt, Germany; the European Astronaut Centre in Cologne, Germany; the European Space Research Institute in Frascati, Italy; the European Space Astronomy Centre in Villanueva de la Cañada, Spain; and liaison offices in Belgium, Russia, and the United States. The ESA operates a launch base in French Guiana.

Each member state funds mandatory ESA activities based on that country's gross national product (the total value of goods and services that are produced by a country over a particular period). Mandatory activities include space science programs and the agency's general budget. In addition, the ESA operates optional projects in which countries may choose to participate and fund.

ESA MISSIONS AND CAPABILITIES. The ESA's past, current, and planned space science missions are described in Chapters 6 through 8. The agency is a partner in the ISS program, which is discussed in Chapter 5. Through this relationship, the ESA plays a key role in NASA's development of the *Orion* spacecraft, which is described in Chapter 2. In "NASA Signs Agreement for a European-Provided Orion Service Module" (January 16, 2013, http://www.nasa.gov/exploration/systems/mpcv/orion_feature_011613.html), NASA explains that the ESA is building the service module for the capsule, which is designed to carry astronauts to Earth's moon and other destinations. NASA notes that the service module will "contain the in-space propulsion capability for orbital transfer, attitude control and high-altitude ascent aborts. It also will generate and store power and provide thermal control, water and air for the astronauts." The first crewed flight of *Orion* is expected during the early 2020s.

In October 2011 the first pair of satellites of the Galileo Positioning System (the ESA's rival to the U.S. GPS) were launched into space from the Guiana Space Centre. The Galileo system will include more than two dozen satellites when it is fully implemented by the early 2020s.

The ESA and some individual European countries may have ASAT capabilities through their missile interceptor systems. Grego indicates that the ESA has under development maneuverable satellites capable of approaching satellites in orbit.

As shown in Table 1.1 and Figure 1.8 in Chapter 1, Europe conducted 11 launches in 2014 to put spacecraft into Earth orbit. Six of the launches were for commercial purposes, and five were for noncommercial purposes. According to the FAA, in *Annual Compendium of Commercial Space Transportation*, all but one of the launches were deemed completely successful. The ESA used three different rocket types for its launches in 2014: Ariane (seven launches), Soyuz (three launches), and Vega (one launch). The Vega rocket was developed by the ESA and the Italian Space Agency for carrying light-weight payloads into orbit. In August 2014 a Russian Soyuz rocket lifted off from the Guiana Space Centre carrying two Galileo satellites. They were inserted into the wrong Earth orbits. However, the ESA notes in "Sixth Galileo Satellite Reaches Corrected Orbit" (March 13, 2015, http://www.esa.int/Our_Activities/Navigation/The_future_-_Galileo/Launching_Galileo/Sixth_Galileo_satellite_reaches_corrected_orbit) that the orbits were corrected over the period of a month.

CHINA

China's space program is overseen by the China National Space Administration and is operated by the China Aerospace Science and Technology Corporation (CASC). The CASC is a state-run enterprise that develops and produces rockets, spacecraft, and related products. It has conducted satellite launches since 1970. CASC launch sites include Jiuquan in the Gobi desert, Taiyuan in northern China, and Xichang in southeastern China.

China's space endeavors are not clearly separated into military and nonmilitary missions as they are in the United States. In *Annual Report to Congress: Military and Security Developments Involving the People's Republic of China 2015* (April 7, 2015, http://www.defense.gov/Portals/1/Documents/pubs/2015_China_Military_Power_Report.pdf), the DOD notes, "China has invested in advanced space capabilities, with particular emphasis on satellite communication (SATCOM), intelligence, surveillance, and reconnaissance (ISR), satellite navigation (SATNAV), and meteorology, as well as manned, unmanned, and interplanetary space exploration."

As previously noted, in 2007 the Chinese government purposely rammed a missile into one of its old weather satellites, purportedly to showcase the nation's military space capabilities. A Chinese satellite launch conducted in 2013 also attracted keen attention. Andrea Shalal-Esa notes in "U.S. Sees China Launch as Test of Anti-satellite Muscle—Source" (Reuters.com, May 15, 2013) that U.S. military observers suspect the satellite was "the first test of a new interceptor that could be used to destroy a satellite in orbit." The Chinese government claims the satellite "carried a science payload to study the earth's magnetosphere."

In addition, China is believed to use satellites as part of its missile defense system. Grego suggests that China has under development maneuverable satellites capable of approaching satellites in orbit.

China has operated an active space science program since 2007, when it launched its first lunar robotic orbiter—*Chang'e 1*—to the moon. That mission and others devoted to scientific purposes (other than advancing human spaceflight) are described in Chapter 6.

As shown in Table 1.1 and Figure 1.8 in Chapter 1, China conducted 16 launches in 2014 to put spacecraft into Earth orbit. All of the missions were for noncommercial purposes. According to the FAA, in *Annual Compendium of Commercial Space Transportation*, all the launches were deemed successful.

Qian Xuesen

The Chinese space program began during the late 1950s under the direction of the rocket engineer Qian Xuesen (c. 1911–2009). Qian was born in China but immigrated to the United States during the 1930s, where he attended the Massachusetts Institute of Technology and the California Institute of Technology (Caltech). He was a key member of the rocketry club at Caltech that evolved into NASA's Jet Propulsion Laboratory. He was also instrumental in the U.S. program to acquire and apply German rocket technology at the end of World War II. In 1950 Qian was accused of being a communist spy and had his security clearance revoked. At the time, he was pursuing U.S. citizenship.

In 1955, after five years under virtual house arrest, Qian was deported to China, where he was put in charge of the nation's budding space program. Under his leadership China developed successful satellite and missile systems. These included the antiship missile called Haiying by the Chinese and Silkworm by the Western media. During the cold war China sold Silkworms to a number of developing countries that were considered to be unfriendly to the United States. Qian also led development of the Chang Zheng (Long March) rockets that became the primary launch vehicle of the Chinese space program.

During the late 1960s Qian fell out of favor with the Chinese leadership and was removed from his post. This disgrace resulted in Qian receiving little credit within China for his accomplishments. However, the rest of the world considers him to be the father of the Chinese space program. In her 1995 biography of Qian, *Thread of the Silkworm*, Iris Chang asserts that deporting the brilliant rocket scientist was "one of the most monumental blunders committed by the United States."

China Reaches Space

On April 24, 1970, *DFH 1*, the first Chinese satellite, was launched into Earth orbit. It was propelled into space by a Long March rocket. Since the 1970s China has conducted many satellite launches using Long March rockets. Figure 3.7 depicts a Long March 3C, one of several configurations in the vehicle family. During the 1990s development began on capsules that were capable of carrying animals, and later humans, into space. In 1999 the first such spacecraft, *Shenzhou 1*, successfully completed 14 orbits around Earth. The word *shenzhou* loosely translates as "divine ship" or "divine vessel." During the first few years of the first decade of the 21st century the Shenzhou series was updated with newer and more powerful Long March rockets. The rockets are designed for orbital missions and when they return to Earth they land with the aid of parachutes in the vast expanse of Inner Siberia.

HUMAN SPACEFLIGHT PROGRAM. On October 15, 2003, China launched its first human spaceflight as part of the Shenzhou program. Chinese astronauts are called *yuhangyuans* or taikonauts. On September 29, 2011, China launched into Earth orbit its first space station, *Tiangong 1* (the word *tiangong* means "heavenly palace"). The article "Tiangong 1" (ChinaDaily.com, September 27, 2011) explains that the station consists of two modules: an experiment module in which humans can "live and work" and a support module that "can provide power supply for orbit maneuvering and spacecraft orbiting."

As of February 2016, the following Shenzhou flights had taken place since 2003:

- *Shenzhou 5*—Yang Liwei (1965–) was launched on October 15, 2003. He spent 21 hours and 23 minutes in space and completed 14 orbits.

FIGURE 3.7

Long March 3C rocket

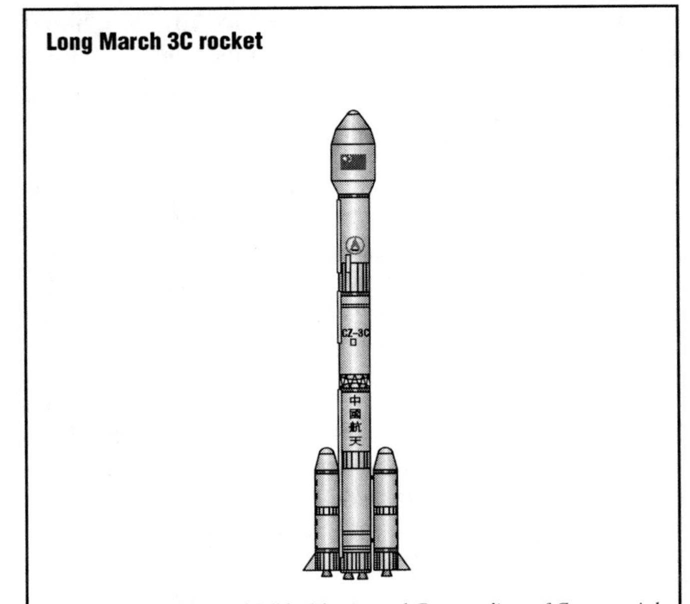

SOURCE: "Long March 3C," in *The Annual Compendium of Commercial Space Transportation: 2014*, Federal Aviation Administration, February 2015, http://www.faa.gov/about/office_org/headquarters_offices/ast/media/FAA_Annual_Compendium_2014.pdf (accessed November 21, 2015)

- *Shenzhou 6*—launched on October 12, 2005, Fei Junlong (1965–) and Nie Haisheng (1964–) spent just over four days orbiting Earth.
- *Shenzhou 7*—Zhai Zhigang (1966–), Liu Buoming (1966–), and Jing Haipeng (1966–) were launched on September 25, 2008. Zhai conducted China's first extravehicular activity during the successful three-day mission.
- *Shenzhou 8*—launched on October 31, 2011, the uncrewed spacecraft was docked with the uncrewed *Tiangong 1* space laboratory, a single module that had been launched into orbit the previous month.
- *Shenzhou 9*—Liu Yang (1978–; the first female taikonaut to visit space), Jing Haipeng (1966–), and Liu Wang (1970–) launched on June 16, 2012, and docked with *Tiangong 1* during their 13-day mission.
- *Shenzhou 10*—launched on June 11, 2013, Nie Haisheng (1964–), Wang Yaping (1980–), and Zhang Xiaoguang (1966–) traveled to *Tiangong 1* during a 15-day mission.

According to Morris Jones, in "China's Space Laboratory Still Cloaked" (SpaceDaily.com, March 10, 2015), sketchy press releases from the Chinese government indicate that a second small space laboratory—*Tiangong 2*—will be put into orbit in 2016. It will likely be visited by taikonauts during a *Shenzhou 11* mission.

In December 2011 the Chinese government announced its latest five-year plan for space exploration. It called for the continuation of earlier plans to build a large space station and to send a taikonaut to the moon. In addition, the nation is developing its own GPS rival called the Beidou Navigation Satellite System that is expected to be completely operational by 2020.

China has shown a keen interest in participating in international space ventures and has agreements with Brazil, Russia, and the ESA. However, under U.S. law NASA is forbidden to collaborate with the Chinese space program because it is run by the Chinese military (i.e., the People's Liberation Army). Andrew Jacobs explains in "China's Space Program Bolstered by First Docking" (NYTimes.com, November 3, 2011) that this U.S. policy dates back to 1989 and was implemented following the Chinese government's violent crackdown on pro-democracy demonstrators. In addition, Chinese scientists and taikonauts are not allowed to attend U.S. space conferences or launches or to visit the *ISS*.

China has stated that it intends to make its future large space station available for visits from astronauts from other nations. In "China's Space Station Planners Put out Welcome Mat" (SpaceNews.com, October 13, 2015), Peter B. de Selding indicates that the station is expected to include a core habitation module that will be launched in 2018. Two "experiment-carrying" modules will then be attached by around 2022. There is room for four additional modules that can be financially sponsored by international partners. Construction of the large space station will require China to have more powerful launch vehicles. China is developing Long March 5 and Long March 7 rockets for this purpose. According to Rina Marie Doctor, in "China Testing Its Largest, Most Powerful Rocket: Long March 5 to Launch by 2016" (TechTimes.com, November 21, 2015), the Long March 5 is a heavy-capacity rocket similar in capabilities to the Delta IV and Atlas V rockets. Both the Long March 5 and Long March 7 launch vehicles are expected to be flight tested in late 2016.

JAPAN

The Japan Aerospace Exploration Agency (JAXA) was created on October 1, 2003, by merging the Institute of Space and Astronautical Science, the National Aerospace Laboratory of Japan, and the National Space Development Agency of Japan. JAXA is headquartered in Tokyo and has more than a dozen field facilities located throughout Japan.

The first Japanese satellite, *Ohsumi*, was launched into space in February 1970 by a Lambda-4S rocket. *Ohsumi* remained in space for more than three decades and was destroyed in 2003, when it reentered Earth's atmosphere. Since 1970 Japan has launched several satellites into space. For example, *Nozomi* was launched in 1998 by an M-5 rocket and was to go into orbit around Mars in 2003. An equipment failure prevented this from happening. Instead, JAXA was forced to put the spacecraft into a solar orbit. Japan has developed several launch vehicles as part of its H-II series including the H-IIA and H-IIB rockets, which were in use as of 2016.

As explained previously, in 1990 the journalist Toyohiro Akiyama became the first Japanese citizen to go into outer space when he flew as a tourist to the Russian space station *Mir*. In 1992 the Japanese scientist Mamoru Mohri (1948–) became the first JAXA astronaut in space when he traveled aboard the U.S. space shuttle as a payload specialist. In 1994 Chiaki Mukai (1952–), the first Japanese woman, went into space as a member of a U.S. space shuttle crew. In 1997, while on a space shuttle mission, Takao Doi (1954–) became the first Japanese astronaut to conduct an extravehicular activity when, as reported by JAXA (October 6, 2011, http://iss.jaxa.jp/en/astro/biographies/doi/index.html), "he manually captured a solar observation satellite and tested a space crane, tools, and techniques being developed for assembly of the International Space Station."

Japan is one of the original partner nations in the *ISS* program and operates the Kibo laboratory aboard the station, which is described in Chapter 5. JAXA has a series of unmanned cargo transport vehicles called H-II transfer vehicles (HTVs) that have visited the station since 2009.

As shown in Table 1.1 and Figure 1.8 in Chapter 1, Japan conducted four launches in 2014 to put spacecraft into Earth orbit. All four flights were noncommercial missions. According to the FAA, in *Annual Compendium of Commercial Space Transportation*, all the launches were deemed successful. Chapter 6 describes Japan's space science endeavors.

Japan may have ASAT capabilities through its missile defense system. Grego indicates that the nation has under development maneuverable satellites capable of approaching satellites in orbit.

OTHER SPACE-ACTIVE NATIONS

Numerous other countries have developed their own satellites. Spacecraft payloads are provided by universities, companies, and space agencies around the world. Most satellites are designated to purposes such as communications, remote sensing, technology testing, navigation, or various scientific pursuits. National security, missile defense, and perhaps ASAT ambitions also drive space technology developments around the world. This is especially true for North Korea and Iran.

Very few countries have developed their own launch vehicles. As of 2016, the vast majority of spacecraft launches relied on rockets sourced from Russia, China, the United States, and Europe.

India has active commercial and space science programs that are operated by the Indian Space Research Organisation (ISRO), an agency of the Indian government. According to the ISRO (2016, http://www.isro.org/about-isro/genesis), the country's space activities began during the 1960s under the direction of Vikram Sarabhai (1919–1971). He is considered to be the father of India's space program. For decades India relied on foreign launch vehicles for its spacecraft. During the early 1990s it developed the Polar Satellite Launch Vehicle and, later, the GeoSynchronous Satellite Launch Vehicle. As shown in Table 1.1 and Figure 1.8 in Chapter 1, India conducted four orbital launches in 2014—three noncommercial launches and one commercial launch. The FAA indicates in *Annual Compendium of Commercial Space Transportation* that all the launches were successful. India's past, current, and future space science missions are described in Chapters 6 and 7.

Canada is also noted for its contributions to space exploration. In 1984 Marc Garneau (1949–) became the first Canadian to travel into space when he flew a mission aboard a U.S. space shuttle. The Canadian Space Agency was founded during the early 1990s. Subsequently, Canadian astronauts made additional shuttle flights and visited the *ISS*. Chapter 5 describes the nation's role in providing key components for the space station.

CHAPTER 4

PRIVATE SPACE EXPLORATION

Private applications of space technology have achieved a significant level of commercial and economic activity, and offer the potential for growth in the future, particularly in the United States.

—Commercial Space Launch Act, Public Law 98-575 (1984)

Since the dawn of the space age the National Aeronautics and Space Administration (NASA) has relied heavily on private industry to supply equipment, components, and personnel for the nation's space program. Likewise, manufacturing companies have capitalized on commercial demand for space-based assets, such as communications satellites. Hundreds of them have been put into orbit. During the late 20th century the first privately operated launch services appeared around the world. Although some are subsidized (provided with funding by national governments), the companies provide clients with alternatives to the launchpads operated solely by government agencies.

Evolving technologies are facilitating new business opportunities in space, such as tourism. In 2004 the first crewed spacecraft developed and launched by a commercial enterprise traveled just over the edge of space and back to Earth. This short trip was an example of suborbital spaceflight and represents a new realm in private space ventures. In 2010 the first commercially developed robotic spacecraft was put into orbit and then recovered successfully. This was a feat only government space agencies had previously achieved. Privately developed robotic spacecraft now make regular resupply visits to the *International Space Station* (*ISS*). With so many prospects on the horizon, space is becoming a very popular arena for doing business.

PRIVATE ORGANIZATIONS

Private organizations and clubs have played a major role in advancing space exploration. As far back as the 1920s, groups of scientists, hobbyists, and other enthusiasts were gathering together to share their passion for rocket science and space travel. For example, the Verein für Raumschiffahrt (VfR; Society for Spaceship Travel) was formed in 1927 in Berlin, Germany, by a group of scientists and authors interested in rocket research. The VfR included many famous members, including Wernher von Braun (1912–1977), who would later design the rockets for the U.S. Apollo missions for NASA. The British Interplanetary Society (BIS) was founded in 1933 in Liverpool, England. This group of scientists and intellectuals advanced many important theories of space flight, including a design for a lunar landing vehicle that was incorporated into the Apollo Program.

Many of the early groups were absorbed by government and military space organizations or evolved into aerospace manufacturing businesses. Private groups continue to advance space flight by researching and developing new technologies, operating commercial space enterprises, promoting public interest in space, and influencing government decisions on the future of spaceflight.

American Institute of Aeronautics and Astronautics

In April 1930 a group of American scientists, engineers, and writers interested in space exploration formed the American Interplanetary Society. In 1934 the name of the group was changed to the American Rocket Society (ARS). By this time the members were predominantly rocket scientists who specialized in the research, design, and testing of liquid-fueled rockets. The ARS featured many prominent members including Robert H. Goddard (1882–1945), whose theories and experiments were instrumental in the development of rocket science during the early 20th century.

During World War II several ARS members started the company Reaction Motors Inc. to support the war effort. Reaction Motors later developed rocket engines that were used in the famous series of X-planes. Over the decades, the company evolved into Morton-Thiokol Inc., the manufacturer of the space shuttle's rockets. In 2001 Alliant Techsystems Inc. bought what was then known as Thiokol Inc.

In 1932 a group of American aeronautical engineers and scientists formed the Institute of Aeronautical Sciences (IAS). Although originally focused on Earth-bound aviation, the IAS grew increasingly interested in space flight. In 1963 the IAS merged with the ARS to become the American Institute of Aeronautics and Astronautics (AIAA).

As of 2016 the AIAA (https://www.aiaa.org/AboutAIAA/) had more than 30,000 members and was the largest professional society in the world devoted to aviation and spaceflight. Its stated mission is to "inspire and advance the future of aerospace for the benefit of humanity." The AIAA has published hundreds of books and hundreds of thousands of technical papers throughout its history.

Planetary Society

The Planetary Society is a nonprofit space advocacy group based in Pasadena, California. It was founded in 1980 by the American scientists Carl Sagan (1934–1996), Bruce Churchill Murray (1931–2013), and Louis Friedman (1940–). The society's stated mission (2016, http://www.planetary.org/about/) is to "Empower the world's citizens to advance space science and exploration."

The society funds projects that support its goals and educate the public about space travel. It also encourages its members and the public to contact government leaders regarding space exploration projects. During the 1980s the society waged a campaign to encourage Congress to restore funding for NASA's Search for Extra-Terrestrial Intelligence (SETI) project. During the early 1990s the battle was over NASA's planned postponement of the Mars Observer mission. In late 2003 and early 2004, Planetary Society members sent thousands of postcards to congressional leaders to protest funding cuts that affected NASA's planned mission to Pluto. According to the Planetary Society, all three of these campaigns were successful in that government funding was restored to the projects.

In 1999 the society started the SETI@home project in which private citizens could allow their home computers to be used to analyze data recorded by a giant radio telescope as part of SETI. By the time the SETI@home project ended in December 2005, more than 5 million people had participated. The project was turned over to the University of California, Berkeley, Space Sciences Laboratory, which operates it under the Berkeley Online Infrastructure for Network Computing.

At the dawn of the 21st century the society launched the project Red Rover Goes to Mars to coincide with NASA's Mars Exploration missions. The project included an essay contest for students that resulted in the names used for the Mars rovers, *Spirit* and *Opportunity*. The contest was sponsored by the Planetary Society and the LEGO Company.

In partnership with LEGO and NASA, the Planetary Society also provided for the creation of DVDs that were mounted to the rovers for the missions. The DVDs were specially crafted out of silica glass (instead of plastic) and contain the names of nearly 4 million people who asked to be listed. Each DVD surface features a drawing of an "astrobot" saying "Hello" to Mars. The rovers safely landed on Mars in January 2004. Photos transmitted to NASA by the rovers after landing showed that the DVDs survived the journey. The rovers remained on Mars and did not return to Earth.

Other components of the Red Rover Goes to Mars project included a contest in which the winning students visited mission control during the Mars Exploration missions and a classroom project in which students built models of the Mars rovers and the Martian landscape.

On June 21, 2005, the Planetary Society launched its first spacecraft, *Cosmos 1*, to test a solar sail in orbit around Earth. A solar sail is made of very thin and highly reflective material that is propelled by photons (packets of light) from the sun. *Cosmos 1* was built in Russia with funding and technical support from the Planetary Society. The mission was cosponsored by the media company Cosmos Studios through a contract with Roscosmos. It was the first space mission ever funded by a private space interest organization. *Cosmos 1* was lost soon after launch when its Russian-supplied Volna rocket failed to fire properly.

The Planetary Society began raising money to fund a second solar sail. In May 2015 the organization conducted a test flight of its first *LightSail*. It was lofted into space inside three CubeSats—small cube-shaped satellites with each side measuring nearly 4 inches (10 cm) in length. They were launched aboard a U.S. military space plane, the X-37B, which is described in Chapter 3.

In "LightSail Spacecraft Wakes Up, Deploys Solar Sail" (June 8, 2015, CBSnews.com), Mike Wall indicates that the mission was plagued by technical glitches including battery problems. Nevertheless, the sail was successfully deployed. Wall notes that *LightSail* was put into a "relatively low orbit" where it was expected to stay for only a few days. According to the Planetary Society (http://sail.planetary.org/missioncontrol), the mission ended after 25 days when the spacecraft was destroyed upon reentry to Earth's atmosphere. In *LightSail: A Solar Sailing Spacecraft from the Planetary Society* (2015, http://sail.planetary.org/), the organization notes that a second *LightSail* is under development. It is expected to fly in 2016 aboard *Prox-1*, a satellite being developed by the Georgia Institute of Technology. The launch vehicle will be a Falcon 9 Heavy rocket, which was still under development as of February 2016.

Space Foundation

The Space Foundation was created in 1983 in Colorado Springs, Colorado. It is a nonprofit organization that advocates for spaceflight for national security, civil, and commercial purposes. It operates daily stock indices that track the performance of space industry companies in the stock markets and sponsors educational activities that are designed to encourage student interest in science, technology, engineering, and math courses and careers. It also operates several award programs under which it recognizes innovators in space exploration. For example, the General James E. Hill Lifetime Space Achievement Award (2016, https://www.spacefoundation.org/about/awards/general-james-e-hill-lifetime-space-achievement-awards) honors individuals for their "lifetime contributions" to the field. The 2016 winner was Richard H. Truly (1937–), a former astronaut and former NASA administrator.

The Space Foundation holds the annual Space Symposium that includes speakers, panel discussions, exhibits, and award ceremonies. According to the Space Foundation, in "About the Space Symposium" (2016, http://www.nationalspacesymposium.org/about/about-national-space-symposium), the first symposium in 1984 had about 250 attendees; in recent years the symposium has hosted more than 11,000 attendees.

X Prize Foundation

The X Prize Foundation is a group of private investors who offer large cash prizes to innovators who successfully compete in various competitions. The organization was founded in 1994 by the aerospace engineer Peter H. Diamandis (1961–).

Prize competitions for flying feats have a long history. In 1919 the New York City hotel owner Raymond Orteig (1870–1939) offered a $25,000 prize to the first aviator who could fly nonstop from New York to Paris or from Paris to New York. Charles A. Lindbergh Jr. (1902–1974) won the Orteig Prize in 1927.

The X Prize Foundation holds competitions in a variety of fields, including space travel. In 1996 Diamandis offered a $10 million prize to the first nongovernmental group that could take a crewed spacecraft into space, return safely to Earth, and repeat the feat with the same spacecraft within two weeks. The prize money was put up by members of the Iranian American Ansari family. As explained later in this chapter, the Ansari X Prize, as it was called, was won in 2004 by the creators of *SpaceShipOne*. Since that time the X Prize Foundation has sponsored additional competitions, including one to put a privately operated robotic lander on Earth's moon. The Google Lunar X Prize includes $30 million in prize money. According to the X Prize Foundation (http://lunar.xprize.org/), the first lander to travel 500 meters (1,640.4 feet) and transmit high-definition video and images back to Earth will win $20 million. Other prizes totaling $10 million are available for achieving other feats.

PRIVATE SPACE VENTURES

The rockets, spacecraft, and technologies associated with space travel are inordinately expensive to develop. Thus, they have historically been heavily subsidized by national governments. As explained in Chapters 2 and 3, civil space agencies and military services around the world have pumped billions of dollars into space endeavors.

In some countries, such as China, the overall economy is controlled tightly by central authorities, which means there is little to no distinction between government and commercial enterprises. For example, the China Aerospace Science and Technology Corporation (CASC) produces the rockets, spacecraft, and related products for China's space program. However, CASC is a state-run business. The storied space program of the Soviet Union was also nationalized (under government ownership and control). Following the fall of the Soviet Union in the early 1990s, privately owned aerospace companies emerged in Russia. However, the federal government often wielded substantial clout in these firms. For example, the Russian government is a partial owner of RSC Energia, the nation's largest aerospace company. During the early 2010s the Russian government began taking steps to renationalize its space industry. The article "Russia's Roscosmos Centralizes Rocket Engine Production" (MoscowTimes.com, June 24, 2015) indicates that in 2015 companies specializing in rocket engine production were being "merged under one roof to create a consolidated rocket engine building company."

The United States has a freer economy with much greater private ownership of resources. Thus, it is easier to distinguish between governmental and nongovernmental U.S. space enterprises. Nonetheless, there has been substantial interaction between the two sectors. Many public-private partnerships have helped advance the U.S. space program. The primary way in which the two sectors interact is through government contracts. Federal agencies, such as the U.S. Department of Defense (DOD) and NASA, pay companies to design, develop, and manufacture rockets and other spacecraft. The U.S. military space program described in Chapter 3 is largely supported by companies working under federal contracts. For example, the United Launch Alliance (ULA), a collaboration between Boeing and Lockheed Martin, supplies the Atlas and Delta rockets used to launch most national security satellites into orbit.

The original National Aeronautics and Space Act of 1958 did not specifically mention commercial goals for the U.S. space program. However, innovators in private industry were already theorizing about using satellites for telecommunications. They turned to NASA for its launch services. In addition, the agency relied heavily on contractors in the private sector to supply labor and hardware. According to Roger D. Launius, in *Apollo: A Retrospective Analysis* (July 1994, http://history.nasa.gov/Apollomon/Apollo.html), "During the 1960s somewhere between 80 and 90 percent of NASA's overall budget went for contracts to purchase goods and services from others." Over the decades much of the money poured into the agency has flowed to businesses supporting the civil space program.

Besides, traditional industry contracts, NASA also engages the private sector through prize competitions, similar to those held by the X Prize Foundation. For example, in 2005 NASA launched Centennial Challenges (http://www.nasa.gov/directorates/spacetech/centennial_challenges/overview.html). The program offers cash prizes for the completion of certain tasks, many of which are related to air or space flight. The competitors cannot use government funding but can include businesses, private organizations, student groups, and even individuals. As of February 2016, the program offered a $5 million prize to design and build a moon satellite for the CubeQuest challenge and a $1.5 million prize for a robotic sampler under the Sample Return Robot Challenge.

During the early decades of the space age, NASA and the military were the only U.S. entities with launch facilities. Only they could provide U.S. companies with access to space. Over time, new opportunities have arisen for space access, including nongovernmental launch services. Technological advances and a new attitude in the U.S. government are increasingly giving U.S. businesses greater leeway to space.

OFFICE OF COMMERCIAL SPACE TRANSPORTATION

The Commercial Space Launch Act was passed in 1984. It established the Office of Commercial Space Transportation within the U.S. Department of Transportation. The office was later moved to the Federal Aviation Administration (FAA). According to John Sloan, in "Introduction to FAA Office of Commercial Space Transportation (AST) and International Outreach" (October 18, 2012, http://www.gwu.edu/~spi/assets/docs/John%20Sloan%20charts.pdf), the office's licensing authority was delegated to the FAA's Associate Administrator for Commercial Space Transportation (AST); this acronym came to be used for the office as a whole.

Sloan indicates that the AST's authority originally applied only to commercial launches. Over time, responsibilities were added over the reentry of commercial spacecraft and over commercial human spaceflight. On its website (2016, http://www.faa.gov/about/office_org/headquarters_offices/ast/about/), the AST lists its roles as follows:

- Regulate the U.S. commercial space transportation industry, to ensure compliance with international obligations of the United States, and to protect the public health and safety, safety of property, and national security and foreign policy interests of the United States;
- Encourage, facilitate, and promote commercial space launches and reentries by the private sector;
- Recommend appropriate changes in Federal statutes, treaties, regulations, policies, plans, and procedures; and
- Facilitate the strengthening and expansion of the United States space transportation infrastructure.

In "About the Office" (2016, http://www.faa.gov/about/office_org/headquarters_offices/ast/about/faq/#l11), the FAA notes that it requires a license or permit for commercial launches that take place within U.S. borders or are conducted abroad by U.S. entities. The agency indicates that it generally "does not license launches by U.S. government organizations." In addition, the FAA does not license the launching of certain small rockets, such as those built by amateur rocket enthusiasts.

The AST explains in "Licenses, Permits & Approvals" (July 10, 2014, http://www.faa.gov/about/office_org/headquarters_offices/ast/licenses_permits/) that it has the authority to issue experimental permits for reusable suborbital vehicles (RSVs). "Suborbital" means less than a complete orbit; thus, a suborbital flight does not make a complete orbital revolution around Earth. Suborbital flights and RSVs are described in detail later in this chapter.

Another responsibility of the AST is to administer the U.S. government's indemnification program for commercial launch and reentry operators. The U.S. Government Accountability Office (GAO) indicates in *Federal Aviation Administration: Commercial Space Launch Industry Developments Present Multiple Challenges* (August 2015, http://www.gao.gov/assets/680/672144.pdf) that federal law requires launch companies to purchase insurance from private insurers to cover claims up to the "maximum probable loss." The GAO notes that "the FAA's determination of the maximum probable loss will not exceed the lesser of $500 million or the maximum third-party liability insurance available on the world market at a reasonable cost, as determined by FAA." The U.S. government covers claims above that amount up to a cap, which was $3.1 billion as of 2015.

According to the GAO, in *Commercial Space Transportation: Industry Trends, Government Challenges, and*

International Competitiveness Issues (June 20, 2012, http://www.gao.gov/assets/600/591728.pdf), when the Commercial Space Launch Amendments Act was passed in 2004, it charged the FAA with overseeing space tourism safety. Nevertheless, the FAA was prohibited from regulating crew and passenger safety "before October 2015, except in response to high-risk incidents, serious injuries, or fatalities." In November 2015 the U.S. Commercial Space Launch Competitiveness Act (https://www.faa.gov/about/office_org/headquarters_offices/ast/media/US-Commercial-Space-Launch-Competitiveness-Act-2015.pdf) was passed. It extends the prohibition period to October 2023.

As of February 2016, no commercial human spaceflights had taken place; thus, the FAA had no practical data on which to base detailed regulations for this burgeoning industry.

Spaceports

As of February 2016, the FAA (http://www.faa.gov/data_research/commercial_space_data/licenses/) listed 10 licensed spaceports in the United States. The commonly used names of the spaceports, their locations, and operators are as follows:

- California Spaceport at the Vandenberg Air Force Base in California—operated by Spaceport Systems International
- Cecil Field in Florida—operated by the Jacksonville Aviation Authority
- Houston Spaceport at Ellington Airport in Texas—operated by the Houston Airport System
- Mid-Atlantic Regional Spaceport at NASA's Wallops Flight Facility in Virginia—operated by the Virginia Commercial Space Flight Authority
- Midland International Airport in Texas—operated by the Midland International Airport
- Mojave Air and Spaceport at the Mojave Airport in California—operated by the Mojave Air & Space Port
- Oklahoma Spaceport at the Clinton-Sherman Airport in Oklahoma—operated by the Oklahoma Space Industry Development Authority
- Pacific Spaceport Complex (formerly the Kodiak Launch Complex) in Alaska—operated by the Alaska Aerospace Development Corporation
- Spaceport America in New Mexico—operated by the New Mexico Spaceflight Authority
- Spaceport Florida at the Cape Canaveral Air Force Station (CCAFS) in Florida—operated by Space Florida

The locations of the spaceports are shown in Figure 4.1, along with the locations of U.S. federal launch sites, sole site operator launch sites, and proposed launch sites (as of July 2015). Note that sole site operator launch sites, such as the Blue Origin launch site in Texas, do not require FAA licensing as spaceports. The sites are exempt because they are used exclusively by the companies that own and operate them.

Licensed spaceports operate with varying levels of government support. The California Spaceport (2016, http://www.calspace.com/SLC-8/SLC-8_Overview.html) claims to be the only spaceport operated solely on private funds—that is, without taxpayer funding. By contrast, the Alaska Aerospace Development Corporation is a public corporation of the state of Alaska. Some of the spaceports use existing launch infrastructure at U.S. military or NASA facilities. Others, such as Spaceport America, were built from scratch. The runway-based spaceports are equipped only for spacecraft that take off and land horizontally like airplanes.

The spaceports cater to different types of paying customers based on their launch capabilities. Some depend on federal government clients, such as the U.S. military, whereas others focus on commercial clients, particularly those in the emerging space tourism business.

Other Regulatory Bodies

The FAA explains in *Annual Compendium of Commercial Space Transportation: 2014* (February 2015, http://www.faa.gov/about/office_org/headquarters_offices/ast/media/FAA_Annual_Compendium_2014.pdf) that it is not the only federal agency that regulates commercial space-related activities in the United States. The Federal Communications Commission (FCC), the National Oceanic and Atmospheric Administration (NOAA), the U.S. Department of Commerce (DOC), and the U.S. Department of State (DOS) are also involved. For example, NOAA licenses commercial remote-sensing satellites. The FCC licenses commercial communications transmissions via satellite to, from, and within the United States. The DOS and the DOC license the export of certain items as required by law. The export of U.S. satellites and related items is particularly sensitive from a national security standpoint because such items could be used for military purposes.

COMMERCIAL SATELLITE LAUNCHES

For decades NASA was the primary launch-service provider for commercial U.S. satellites. According to Daniel R. Glover, in *Beyond the Ionosphere: Fifty Years of Satellite Communication* (1997, http://history.nasa.gov/SP-4217/ch6.htm), the *Telstar 1* satellite was the agency's first privately sponsored space launch in 1962. The satellite was developed by AT&T and Bell Telephone Laboratories to reduce reliance on the undersea

FIGURE 4.1

Commercial space launch sites and proposed sites as of July 2015

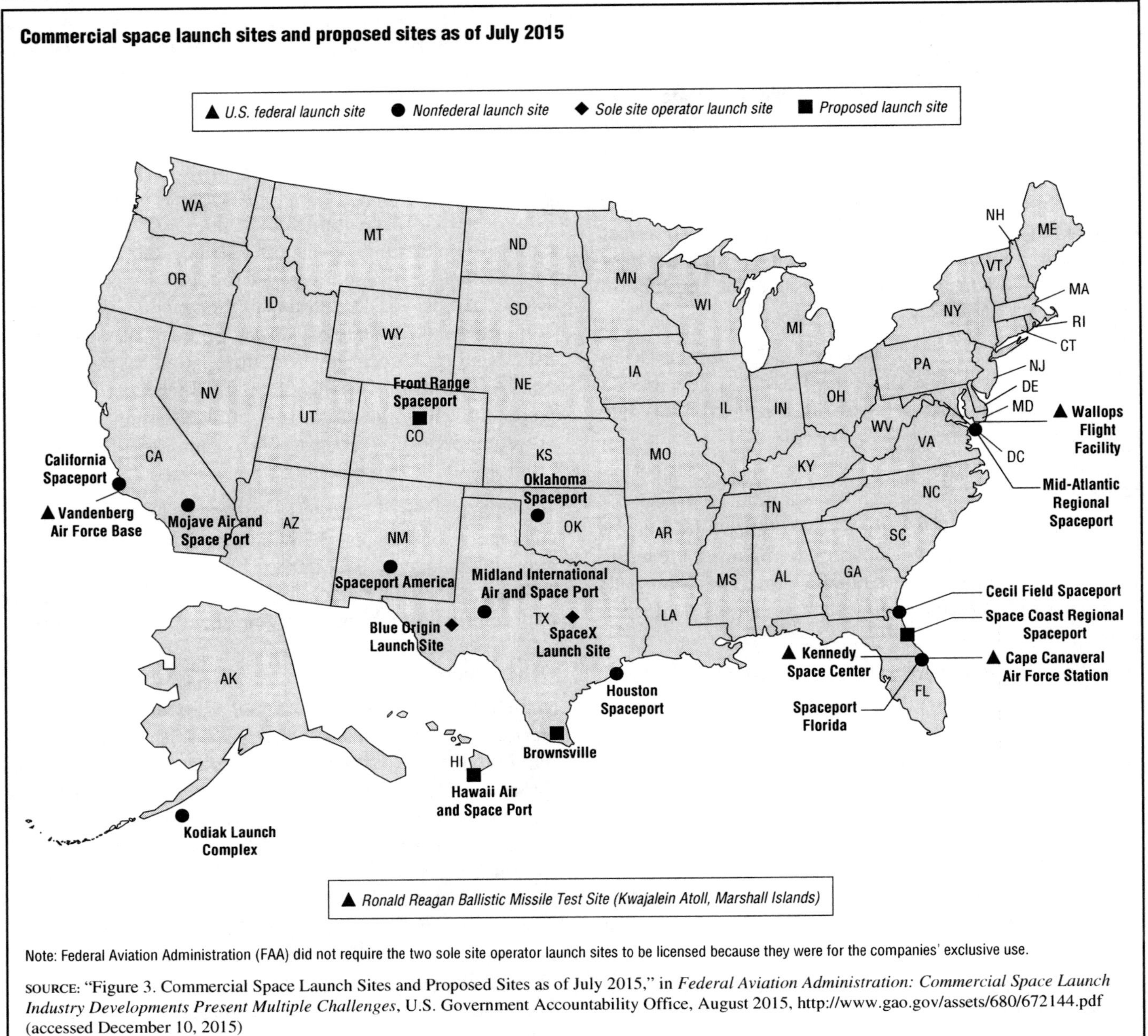

Note: Federal Aviation Administration (FAA) did not require the two sole site operator launch sites to be licensed because they were for the companies' exclusive use.

SOURCE: "Figure 3. Commercial Space Launch Sites and Proposed Sites as of July 2015," in *Federal Aviation Administration: Commercial Space Launch Industry Developments Present Multiple Challenges*, U.S. Government Accountability Office, August 2015, http://www.gao.gov/assets/680/672144.pdf (accessed December 10, 2015)

cables that carried telecommunications across the Atlantic Ocean. Glover notes that *Telstar 1* provided the "first transmission of live television across the Atlantic Ocean from the United States to France" on the day it was launched. The companies paid NASA $6 million for launch, tracking, and telemetry services. Until the 1980s NASA continued to make money from the private sector by launching commercial satellites.

As noted in Chapter 2, the 1986 disaster that destroyed the space shuttle *Challenger* prompted numerous changes in NASA's operation. According to NASA (http://www.hq.nasa.gov/office/pao/History/policy88.html), these key policy changes were laid out by President Ronald Reagan (1911–2004) in the "Presidential Directive on National Space Policy," which was issued on February 11, 1988.

The U.S. military and private sector were forced to return to using expendable launch vehicles (ELVs). Furthermore, shuttles were instructed to cease carrying commercial and military payloads. In *Space Launch Vehicles: Government Activities, Commercial Competition, and Satellite Exports* (March 20, 2006, https://www.fas.org/sgp/crs/space/IB93062.pdf), Carl E. Behrens notes that this restriction "facilitated the emergence of a U.S. commercial space launch industry whose participants had long argued that they could not compete against government-subsidized shuttle launch prices."

In 1995 Boeing formed a satellite-launching business called the Sea Launch Company with Russian, Norwegian, and Ukrainian partners. It was headquartered in Long Beach, California, and operated a rocket launch platform on a modified oil-drilling platform in the South Pacific Ocean. Another major company to enter the space launch industry was International Launch Services (ILS; http://www.ilslaunch.com/about-us/ils-legacy). ILS was formed in 1995 following the merger of two independent launch providers: Martin Marietta and Lockheed.

The FAA (2016, http://www.faa.gov/about/office_org/headquarters_offices/ast/launch_data/historical_launch/) maintains a database of every launch it has licensed or permitted dating back to 1989. Examination of the FAA database shows the United States had a robust commercial launching industry through the 1990s. For example, 22 launches were conducted during 1998. Almost all of them (17) were satellites for commercial clients, such as telecommunications companies. Three launches were for U.S. government payloads—*Lunar Prospector* and *SNOE* for NASA and *GEOSAT F/O* for the U.S. Navy. In addition, satellites were launched in 1998 for the UK military (*Skynet 4D*) and Brazil's space agency (*SCD-2*).

Through the 1990s most U.S. launches were by Delta or Atlas rockets. These are ELVs that were developed to support the DOD and NASA. In addition, the vast majority of U.S. launches took place at DOD and NASA facilities, particularly the Cape Canaveral Air Force Station (CCAFS) in Florida.

In 1997 the FAA licensed its first launch outside the United States. On April 21, 1997, a Spanish satellite was launched into orbit from Spanish territory in the Canary Islands. The launch vehicle was a Pegasus XL rocket developed by Orbital Sciences. (See Figure 4.2.)

FIGURE 4.2

Pegasus XL

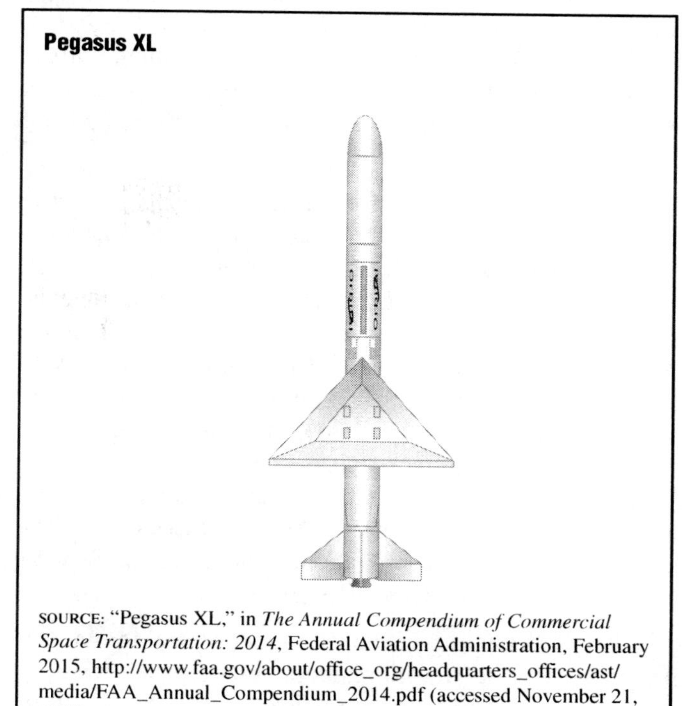

SOURCE: "Pegasus XL," in *The Annual Compendium of Commercial Space Transportation: 2014*, Federal Aviation Administration, February 2015, http://www.faa.gov/about/office_org/headquarters_offices/ast/media/FAA_Annual_Compendium_2014.pdf (accessed November 21, 2015)

Foreign Competition

After the Soviet Union collapsed during the early 1990s, Russian-made rockets became available to commercial clients. As noted earlier, Sea Launch and ILS were both founded in 1995. On March 28, 1999, Sea Launch test-launched a Russian Zenit rocket. Later that year it used a Zenit rocket to launch its first commercial satellite, the U.S.-made *DIRECTV 1R*.

By the dawn of the 20th century, U.S. commercial launch service providers faced fierce foreign competition. A major contender was the French company Arianespace, which in 1979 had become the world's first commercial provider of launch services. As explained in Chapter 3, the company's Ariane rocket family became the primary launch vehicles for European spacecraft. Arianespace launches take place at the Kourou spaceport in French Guiana in South America. (See Figure 3.6 in Chapter 3.) U.S. companies found it increasingly difficult to compete with foreign-made rockets and foreign-provided launch services. In addition, two of the largest U.S. launch consortiums, ILS and the Sea Launch Company, were acquired by foreign investors.

The two original ILS partners, Martin Marietta and Lockheed, were bought by other companies, and ILS (2016, http://www.ilslaunch.com/about-us) became a subsidiary of a Russian company called the Khrunichev State Research and Production Space Center. As of February 2016, ILS was still headquartered in the United States; however, it provided only Russian-made Proton rockets that launched from the Baikonur Cosmodrome in Kazakhstan. As explained in Chapter 3, the Russian government pays the government of neighboring Kazakhstan for use of the cosmodrome.

In January 2007 a rocket carrying a Dutch communications satellite exploded during liftoff from the Sea Launch Company's platform in the Pacific Ocean. The resulting bad publicity drove potential customers to other launch service providers. In 2010 the company filed for bankruptcy. W. J. Hennigan notes in "Sea Launch Looking up Again with New Rocket Mission" (LATimes.com, September 22, 2011) that in late 2010 the Russian company Rocket & Space Corp. Energia invested $155 million in Sea Launch to boost it out of bankruptcy. It also moved the Sea Launch headquarters to Bern, Switzerland, but continued to maintain facilities in California. As a result, Sea Launch launches remained subject to FAA permits. In September 2011 Sea Launch successfully conducted its first launch since the 2007 mishap. However, in "China Eyes Purchase of Sea Launch Assets" (SpaceNews.com, July 17, 2015), Peter B. de Selding notes that a 2013 launch failure seriously hurt the company's business. Sea Launch began selling assets and, as of February 2016, had not conducted a launch since 2014.

The involvement of foreign governments in space launch companies helps their success in the international market. The French and Russian commercial space launch industries are subsidized by their respective governments. Arianespace is also subsidized by the European Space Agency (ESA). The state-run Chinese company CASC began offering commercial satellite launches during the 1990s. According to the GAO, in *Commercial Space Launch Act: Preliminary Information on Issues to Consider for Reauthorization* (June 6, 2012, http://www.gao.gov/assets/600/591391.pdf), the French, Russian, and Chinese governments also provide unlimited indemnification coverage for satellite-launching operations.

By the beginning of the 21st century, foreign launch providers serving the commercial market had severely reduced U.S. competitiveness. The FAA (2016, http://www.faa.gov/about/office_org/headquarters_offices/ast/launch_data/historical_launch/) indicates that from 2000 through 2011 it licensed only 82 U.S. launches, or fewer than seven launches per year on average. Sea Launch and its successor Rocket & Space Corp. Energia had the largest number (29) of launches during this period, but all of them were conducted using Russian Zenit rockets.

Table 4.1 is a list compiled by the FAA of non-U.S. commercial orbital launch vehicles that were operational in 2014. French, Russian, and Chinese operators dominate the list.

The GAO notes in *Commercial Space Transportation: Industry Trends, Government Challenges, and International Competitiveness Issues* that between 2002 and 2011 total worldwide revenue related to commercial space launches was between $1 billion and $2.5 billion annually. The U.S. private sector captured, at most, only a few hundred million dollars each year.

SpaceX

In 2009 an American newcomer entered the launch service business. Space Exploration Technologies Corp. (SpaceX) is a U.S. company founded in 2002 by Elon Musk (1971–), a cofounder of the PayPal online payment service. SpaceX developed the Falcon launch vehicle and launched it at Omelek Island, which is part of the Kwajalein Atoll in the Marshall Islands in the central Pacific Ocean.

After its founding, SpaceX quickly obtained work from U.S. government agencies with payloads for its Falcon 1 rockets. However, the first three launches in

TABLE 4.1

Orbital launch vehicles available for commercial use outside the United States

[There are 13 expendable launch vehicle types available for commercial use outside the United States: Ariane 5, Dnepr, Epsilon, GSLV, H-IIA/B, Long March 2D, Long March 3A, Proton M, PSLV, Rockot, Soyuz 2, Vega, and Zenit 3SL/SLB.]

Operator	Vehicle	Year of first launch	Total/2014 launches	Active launch sites	Mass to GTO kg (lb)	Mass to LEO kg (lb)	Mass to SSO kg (lb)
Antrix/ISRO	GSLV	2001	8/1	Satish Dhawan	2,500 (5,516)	5,000 (11,023)	—
Antrix/ISRO	PSLV	1993	26/3	Satish Dhawan	1,425 (3,142)	3,250 (7,165)	1,750 (3,850)
Arianespace	Ariane 5	1996	77/6	Guiana Space Center	9,500 (20,944)	21,000 (46,297)	10,000 (22,046)
Arianespace	Soyuz 2	2004	38/10	Baikonur, Guiana Space Center, Plesetsk	3,250 (7,165)	4,850 (10,692)	4,400 (9,700)
Arianespace	Vega	2012	3/1	Guiana Space Center	—	1,500 (3,307)	—
CGWIC	Long March 2	2C: 1975 2D: 1992	2C: 38/4 2D:15/2	Jiuquan, Taiyuan, Xichang	2C: 1,250 (2,756)	2C: 3,850 (8,488)	2C: 1,900 (4,189) 2D: 1,300 (2,866)
CGWIC	Long March 3A	A: 1994 B: 1996 BE: 2007 C: 2008	A: 23/0 B: 13/0 BE:13/0 C: 11/1	Xichang	A: 2,600 (5,732) B: 5,100 (11,244) BE: 5,500 (12,125) C: 3,800 (8,378)	—	—
Eurockot	Rockot	1990	22/2	Baikonur, Plesetsk	—	2,140 (4,718)	—
ILS	Proton M	2001	86/8	Baikonur	6,920 (15,256)	23,000 (50,706)	—
ISC Kosmotras	Dnepr	1999	21/2	Baikonur, Dombarovsky	—	3,700 (8,157)	2,300 (5,071)
JAXA	Epsilon	2013	1/0	Uchinoura	—	1,200 (2,646)	700 (1,543)
Mitsubishi	H-IIA/B	A: 2001 B: 2009	A: 26/4 B: 4/0	Tanegashima	A: 4,100–6,000 (9,039–13,228) B: 8,000 (17,637)	A: 10,000–15,000 (22,046–33,069) B: 19,000 (41,888)	—
Sea Launch/ Land Launch	Zenit 3	3SL: 1999 3SLB: 2008	3SL: 36/1 3SLB: 6/0	Baikonur, Sea Launch	3SL: 6,000 (13,228) 3SLB: 3,500 (7,716)	—	—

CGWIC = China Great Wall Industry Corporation. GSLV = Geosynchronous Satellite Launch Vehicle. GTO = geosynchronous transfer orbit. ILS = International Launch Services. JAXA = Japan Aerospace Exploration Agency. kg = kilograms. LEO = Low Earth Orbit. lb = pounds. PSLV = Polar Satellite Launch Vehicle. SSO = Sun-Synchronous Orbit.

SOURCE: "Table 2. Currently Available Non-U.S. Commercial Launch Vehicles," in *The Annual Compendium of Commercial Space Transportation: 2014*, Federal Aviation Administration, February 2015, http://www.faa.gov/about/office_org/headquarters_offices/ast/media/FAA_Annual_Compendium_2014.pdf (accessed November 21, 2015)

2006 through mid-2008 failed. On September 28, 2008, the fourth Falcon 1 launch was successful. The rocket carried a payload mass simulator, which is basically a test load designed to simulate an actual payload. The following year a Falcon 1 carried its first commercial payload, the RazakSAT, for the Astronautic Technology Sdn Bhd, a satellite company controlled by the Malaysian government. The satellite was put into a low Earth orbit (LEO), which lies approximately 100 to 1,200 miles (160 to 1,900 km) above Earth's surface.

Meanwhile SpaceX had developed an enhanced rocket called the Falcon 9, which is capable of carrying much heavier payloads into more diverse orbits. In *Annual Compendium of Commercial Space Transportation: 2014*, the FAA indicates that the rocket's first configuration or version (v) was dubbed Falcon 9 v1.0 and debuted in 2010.

In 2013 SpaceX unveiled a new version called the Falcon 9 v1.1. (See Figure 4.3.) On December 3, 2013, SpaceX used a Falcon 9 v1.1 at the CCAFS to launch the *SES-8*, a commercial satellite operated by SES, a European telecommunications company. The launch was particularly significant because the satellite was put into a geosynchronous transfer orbit (GTO). According to NASA (2016, http://www2.jpl.nasa.gov/basics/bsf5-1.php), a GTO is an elliptical orbit that at its farthest point from Earth is about 22,991 miles (37,000 km) above the planet's surface. From a GTO orbit a satellite can fire its small onboard rockets to maneuver into its final geosynchronous orbit. This is an orbit that duplicates Earth's time period to rotate on its axis. A geostationary orbit is a geosynchronous orbit that is perfectly round. In this orbit a satellite appears to remain "parked" above one point on Earth. It is a very common orbit for communications satellites. Launching satellites into a GTO is more difficult than launching them into LEO. Thus, SpaceX's achievement attracted attention.

As shown in Table 1.1 and Table 1.2 in Chapter 1, there were 11 U.S. commercial orbital launches during 2014. SpaceX launched four private satellites and conducted two launches for NASA. Orbital Sciences Corporation (or Orbital Sciences) conducted four launches for NASA, but only three were successful. The ULA launched a satellite for a private company.

According to the GAO, in *Federal Aviation Administration: Commercial Space Launch Industry Developments Present Multiple Challenges*, U.S. space launch revenue was $1.1 billion in 2014, up dramatically over the previous decade. The GAO notes, "the growth in the U.S. commercial space launch industry is largely being led by SpaceX being more price competitive compared with foreign launch providers."

THE SPACEX COST DIFFERENCE. SpaceX receives a lot of attention for its low prices. In "Is SpaceX Changing the Rocket Equation?" (AirSpaceMag.com, January 2012), Andrew Chaikin describes the major cost-cutting factors that Musk implemented at SpaceX. The company designs and manufactures most of its own rocket components, rather than buying them piece-by-piece from aerospace vendors. This includes small mechanical parts like valves and sophisticated elements, such as flight computers. Chaikin states, "that's something SpaceX didn't originally set out to do, but was driven to by suppliers' high prices."

Chaikin notes that SpaceX engineers learned a great deal about rocket design by studying the "vast technical archive" of information at NASA. In fact, the company modeled its rocket engine, the Merlin, on the engines used on the Apollo modules that landed on Earth's moon. Falcon launch vehicles use multiple Merlin engines, rather than relying on a mixture of engine and booster types, as is common for other launch vehicles, such as the Atlas V.

FIGURE 4.3

Falcon 9

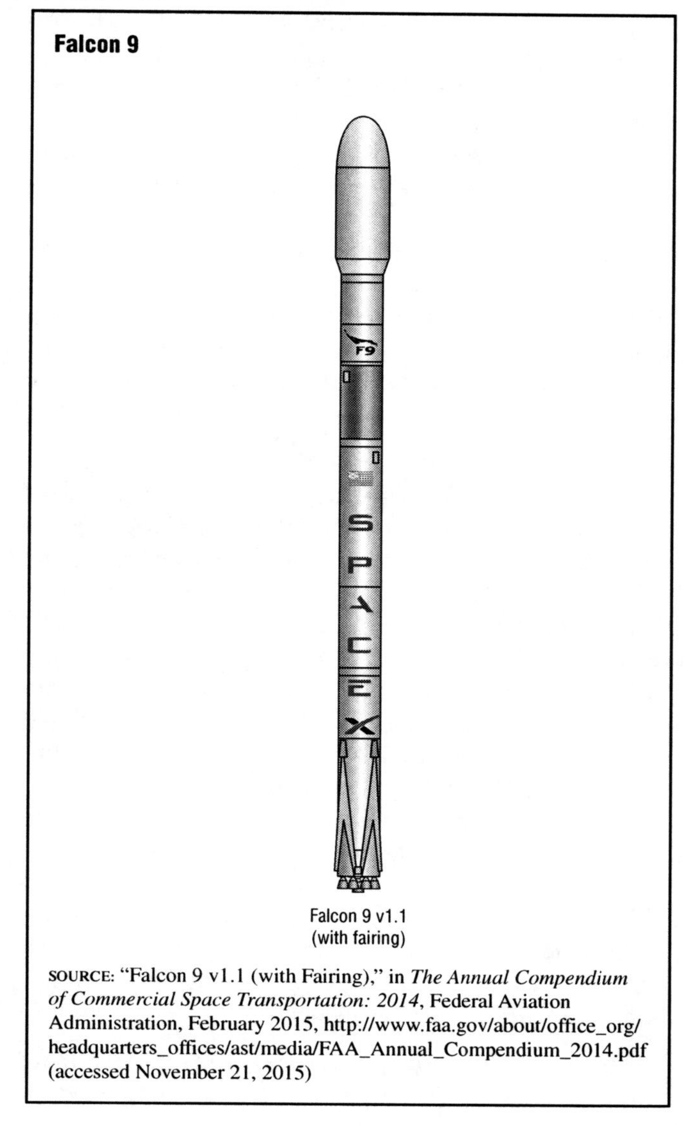

Falcon 9 v1.1
(with fairing)

SOURCE: "Falcon 9 v1.1 (with Fairing)," in *The Annual Compendium of Commercial Space Transportation: 2014*, Federal Aviation Administration, February 2015, http://www.faa.gov/about/office_org/headquarters_offices/ast/media/FAA_Annual_Compendium_2014.pdf (accessed November 21, 2015)

One innovation that could make the Falcon line even cheaper in the future is reusable rocket stages. Historically rocket stages have been expendable. After exhausting their fuel supply they fall away from the launch vehicle during flight and are lost. On December 21, 2015, SpaceX made history by guiding a used rocket stage back to Earth for a soft vertical landing. The first-stage booster from a Falcon 9 returned approximately eight minutes after it helped launch a multi-satellite payload into orbit. SpaceX had previously attempted, but failed, to safely land used Falcon 9 boosters on floating platforms in the ocean. If used rocket stages can be recovered and reused, it would dramatically lower the price of space launches.

2015—A MIXED YEAR. SpaceX had a mixed year in 2015. In February 2015 a Falcon 9 launched the *Deep Space Climate Observatory* into orbit. The NOAA spacecraft is described in Chapter 6. The launch was called the DSCVR mission and helped SpaceX become certified by the U.S. Air Force to carry national security payloads. (See Chapter 3.) In November 2015 SpaceX was the only bidder for a 2018 launch of a global positioning system satellite for the U.S. Air Force. As of February 2016, the contract had not been awarded but was expected to go to SpaceX. In addition, the company (http://www.spacex.com/missions) notes that it conducted three launches for commercial clients during March, April, and December 2015, and launches for NASA in January and April 2015. However, the company suffered a serious blow during June 2015 when a Falcon 9 exploded minutes after takeoff. It was carrying cargo bound for the *ISS*, a mission described later in this chapter.

The mishap did not appear to seriously hurt business at SpaceX, and a Falcon 9 launch was completed in January 2016. As of February 2016, the company's flight manifest (http://www.spacex.com/missions) listed dozens of scheduled Falcon 9 launches for 2016 and beyond. Also in 2016, the company plans to conduct the first demonstration flight of its new Falcon Heavy launch vehicle. According to SpaceX (2016, http://www.spacex.com/falcon-heavy), the Falcon Heavy will be able to "lift more than twice the payload of the next closest operational vehicle, the Delta IV Heavy, at one-third the cost." If the new vehicle line passes the required certification tests, it will compete to carry large national security payloads into orbit. The Falcon Heavy launch vehicle has also been chosen by NASA to transport astronauts to and from the *ISS* in the future.

NASA PROVIDES NEW COMMERCIAL OPPORTUNITIES

As described in Chapter 2, the U.S. space shuttle program began winding down during the first decade of the 20th century. However, the United States and its international partners were still committed to operating the *ISS* for years to come. NASA did not have a spacecraft ready to replace the retiring shuttle. The United States had to rely on other ISS program partners, chiefly Russia, Japan, and the ESA, to provide transport to and from the station until new U.S. spacecraft became available.

NASA was directed to forge public-private partnerships to facilitate U.S. participation in the ISS program. In *Commercial Launch Vehicles: NASA Taking Measures to Manage Delays and Risks* (May 26, 2011, http://www.gao.gov/assets/130/126310.pdf), the GAO indicates that one impetus was provided by President George W. Bush (1946–). In February 2004 he laid out his Vision for Space Exploration (http://www.nasa.gov/pdf/55583main_vision_space_exploration2.pdf), which directed NASA to work with the U.S. private sector to find new options for transporting American goods and astronauts to the *ISS*. In response, NASA formed the Commercial Orbital Transportation Services (COTS) program.

ISS Cargo Transportation

NASA's first COTS goal was to demonstrate that private spacecraft could successfully transport cargo to the *ISS*. According to the agency (March 2, 2012, http://www.nasa.gov/offices/c3po/about/c3po.html), in 2006 two private companies were selected for the initial demonstration phase: SpaceX and Rocketplane-Kistler in Oklahoma. Rocketplane-Kistler was dropped from the program after it failed to meet designated milestones. It was replaced in 2008 with Orbital Sciences.

SpaceX developed the Dragon, which is depicted in Figure 4.4. Its launch vehicle is the company's Falcon 9 rocket. Figure 4.5 shows a Falcon 9 v1.1 rocket loaded with a Dragon capsule. Orbital Sciences developed a spacecraft called the Cygnus. (See Figure 4.6.) During its research and development phase, the Cygnus was launched by the company's Taurus II rocket. In 2011 Orbital Sciences changed the launch vehicle name to Antares. Figure 4.7 shows an Antares 120, one of several configurations of the rocket. In 2015 Orbital Sciences merged with divisions of Alliant Techsystems to form the new company Orbital ATK. (Alliant Techsystems manufactured the solid rocket boosters that NASA used in the space shuttle system.)

Both the Dragon and the Cygnus can go into Earth orbit and dock with the *ISS*. They can be loaded with trash and other items at the station before their return flights. The Dragon has a heat shield and parachute system and splashes down in the ocean. Thus, the Dragon capsules are reusable. By contrast, the Cygnus spacecraft and their contents are destroyed during fiery reentries into Earth's atmosphere.

DRAGON FLIGHTS. On June 4, 2010, a Falcon 9 rocket carried a mock-up Dragon into orbit. The test flight was

FIGURE 4.4

Dragon (cargo only)

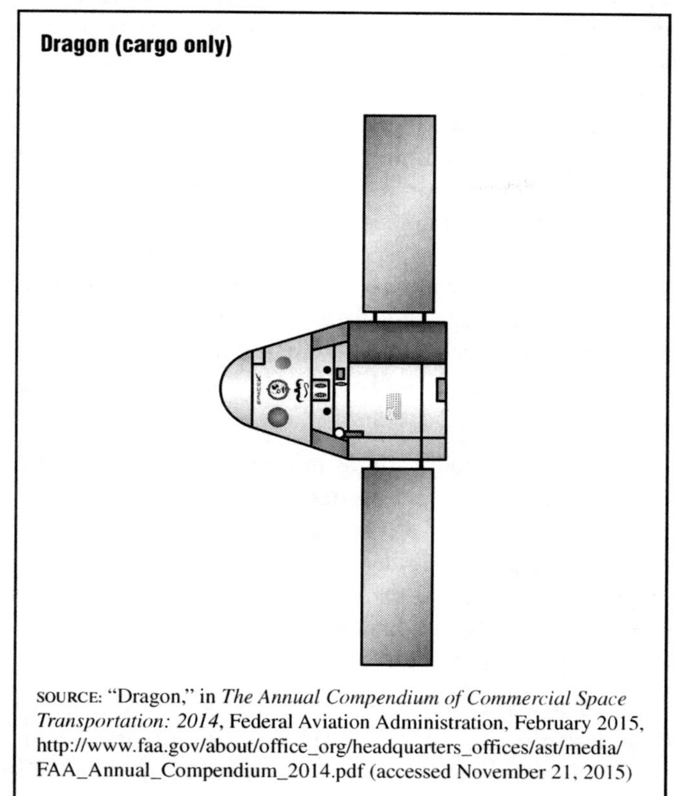

SOURCE: "Dragon," in *The Annual Compendium of Commercial Space Transportation: 2014*, Federal Aviation Administration, February 2015, http://www.faa.gov/about/office_org/headquarters_offices/ast/media/FAA_Annual_Compendium_2014.pdf (accessed November 21, 2015)

successful. On December 8, 2010, an actual Dragon spacecraft successfully completed its first demonstration flight under the COTS program. NASA states in *COTS 2 Mission Press Kit* (May 14, 2012, http://www.nasa.gov/pdf/649910main_cots2_presskit_051412.pdf) that "SpaceX became the first commercial company to send a spacecraft to low-Earth orbit and recover it successfully, something that only three governments—the United States, Russia and China—have ever done."

A second COTS demonstration flight of the Dragon took place in May 2012. It actually included activities that had been planned for a third COTS demonstration flight. Thus, it was a combined second and third demonstration mission. On May 22, 2012, the Dragon was launched into orbit. It was maneuvered near the *ISS* and, after considerable systems testing, was docked to the station. It was the first commercial spacecraft in history to dock with the *ISS*. The Dragon returned safely to Earth. Later that year NASA declared the COTS demonstration phase completed for the Dragon spacecraft.

In *Commercial Cargo: NASA's Management of Commercial Orbital Transportation Services and ISS Commercial Resupply Contracts* (June 13, 2013, http://oig.nasa.gov/audits/reports/FY13/IG-13-016.pdf), NASA's Office of Inspector General (OIG) notes that the agency agreed to pay SpaceX $396 million for successful completion of the COTS program. However, SpaceX "shared

FIGURE 4.5

Falcon 9 with Dragon

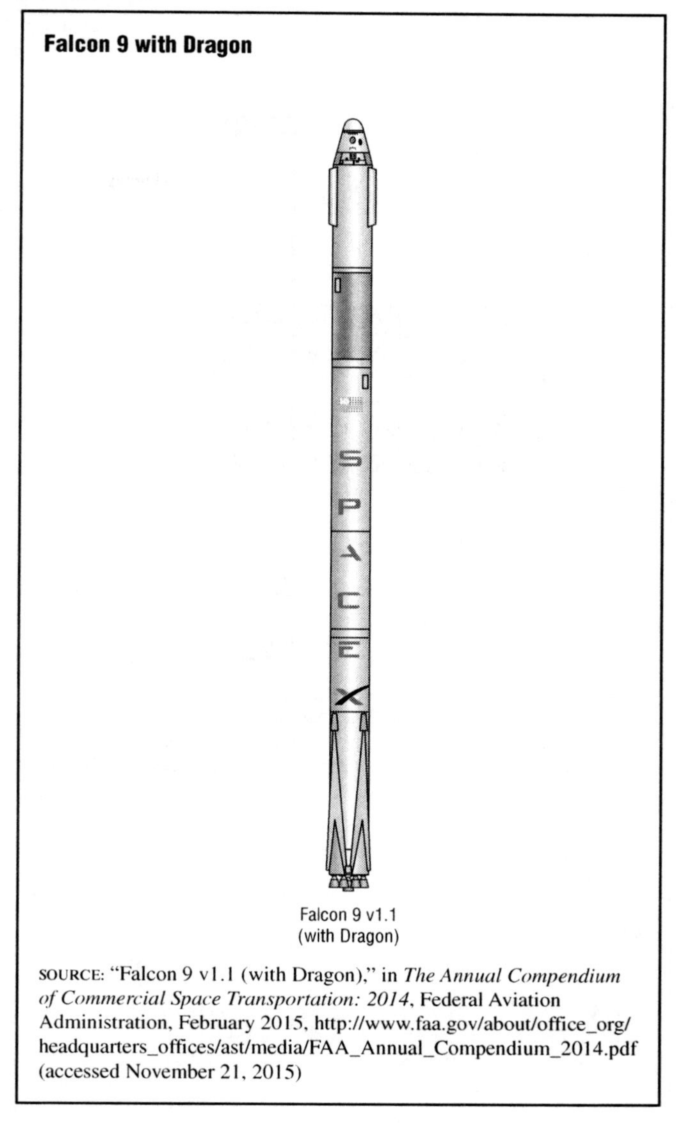

SOURCE: "Falcon 9 v1.1 (with Dragon)," in *The Annual Compendium of Commercial Space Transportation: 2014*, Federal Aviation Administration, February 2015, http://www.faa.gov/about/office_org/headquarters_offices/ast/media/FAA_Annual_Compendium_2014.pdf (accessed November 21, 2015)

costs" with NASA by contributing more than half of the funds needed for spacecraft development. In 2008 the two entered into a contract calling for NASA to pay SpaceX $1.6 billion for 12 cargo missions to the *ISS*. According to Stephen Clark, in "NASA Orders Missions to Resupply Space Station in 2017" (SpaceFlightNow.com, March 7, 2015), in March 2015 NASA added three additional missions under an extension provision of the original contract.

On October 7, 2012, a Dragon was launched on its first official commercial resupply (CRS) mission to the *ISS*. NASA dubbed the mission SpaceX-1 or CRS-1. SpaceX followed up with one cargo resupply mission in 2013, two missions in 2014, and one mission in early 2015. All the flights were successful. On June 28, 2015, a Falcon 9/Dragon launch to the *ISS* ended in disaster when the spacecraft exploded approximately two minutes after lifting off from the CCAFS. In "NASA Vows to Press on Despite SpaceX Rocket Explosion" (USAToday.com,

FIGURE 4.6

Cygnus

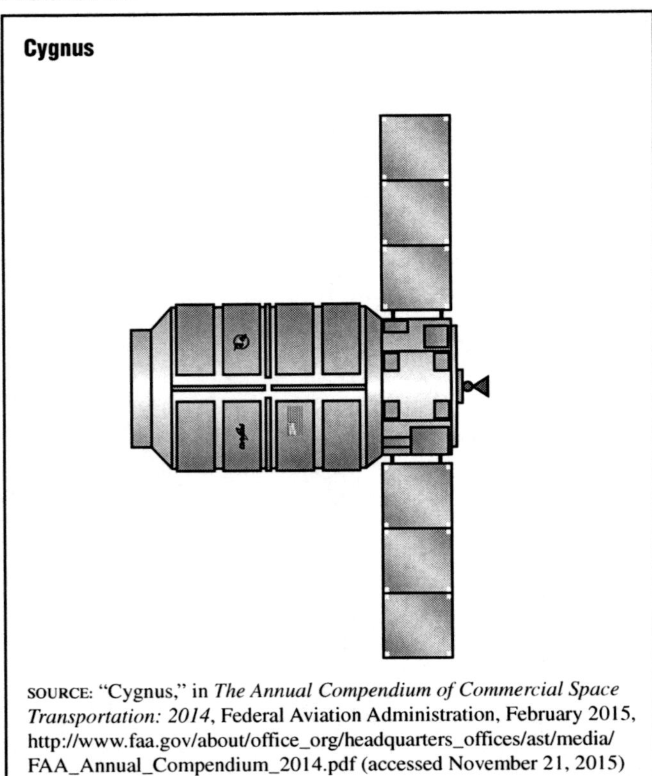

SOURCE: "Cygnus," in *The Annual Compendium of Commercial Space Transportation: 2014*, Federal Aviation Administration, February 2015, http://www.faa.gov/about/office_org/headquarters_offices/ast/media/FAA_Annual_Compendium_2014.pdf (accessed November 21, 2015)

FIGURE 4.7

Antares 120

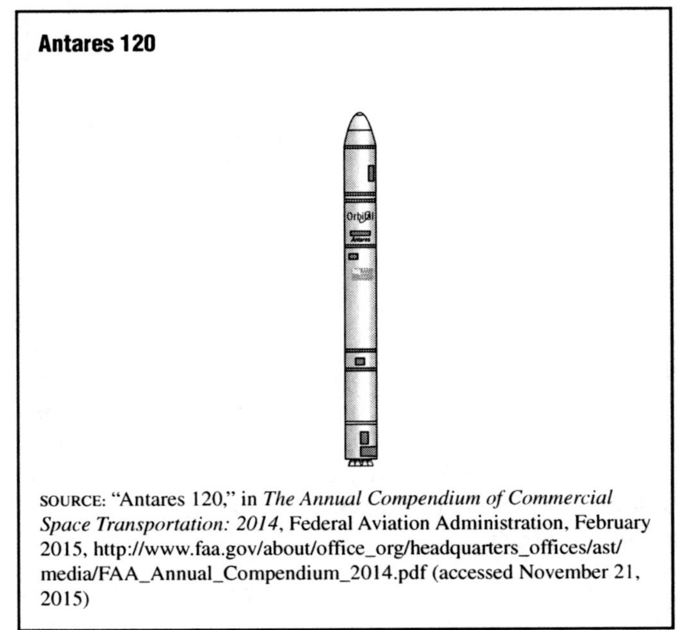

SOURCE: "Antares 120," in *The Annual Compendium of Commercial Space Transportation: 2014*, Federal Aviation Administration, February 2015, http://www.faa.gov/about/office_org/headquarters_offices/ast/media/FAA_Annual_Compendium_2014.pdf (accessed November 21, 2015)

June 29, 2015), James Dean indicates that the mishap occurred when the spacecraft was about 29 miles (46.7 km) above Earth's surface. Pieces of the vehicle and its cargo fell into the Atlantic Ocean.

SpaceX began an accident investigation under the oversight of the FAA. As of February 2016, an official report had not been released to the public. SpaceX statements indicate that the company suspects failure of a bracket-like piece of metal called a strut. According to Andy Pasztor, in "SpaceX Reviewing Suppliers' Quality Control before Renewing Launches" (WSJ.com, September 1, 2015), the preliminary investigation did not reveal "any major design or assembly problems beyond the failure of a commonplace metal strut." SpaceX (July 20, 2015, http://www.spacex.com/news/2015/07/20/crs-7-investigation-update) states, "Despite the fact that these struts have been used on all previous Falcon 9 flights and are certified to withstand well beyond the expected loads during flight, SpaceX will no longer use these particular struts for flight applications."

SpaceX returned to flight in December 2015, when a Falcon 9 launch vehicle carried multiple small satellites into orbit for the private company Orbcomm. As described earlier, that launch was successful, and the first stage of the rocket was recovered safely at CCAFS. As of February 2016, NASA had not announced when SpaceX would resume cargo deliveries to the *ISS*.

CYGNUS FLIGHTS. According to NASA's OIG, in *Commercial Cargo*, NASA agreed to pay Orbital Sciences $288 million for successful completion of the COTS program. The company contributed more than half of the funds needed for spacecraft development. In 2008 NASA signed a contract with Orbital Sciences for $1.9 billion for eight Cygnus cargo missions to the *ISS*. Clark notes in "NASA Orders Missions to Resupply Space Station in 2017" that in March 2015 NASA added one additional mission under an extension provision of the original contract.

Cygnus underwent its first and only COTS demonstration flight on September 18, 2013, when a Cygnus spacecraft was launched by an Antares rocket into orbit. The Cygnus was successfully docked with the *ISS* and released several weeks later. On January 9, 2014, a Cygnus was launched on its first official CRS mission to the space station. Orbital Sciences followed up with another cargo resupply mission in early 2014. Both flights were successful. On October 28, 2014, an Antares/Cygnus bound for the *ISS* failed only seconds after lifting off from the Mid-Atlantic Regional Spaceport in Virginia. As it began to fall back toward the ground it was blown up by controllers. The spacecraft and its payload were largely destroyed, and there was some damage to the launch facilities on the ground.

After an investigation that lasted nearly a near, NASA reported its findings on the mishap in *NASA Independent Review Team: Or–3 Accident Investigation Report* (October 9, 2015, http://www.nasa.gov/sites/default/files/atoms/files/orb3_irt_execsumm_0.pdf). The agency believes that an explosion occurred in an oxygen turbopump in one of the rocket's AJ26 engines. This caused the launch vehicle to lose thrust. NASA notes that AJ26 engines are modified Soviet rocket engines manufactured during the 1970s for the N-1 rocket. (As explained in Chapter 3, the N-1 rocket program was a failed Soviet attempt to develop a spacecraft capable of traveling to Earth's moon.) NASA was unable to determine the exact cause of the turbopump explosion. The agency suspects that design flaws, inadequate testing, and manufacturing defects played a role. Particles of titanium and silica debris were also detected in the engine that initially exploded.

NASA states that during its investigation, Orbital ATK began testing RD-181 engines as alternatives to AJ26 engines for its Antares rockets. In "RD-181 Engines Prepared for Shipment to U.S." (SpaceFlight Now.com, June 1, 2015), Clark explains that Russian-made RD-181 engines are modeled after the engines used on the ULA's Atlas V and Russia's Zenit launch vehicles.

On December 6, 2015, Orbital ATK resumed cargo shipments to the *ISS*. A Cygnus capsule was launched aboard an Atlas V launch vehicle at the CCAFS. The mission was successful.

CARGO MISSIONS THROUGH 2020. In 2014 NASA opened a bidding period for a second round of CRS contracts covering *ISS* cargo deliveries through 2020. In "NASA Delays Cargo Award Again, Eliminates Boeing from Contention" (SpaceFlightNow.com, November 5, 2015), Clark indicates that SpaceX, Orbital ATK, Boeing, Lockheed Martin, and Sierra Nevada all submitted proposals. Boeing developed a capsule called the Crew Space Transportation (CST)-100. (See Figure 4.8.) It won a contract under NASA's *ISS* crew transport program, which is described later in this chapter. Boeing says the CST-100 could be modified to carry cargo only. Lockheed Martin's cargo capsule is called the Jupiter/Exoliner. Sierra Nevada's spacecraft—the Dream Chaser—is a space plane. As shown in Table 4.2 its launch vehicle would be the Atlas V. In January 2016 NASA announced it had picked Orbital ATK, Sierra Nevada, and SpaceX for the second CRS contract.

ISS Crew Transportation

NASA's second priority for its ISS program was development by private partners of crewed vehicles capable of traveling to and from the station. This need became even more pressing after the space shuttle program's last flight occurred in July 2011. Ever since, NASA has relied on Russian Soyuz rockets and spacecraft to transport American astronauts to and from the *ISS*.

In 2010 the agency entered into agreements with private companies to begin developing and testing vehicles and technologies. According to the FAA, in *Annual Compendium of Commercial Space Transportation: 2014*, the initial financial awards for Commercial Crew Development (CCDev1) totaled $49.8 million and went to five companies: Blue Origin of Washington, Boeing, the Paragon Space Development Corporation of Arizona, Sierra Nevada (which is based in Nevada), and the ULA.

FIGURE 4.8

Crew Space Transportation (CST)-100 capsule

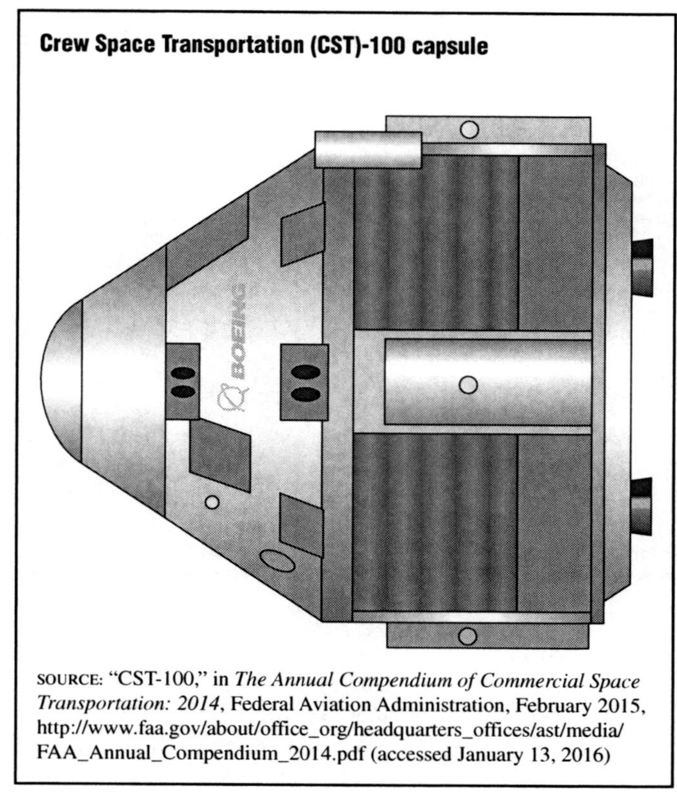

SOURCE: "CST-100," in *The Annual Compendium of Commercial Space Transportation: 2014*, Federal Aviation Administration, February 2015, http://www.faa.gov/about/office_org/headquarters_offices/ast/media/FAA_Annual_Compendium_2014.pdf (accessed January 13, 2016)

In 2011 a second set of awards (CCDev2) was established. The FAA notes that a total of $315.5 million went to Boeing, Blue Origin, Sierra Nevada, and SpaceX. Unfunded CCDev2 agreements were made with the ULA, Excalibur Almaz, Inc., of Texas, and Alliant Techsystems Inc. of Utah. In 2012 NASA announced its Commercial Crew Integrated Capability (CCiCAP) funding levels. A total of just over $1.1 billion was earmarked for three companies—Boeing, Sierra Nevada, and SpaceX.

On September 16, 2014, NASA made its final selection. Jeff Foust indicates in "NASA Selects Boeing and SpaceX for Commercial Crew Contracts" (SpaceNews.com, September 16, 2014) that Commercial Crew Transportation Capability (CCtCap) contracts worth $6.8 billion were awarded to Boeing ($4.2 billion) and SpaceX ($2.6 billion). NASA hopes the companies' spacecraft will be certified by the end of 2017 to transport astronauts to the *ISS*. The crew capsules are as follows:

- Boeing's CST-100 spacecraft is shown in Figure 4.8. It is also called the Starliner. In *Annual Compendium of Commercial Space Transportation: 2014*, the FAA notes the reusable capsule is designed to transport up

TABLE 4.2

Orbital vehicles and platforms for transport of cargo and/or crew in low Earth orbit

Operator	Vehicle	Launch vehicle	Maximum cargo kg (lb)	Maximum crew size	First flight
SpaceX	Dragon (cargo)	Falcon 9	6,000 (13,228)	0	2010
SpaceX	Dragon V2 (crew)	Falcon 9	TBD	7	2017
Orbital	Cygnus (Standard)	Antares	2,000 (4,409)	0	2013
Orbital	Cygnus (Enhanced)	Antares	2,700 (5,952)	0	2015
Boeing	CST-100	Atlas V Delta IV Falcon 9	TBD	7	2017
SNC	Dream Chaser	Atlas V	TBD	7	TBD
Blue Origin	Space Vehicle	Atlas V Blue Origin RBS	TBD	7	TBD

Operator	Platform	On-orbit vehicle	Maximum volume m^3 (ft^3)	Maximum crew size	First flight
Bigelow Aerospace	BA 330	Dragon CST-100	330 (11,653)	6	TBD
Bigelow Aerospace	BEAM	Dragon	32 (1,125)	TBD	2015

Notes: BA = Bigelow Aerospace. BEAM = Bigelow Expandable Activity Module. CST = Crew Space Transportation. ft^3 = cubic feet. kg = kilograms. lb = pounds. m^3 = cubic meters. RBS = Reusable Booster System. SNC = Sierra Nevada Corporation. TBD = To Be Determined.

SOURCE: "Table 7. On-Orbit Vehicles and Platforms," in *The Annual Compendium of Commercial Space Transportation: 2014*, Federal Aviation Administration, February 2015, http://www.faa.gov/about/office_org/headquarters_offices/ast/media/FAA_Annual_Compendium_2014.pdf (accessed November 21, 2015)

to seven crew members or a combination of people and cargo to the *ISS*. Each CST-100 can be reused up to 10 times. The spacecraft includes flight components modeled on NASA's Apollo and Orion systems, which are described in Chapter 2. As shown in Table 4.2, the spacecraft can be launched by Atlas V, Delta IV, or Falcon 9 launch vehicles. The capsule would splash down to Earth using a parachute system.

- SpaceX's Dragon V2 spacecraft is shown in Figure 4.9. In "SpaceX Completes Crewed Orbit and Entry Review" (August 15, 2013, http://www.nasa.gov/content/spacex-completes-crewed-orbit-and-entry-review/#.Uv5sfZ5dWrY), Rebecca Regan quotes a SpaceX spokesperson as saying, "SpaceX's Dragon spacecraft was designed from the outset to accommodate the upgrades necessary to safely carry people." According to the FAA, the crewed capsule will carry up to seven people and launch atop a Falcon 9 rocket. It features a parachute system for Earth splashdowns.

Table 4.2 compares design data for the two crewed spacecraft. It also includes information about other U.S. commercial vehicles and platforms associated with the ISS program. One is an unnamed crewed vehicle under development by Blue Origin, a company described in detail later in this chapter. Bigelow Aerospace of Nevada is developing inflatable orbital platforms (or modules) that can be used on the *ISS*. The two designs are called the BA 330 and the Bigelow Expandable Activity Module. They build on expandable space habitats that the company developed under its Genesis program during the first decade of the 21st century. Genesis I and Genesis II were launched into orbit in 2006 and 2007, and according to the company (http://bigelowaerospace.com/bigelow-aerospace-spacecraft/genesis/) were still operational as of February 2016.

FIGURE 4.9

Dragon V2 capsule for astronauts

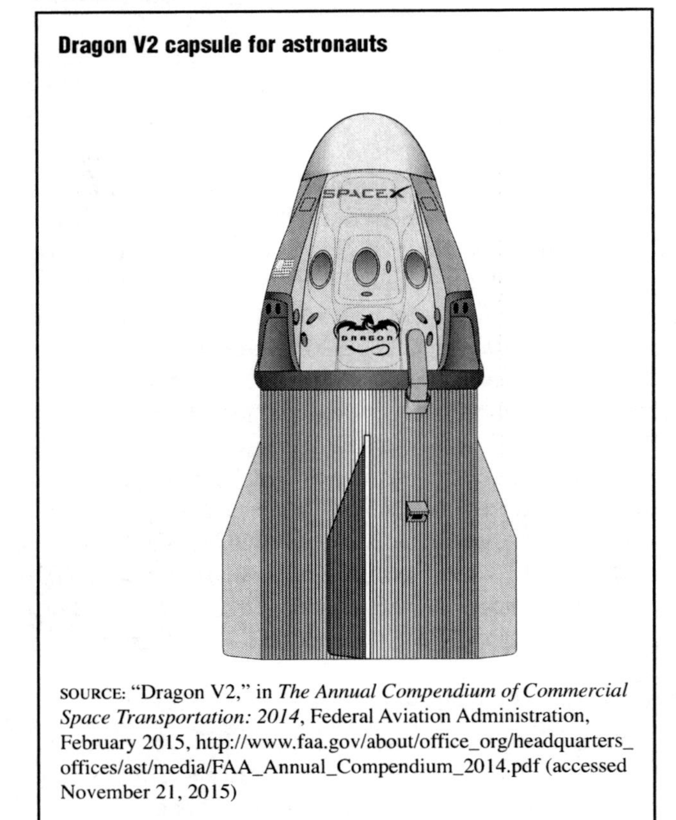

SOURCE: "Dragon V2," in *The Annual Compendium of Commercial Space Transportation: 2014*, Federal Aviation Administration, February 2015, http://www.faa.gov/about/office_org/headquarters_offices/ast/media/FAA_Annual_Compendium_2014.pdf (accessed November 21, 2015)

COMMERCIAL SUBORBITAL SPACEFLIGHT

A suborbital flight does not make a complete orbit around Earth. It is easier for private companies to develop suborbital vehicles, rather than orbital vehicles, because it takes much more force to boost a spacecraft to sufficient velocity and altitude to enter a sustainable orbit. Suborbital space flight is also much cheaper because it can be achieved through suborbital reusable launch vehicles (sRLVs). The vehicles can be reused numerous times. They may leave and return to Earth in a horizontal flight profile (like an airplane) or in a vertical flight profile. An example of the latter is a capsule that is launched by a rocket and later lands with the aid of parachutes.

In *The U.S. Commercial Suborbital Industry: A Space Renaissance in the Making* (October 2011, http://www.faa.gov/about/office_org/headquarters_offices/ast/media/111460.pdf), the FAA sees seven potential markets for the commercial suborbital spaceflight industry:

- Human spaceflight—mainly tourists, but also researchers studying the effects of microgravity and scientists accompanying payloads
- Basic and applied research—experiments in various disciplines, such as biology, physics, and space and earth science
- Aerospace technology testing and demonstration—payloads from aerospace companies desiring to test or demonstrate their products under suborbital conditions
- Remote sensing—applications that image Earth for commercial, civil, government, and military clients
- Education—researchers or payloads from educational institutions and organizations
- Media and public relations—clients who desire photographs or video taken from space
- Point-to-point transportation—passengers who wish to travel to points around the globe in far less time than it takes by airplane

Space Tourism

Private development of sRLVs has been driven in large part by the prospects for space tourism. During the 1990s the Russian Space Agency allowed private citizens to visit the space station *Mir* and the *ISS* for fees ranging from $15 million to $40 million per tourist. Most of the trips were arranged through the private U.S. company Space Adventures. Formed in 1998, the company offers customers opportunities in space tourism and related entertainment areas, such as "zero gravity" experiences. Other companies have also entered the industry. The demand by private citizens for space travel is expected to grow substantially during the 21st century.

SpaceShipOne

In 2004 a major milestone in space exploration was achieved when the first nongovernmental crewed spacecraft traveled to space and back. The spacecraft was called *SpaceShipOne*, and it was funded by the private investor Paul Allen (1953–), a cofounder of the Microsoft Corporation. In 2001 Allen contracted the California design firm Scaled Composites to develop an sRLV that was capable of carrying at least one passenger on a suborbital flight.

Aside from reengaging the public's interest and passion in space exploration, Allen set out to win the Ansari X Prize (http://www.xprize.org/). This prize was offered by a group of private investors called the X Prize Foundation, which was created by Diamandis. The Ansari family was the prime funder of the $10 million prize, which was available to any nongovernmental group that could achieve the following:

- Build a spaceship and fly three people (or at least one person plus the equivalent weight of two people) into space (defined as an altitude of at least 62 miles [100 km] above Earth's surface)
- Return safely to Earth
- Repeat the feat with the same spaceship within two weeks

On June 21, 2004, the U.S. test pilot Mike Melvill (1940–) of Scaled Composites became the first person to pilot a privately built plane into space when he took *SpaceShipOne* to an altitude of 62.2 miles (100.1 km) during a test flight. On September 29, 2004, he achieved an altitude of 63.9 miles (102.8 km). Five days later the U.S. pilot Brian Binnie (1953–) took the same plane to an altitude of 69.6 miles (112 km) to win the Ansari X Prize.

The flights were conducted from an airstrip in Mojave, California, using a carrier plane called *White Knight* (it would later be dubbed *WhiteKnightOne*). The carrier plane transported *SpaceShipOne* to an altitude of approximately 47,000 feet (14,300 m) and released it. A rocket motor aboard the spaceship was fired to propel it vertically into space. The pilots experienced about three minutes of weightlessness at the height of their journeys. During reentry, the wings of *SpaceShipOne* were maneuvered to provide maximum drag and slow its descent. The spacecraft glided back to the airstrip and landed like a plane. (See Figure 4.10, but note that weightlessness occurred only at the uppermost reach of the spacecraft.)

The *White Knight* was a manned twin-turbojet carrier aircraft that was designed to fly to high altitudes carrying a payload of up to 4 tons (3.6 t). It was named after two U.S. Air Force pilots (Robert Michael White [1924–2010] and William J. Knight [1929–2004]) who earned their astronaut wings flying the experimental X-15 aircraft during the early 1960s.

SpaceShipOne used a unique hybrid rocket motor that was fueled by liquid nitrous oxide (laughing gas) and solid hydroxy-terminated polybutadiene (a major constituent of the rubber that is used in tires). The individual fuel components are nontoxic and are not hazardous to transport or store. They do not react when mixed

FIGURE 4.10

Flight sequence for *SpaceShipOne*

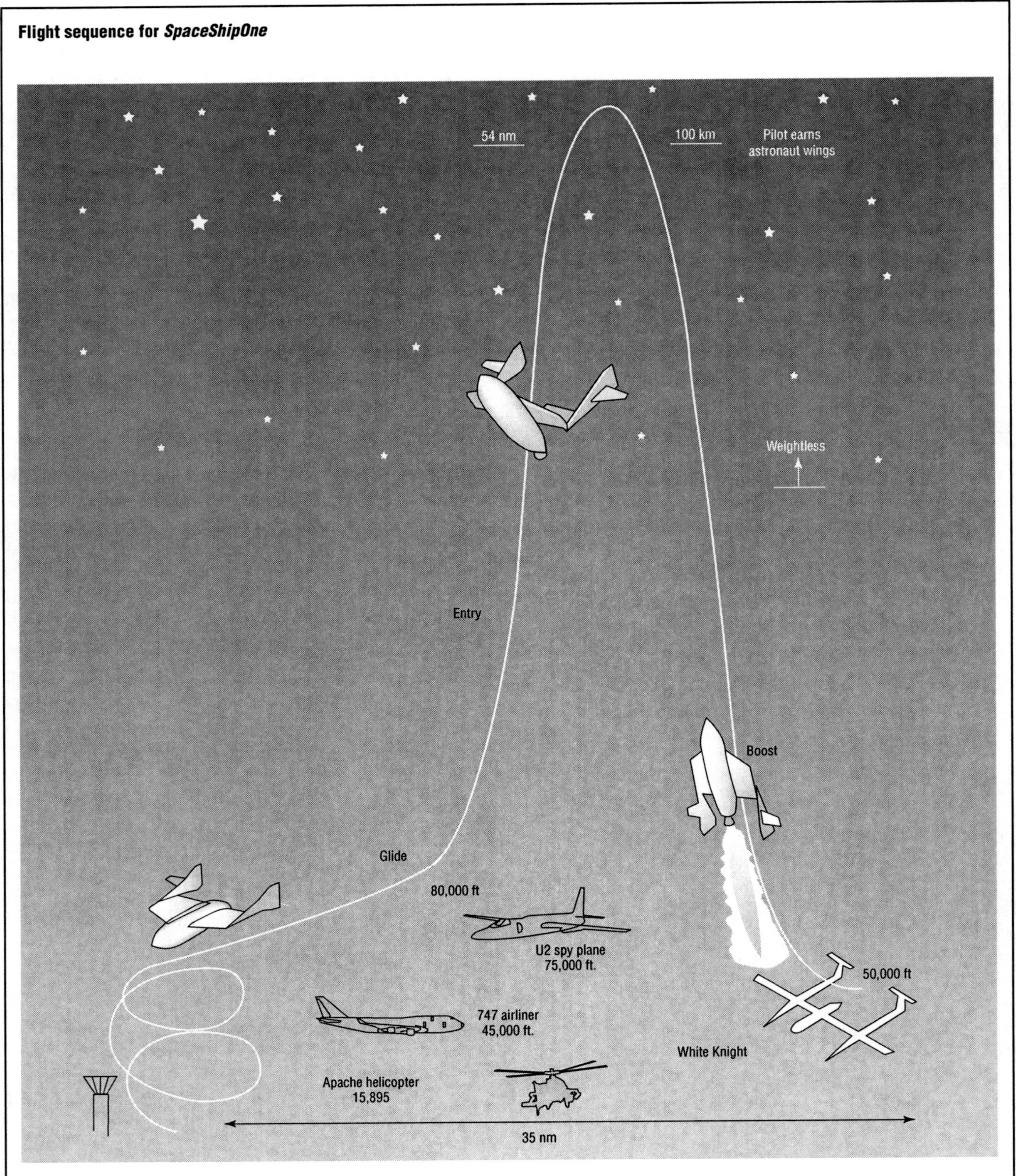

Note: ft. = feet; km = kilometers; nm = nautical mile.

SOURCE: "Untitled," in *SpaceShipOne…The First Private Manned Space Program*, Scaled Composites, LLC, undated, http://www.scaled.com/images/uploads/pdf/Lithograph_-_SpaceShipOne_p2.pdf (accessed December 11, 2015). © 2004 Mojave Aerospace Ventures LLC; SpaceShipOne is a Paul G. Allen Project.

together unless a flame is supplied. In *SpaceShipOne*, the nitrous oxide was gasified before combustion.

SpaceShipOne has been on display at the Smithsonian Institution's National Air and Space Museum in Washington, D.C., since 2005. It is part of the *Milestones of Flight* exhibit, along with other legendary air- and spacecraft from the beginning of aviation through the space age.

Virgin Galactic and the Spaceship Company

In 2004 Richard Branson (1950–), the founder of the Virgin Group, created Virgin Galactic, which bills itself as the world's first commercial "spaceline." The following year Branson and Burt Rutan (1943–), the president of Scaled Composites, announced the formation of a new aerospace production company: The Spaceship Company (TSC). In 2012 Virgin Galactic (2016, http://thespaceshipcompany.com/who-we-are/) acquired full ownership of TSC.

Using original technology licensed from Allen, TSC plans to build a small fleet of spacecraft based on the designs of the *White Knight* and *SpaceShipOne*. On August 2, 2007, an explosion during a rocket ground test killed three workers at a Scaled Composites facility in the California desert. The explosion involved the firing of the nitrous oxide delivery system. However, work continued on the development of the new SpaceShipTwo (SS2) and WhiteKnightTwo (WK2) spacecraft. It should be noted that SS2 and WK2 are the names for a series of crafts that TSC plans to develop.

In December 2009 Virgin Galactic revealed the first SS2 to the public at a ceremony in California. Like its predecessor, the new vehicle is carried aloft by a carrier plane or "mothership." The first WK2 was named Virgin Mothership *Eve*, after Branson's mother. At the unveiling ceremony for the first SS2, Governor Arnold Schwarzenegger (1947–) of California and Governor Bill Richardson (1947–) of New Mexico christened the craft with the name Virgin Space Ship (VSS) *Enterprise*. (The use of the initials VSS is based on the tradition that the names of U.S. and British military naval vessels are preceded by the initials USS, which stands for United States Ship, and HMS, which stands for Her Majesty's Ship, respectively.)

Soon after its founding, Virgin Galactic (2016, http://www.virgingalactic.com/human-spaceflight/fly-with-us/) began selling seats on its future suborbital space flights to tourists. As of February 2016, hundreds of people were believed to have signed up for a trip aboard an SS2. The price per ticket at the time was $250,000 per flight.

The SS2 is designed to carry two crew members and six passengers. In March 2010 the spacecraft underwent its first "captive carry" flight while attached to its mothership. Later that year the *Enterprise* conducted its first glide flight. The spacecraft features a unique wing design, including a "feathering system" that facilitates smooth travel through the upper atmosphere as it returns to Earth from the edge of outer space. TSC operates ground facilities at the Mojave Air and Space Port in Mojave, California.

On October 31, 2014, the *Enterprise* was conducting a test flight at a high altitude (but not in space) when disaster struck. The SpaceShipTwo broke apart during flight. The pilot, Peter Siebold (1971–), was ejected from the aircraft and parachuted to safety, but was seriously injured. The copilot, Michael Alsbury (1975–2014), was killed. The National Transportation Safety Board (NTSB) investigated the accident and reported its findings in *In-Flight Breakup during Test Flight: Scaled Composites SpaceShipTwo, N339SS, near Koehn Dry Lake, California, October 31, 2014* (July 28, 2015, http://www.ntsb.gov/investigations/AccidentReports/Reports/AAR1502.pdf). The agency concludes that human error contributed to the crash noting,

> the probable cause of this accident was Scaled Composites' failure to consider and protect against the possibility that a single human error could result in a catastrophic hazard to the SpaceShipTwo vehicle. This failure set the stage for the copilot's premature unlocking of the feather system as a result of time pressure and vibration and loads that he had not recently experienced, which led to uncommanded feather extension and the subsequent aerodynamic overload and in-flight breakup of the vehicle.

In addition, the NTSB identified shortcomings in the experimental permitting process and safety inspections conducted by the AST prior to the mishap. The NTSB provided recommendations for improving permitting and inspection procedures.

In February 2016 Virgin Galactic announced the completion of a second SpaceShipTwo, which was named VSS *Unity*. At that time Virgin Galactic had not publicly stated when its first tourist flight would take place; however, media sources estimate it could occur in 2017 or 2018.

Other Suborbital Space Flight Companies

Numerous other companies are planning or developing sRLVs. The tourism trade is one driving force behind the growth in the industry. In addition, NASA is spurring the commercial market through its Flight Opportunities Program (FOP). According to the agency (July 30, 2015, http://www.nasa.gov/centers/dryden/news/FactSheets/FS-102-DFRC.html#.Uv-A6J5dWrY), it is "purchasing flight services from emerging, commercial suborbital platform providers and facilitating access to reduced gravity environments, brief periods of weightlessness and high-altitude atmospheric conditions." The arrangements allow NASA to conduct research, develop technologies, and support the commercial sector. As of February 2016, the

TABLE 4.3

U.S. commercial suborbital reusable launch vehicles

Operator	SRV	Seats[a]	Maximum cargo kg (lb)	Price	Announced operational date
Blue Origin	New Shepard	3+	120[b] (265)	Not announced	Not announced
Masten Space Systems	Xaero	—	12 (26)	Not announced	Not announced
	Xombie	—	20 (44)	Not Announced	Not announced
	Xogdor	—	25 (55)	Not announced	Not Announced
UP Aerospace	SpaceLoft XL	—	36 (79)	$350,000 per launch	2006 (actual)
Virgin Galactic	SpaceShipTwo	6	600 (1,323)	$200,000 per seat	2015–2016
XCOR Aerospace	Lynx Mark I	1	120 (265)	$95,000/seat	2014
	Lynx Mark II	1	120 (265)	$95,000/seat	2014
	Lynx Mark III	1	770 (1,698)	$95,000/seat	2015–2016

[a]Passengers only; several vehicles are piloted.
[b]Net of payload infrastructure.
Notes: kg = kilograms. lb = pounds.

SOURCE: "Table 5. SRVs and Providers," in *The Annual Compendium of Commercial Space Transportation: 2014*, Federal Aviation Administration, February 2015, http://www.faa.gov/about/office_org/headquarters_offices/ast/media/FAA_Annual_Compendium_2014.pdf (accessed November 21, 2015)

agency (https://flightopportunities.nasa.gov/technologies/) indicated that 151 technologies were being investigated.

The major companies involved in sRLV development as of February 2016 are described in the following sections. A summary compiled by the FAA compares their capabilities in Table 4.3.

BLUE ORIGIN. Blue Origin was founded in 2000 by Jeffrey Bezos (1964–), who also founded the online shopping site Amazon.com. Bezos is famously tight-lipped about the aerospace company's operations. As noted earlier, Blue Origin won financial awards from NASA in 2010 and 2011 as part of the agency's CCDev1 and CCDev2 programs. NASA has provided technical support to Blue Origin on an engine development project and has helped the company review and test an unnamed crewed spacecraft.

As shown in Table 4.3, Blue Origin has also developed an sRLV called New Shepard that seats at least three passengers. (See Figure 4.11.) In *Annual Compendium of Commercial Space Transportation: 2014*, the FAA notes that the vehicle takes off and lands vertically. It features a crew capsule and propulsion module that separate from each other during flight. The propulsion module is designed to return to the launchpad to be recovered intact. The crew module lands with the assistance of parachutes.

FIGURE 4.11

New Shepard

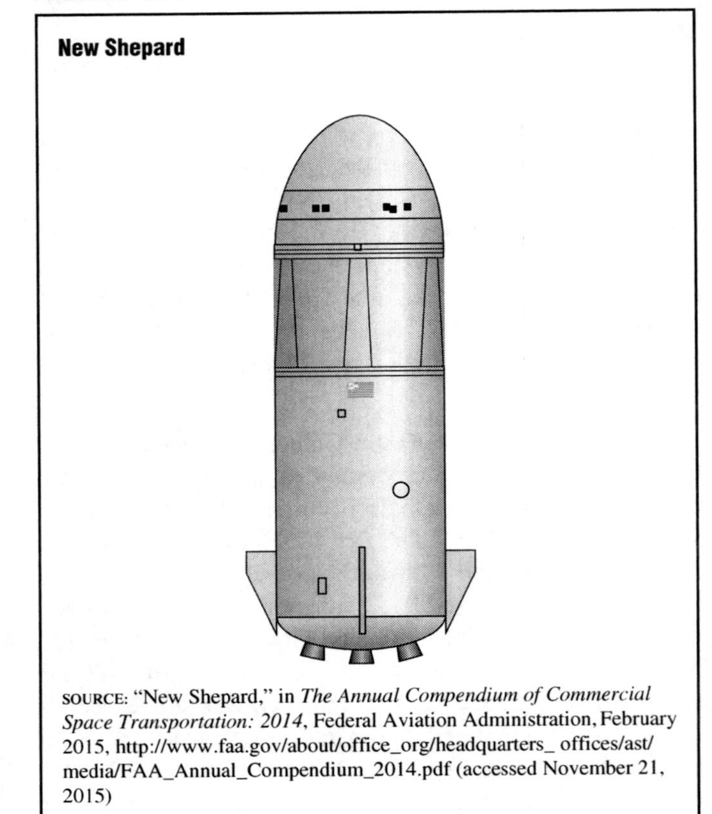

SOURCE: "New Shepard," in *The Annual Compendium of Commercial Space Transportation: 2014*, Federal Aviation Administration, February 2015, http://www.faa.gov/about/office_org/headquarters_ offices/ast/media/FAA_Annual_Compendium_2014.pdf (accessed November 21, 2015)

Blue Origin received an FAA AST Experimental Permit in 2014 authorizing flight testing of the new spacecraft. The first test flight took place on April 29, 2015. According to a Blue Origin blog post that day (https://www.blueorigin.com/news/blog/first-developmental-test-flight-of-new-shepard), the unmanned spacecraft soared to 307,000 feet (93.6 km). The company lost control of the propulsion module during descent and was unable to recover it; however, the crew module landed safely. On November 24, 2015, Blue Origin was able to successfully recover the propulsion module following a test flight. The company (https://www.blueorigin.com/news/news/blue-origin-makes-historic-rocket-landing) notes that its New Shepard flew to 329,839 feet (100.5 km) and the modules separated. The propulsion module was guided back to Earth and landed approximately 4.5 feet (1.4 m) from the center of the launchpad. The crew capsule also landed safely.

Blue Origin's feat was remarkable because it was the first successful recovery of a suborbital rocket. If the company can reuse its propulsion module, it can substantially lower its launch costs in the future.

MASTEN SPACE SYSTEMS. David Masten (1968–) spent a long career in information technology and software development before founding Masten Space Systems in 2004. In 2009 the company won $1 million in NASA's Lunar Lander Challenge.

According to the FAA, in *Annual Compendium of Commercial Space Transportation: 2014*, the company's Xaero and Xogdor line of suborbital spacecraft will have vertical takeoff and vertical land capabilities and will make "soft" landings" without the need for parachutes. The sRLVs are not designed for human spaceflight, but for payload flights. (Note that the Xombie is actually a technology testing vehicle and is not intended to carry payloads.)

As of February 2016, Masten Space Systems did not have any active launch licenses with the FAA (http://www.faa.gov/data_research/commercial_space_data/licenses/).

UP AEROSPACE. UP Aerospace was founded in 1998 by Jerry Larson (1960?–). In "Larson's Spent a Decade Going UP" (BizJournals.com, June 14, 2013), Dan Mayfield indicates that Larson was formerly an engineer at Lockheed. Since 2006 UP Aerospace has launched sounding rockets, which are payload-carrying rockets that travel into the atmosphere, typically to conduct measurements or collect data. Suborbital sounding rockets travel briefly into space before returning to Earth. David Leonard indicates in "Suborbital Rocket Launches Human Remains, Wedding Rings into Space" (Space.com, May 23, 2011) that customers pay for cremated remains, wedding rings, and other objects to be flown into space aboard the sounding rockets.

The SpaceLoft XL is UP Aerospace's primary sRLV. As shown in Table 4.3, it can carry payloads weighing up to 79 pounds (36 kg).

As of February 2016, UP Aerospace did not have any active launch licenses with the FAA (http://www.faa.gov/data_research/commercial_space_data/licenses/).

XCOR AEROSPACE. XCOR Aerospace (2016, http://xcor.com/about-us/company-overview/?p=aerospace) was founded in 1999 by four innovators devoted to space technology. The company's Lynx spacecraft is designed to carry only a pilot and a passenger. The Lynx takes off and lands horizontally like an airplane. Three versions are under development, as shown in Table 4.3. All three sRLVs will be marketed to tourists and for carrying small payloads.

XCOR Space Expeditions (2016, http://spaceexpeditions.xcor.com/spacecraft/) sells tickets for flights aboard the Lynx spacecraft. As of February 2016, the first flights were expected to take place in late 2016. The company, however, did not have any active licenses with the FAA (http://www.faa.gov/data_research/commercial_space_data/licenses/).

RESOURCE EXTRACTION IN OUTER SPACE?

Most commercial space ventures involve satellite transport or the tourist trade. There is growing interest in extracting resources, such as metals and other substances, from asteroids, Earth's moon, and other heavenly bodies.

Asteroids are particularly tempting targets. In "New NASA Mission to Help Us Learn How to Mine Asteroids (July 30, 2015, https://www.nasa.gov/content/goddard/new-nasa-mission-to-help-us-learn-how-to-mine-asteroids), William Steigerwald notes, "even a little, house-sized asteroid should contain metals possibly worth millions of dollars." Some asteroids also contain large amounts of water bound up in clay deposits. This water could be extracted and used as-is or separated into its hydrogen and oxygen components for rocket fuel. Space-based water and fuel, in particular, will be important resources when humans begin spending extended periods off of Earth.

U.S. companies, such as Planetary Resources and Moon Express, are seriously considering resource extraction from outer space bodies. In April 2015 a Dragon capsule launched by a Falcon 9 rocket carried a Planetary Resources satellite named *Arkyd 3 Reflight* (*A3R*) to the *ISS*. In "Planetary Resources' First Spacecraft Successfully Deployed, Testing Asteroid Prospecting Technology on Orbit" (July 16, 2015, http://www.planetaryresources.com/2015/07/planetary-resources-first-spacecraft-deployed/), the company notes that the satellite was released from the space station in July 2015 to begin a 90-day mission. *A3R* is testing various technologies that Planetary Resources plans to use in future spacecraft that will robotically search for near-Earth asteroids suitable for mining. Moon Express was competing, as of February 2016, for the Google Lunar X Prize. As noted earlier, this is a $30 million competition to land a privately funded robot on Earth's moon by year-end 2017. Moon Express (http://www.moonexpress.com/) hopes to one day extract valuable substances, such as platinum and helium, from Earth's moon.

The U.S. Commercial Space Launch Competitiveness Act was passed in November 2015. Title IV of the law specifies that U.S. citizens can "possess, own, transport, use, and sell" resources obtained from outer space so long as the acquisition is conducted in accordance with U.S. law and the nation's international obligations. The United States is a party to the Treaty on Principles Governing the Activities of States in the Exploration and Use of Outer Space, including the

Moon and Other Celestial Bodies, which is described in Chapter 1. The treaty dictates that outer space cannot be appropriated or claimed for ownership by any nation. U.S. advocates of off-Earth resource extraction argue that outer space is similar legally to international waters on this planet. The fish and other aquatic creatures in international waters are not owned by any one nation but can be extracted, possessed, and sold by commercial enterprises from any nation. It remains to be seen if a similar international consensus will develop on how to handle the controversial issues surrounding outer space resource extraction.

CHAPTER 5

THE *INTERNATIONAL SPACE STATION*

I am directing NASA to develop a permanently manned space station and to do it within a decade.

—President Ronald Reagan, State of the Union Address (January 25, 1984)

A space station is an orbiting structure that is designed to accommodate visiting crew members for an extended period. In 1984 the U.S. government envisioned building a continuously manned space station in which scientists would conduct long-term research. The station was to be large and spacious, with room for up to 10 crew members at a time. The U.S. space shuttle was going to be the workhorse that carried cargo and astronauts to the station and back on a routine basis.

Because it was an expensive undertaking, the United States invited other countries to participate. Eventually, 15 countries did so, including Russia, which assumed a major role in the project. The space station became an international venture. It also became extremely expensive. The design was changed several times to bring costs down. The development phase alone dragged on for more than a decade. In 1998 construction finally began on the *International Space Station* (*ISS*). Cost overruns and delays, including a long one after the 2003 loss of the space shuttle *Columbia*, were problematic. The U.S. government decided to phase out the shuttle program and cease *ISS* construction at a newly defined "core complete" configuration.

The resulting *ISS* is much smaller than the space station that was originally envisioned and has fewer research capabilities. This is disappointing to many scientists around the world. Others, however, believe that even a downsized *ISS* is a major step on humanity's journey into outer space.

EARLY SPACE STATIONS

Many early space scientists and engineers developed proposals for Earth-orbiting space stations. For example, during the 1920s Hermann Oberth (1894–1989) proposed an orbiting structure called a *weltraumstation* (space station) that would serve as a launching and refueling station for spacecraft engaged in deep space travel. Oberth's writings had a profound effect on the young Wernher von Braun (1912–1977), who later played a key role in the Apollo program in the United States. National Aeronautics and Space Administration (NASA) historians indicate that a space station was always considered the next step after Apollo. When the Soviets realized that they could not beat the Americans to the moon during the 1960s, they turned their attention to building a space station.

Early space stations operated by NASA and the Soviet Union/Russian Republic are described in Chapters 2 and 3, respectively. They were all placed in low Earth orbit (LEO), which is approximately 100 to 1,200 miles (160 to 1,900 km) above Earth's surface. Below the bottom edge of LEO the air drag from Earth's atmosphere is dense enough to pull spacecraft downward quickly. Above the top edge of LEO lies a thick region of radiation called the inner Van Allen radiation belt, which poses a hazard to human life and sensitive electronic equipment. Spacecraft in LEO travel at about 17,000 miles per hour (27,400 kph) and orbit Earth once every 90 minutes or so.

AN INTERNATIONAL COALITION FORMS

In his State of the Union Address (January 25, 1984, http://reagan2020.us/speeches/state_of_the_union_1984.asp), President Ronald Reagan (1911–2004) directed NASA to develop a space station before the end of a decade. He later named the new space station *Freedom*.

Marcia S. Smith of the Congressional Research Service describes the early history of the space station in *NASA's Space Station Program: Evolution of Its Rationale and Expected Uses, Testimony before the Subcommittee on*

Science and Space Committee on Commerce, Science, and Transportation, United States Senate (April 20, 2005, http://history.nasa.gov/isstestimonysmith.pdf). At that time, designers envisioned a station that could accommodate a crew of up to 10 people.

The project was initially expected to cost the United States only $8 billion because several foreign governments intended to participate. By 1988 Canada, Japan, and nine European countries had signed formal agreements to participate in the project. However, the cost of the venture escalated as design continued year after year. By 1991 NASA's cost estimate for *Freedom* was $30 billion. Congress demanded several redesigns to save money and threatened to cancel the project altogether. Each redesign resulted in a smaller station which frustrated NASA's foreign partners. Smith reports that by 1993 NASA had already spent $11.4 billion on design costs alone. However, not one piece of hardware had been launched into space.

Russia Comes on Board

By the early 1990s the Soviet Union had collapsed. The administration of U.S. President Bill Clinton (1946–) began talks with the new Russian government and welcomed Russia's eagerness to participate in the space station project. According to Smith, President Clinton saw Russian participation as a way to improve foreign relations and put pressure on the Russians to abide by newly signed ballistic missile agreements. He also wanted to provide Russian scientists with jobs to keep them from selling valuable information to enemies of the United States. On September 2, 1993, the two countries signed the Joint Declaration on Cooperation in Space. By the end of the year, NASA and the Russian Space Agency had ironed out a detailed work plan for what was now called the *International Space Station*.

The collaboration with Russia was remarkable from a political standpoint. Only a few years earlier the two sides had been cold war enemies. The new Russian Republic was struggling financially. This raised some concerns among U.S. politicians and scientists. They publicly expressed fears that the cash-strapped Russian government would not be able to fulfill its obligations to the program. U.S. fears grew even greater during the late 1990s as the Russian space station *Mir* experienced numerous problems, as described in Chapter 3. However, the partnership between the two countries survived, and they negotiated deals for *ISS* construction and operation.

The Agreement

On January 29, 1998, the U.S. government signed the Agreement among the Government of Canada, Governments of Member States of the European Space Agency, the Government of Japan, the Government of the Russian Federation, and the Government of the United States of America Concerning Cooperation on the Civil International Space Station (http://www.state.gov/documents/organization/107683.pdf). It is known as the intergovernmental agreement (IGA). The document outlined the agreement among the partners for design, development, operation, and utilization of the *ISS*. The previous year the United States had signed a separate agreement with Brazil giving it utilization rights in exchange for supplying *ISS* parts.

In early 1998 it was expected that at least 40 spacecraft launches spread over five to seven years would be required to assemble more than 100 components into the *ISS*. At that time, the station was designed for a crew of seven people. The space station was expected to be completed by 2005 or 2006 and cost the United States $26 billion to complete. Both estimates would prove overly optimistic.

Downsizing

According to Smith, in 2001 the projected U.S. cost to complete the *ISS* was $4 billion more than expected. NASA reassessed the project and called for seriously downsizing it. By this time the station had fallen under the terms of four U.S. presidents. The administration of George W. Bush (1946–) agreed with the downsizing plan. It was decided to configure the station for only a three-person crew. Previous plans had called for a seven-person crew. This was disappointing news for space scientists. NASA also recommended that a specific "end state" for station construction be set that could be achieved within NASA's budget expectations. This meant elimination of some components that had been planned for the station, including a crew return vehicle, a propulsion module, and a habitation module.

The United States' international partners were displeased with the plan for a smaller crew size because it meant fewer chances for their personnel to visit the station. Furthermore, many scientists were disappointed with the reduction in research potential afforded by the smaller station.

In January 2004 President Bush announced his plan for the U.S. space program, one that focused on crewed missions to Earth's moon and Mars. Because this plan required a new type of spacecraft, it called for the retirement of the space shuttle fleet by 2010. The president also wanted to end *ISS* assembly as soon as the core complete configuration was obtained and eliminate all ISS research projects that did not support the new plans for space travel.

In November 2005 the NASA administrator Michael D. Griffin (1949–; http://www.gpo.gov/fdsys/pkg/CHRG-109hhrg24151/html/CHRG-109hhrg24151.htm) appeared before the U.S. House of Representatives' Committee on

Science to provide an update on NASA's plans for the future, including assembly of the *ISS*. Griffin stated that the agency planned to assemble enough infrastructure on the station "to enable a potential six person crew and meaningful utilization of the *ISS*."

Ownership and Cost-Sharing

The international partners involved in the ISS project as of 2016 were the United States, Russia, Canada, Japan, and 11 members of the European Space Agency (ESA)—Belgium, Denmark, France, Germany, Italy, the Netherlands, Norway, Spain, Sweden, Switzerland, and the United Kingdom. As noted earlier, Brazil obtained utilization rights for the station through an agreement with NASA.

The 1998 IGA recognizes five partners in the ISS: NASA, the Russian Space Agency (now called the Roscosmos State Corporation, or Roscosmos), the ESA, the Canadian Space Agency (CSA), and the Japanese Space Agency (now called the Japan Aerospace Exploration Agency or JAXA). Each partner retains ownership of any elements, equipment, and materials it provides to the station. In addition, each partner has jurisdiction "over personnel in or on the Space Station who are its nationals." The IGA notes that NASA worked out memoranda of understanding (MOUs) with each of the participating space agencies specifying the arrangements for coordinating *ISS* construction and operation.

In *Space Station: Impact of the Grounding of the Shuttle Fleet* (September 2003, http://www.gao.gov/new.items/d031107.pdf), the U.S. General Accounting Office (GAO; now the U.S. Government Accountability Office) discusses the cost-sharing agreement for the *ISS*. NASA is responsible for the entire cost for ground operations and common supplies for the station. NASA is then reimbursed by the partner countries for their share, depending on their level of participation. Partner countries also fund operations and maintenance for any elements they contribute to the *ISS* and for any research activities they conduct. The IGA and the MOUs permit bartering arrangements between partners in lieu of cash payments. For example, the ESA (November 2, 2013, http://www.esa.int/Our_Activities/Human_Spaceflight/ATV/A_fiery_end_to_a_perfect_mission_ATV_Albert_Einstein) indicates that it paid its "dues" through 2017 to NASA by making cargo deliveries to the station using ESA spacecraft.

LOGISTICAL SUPPORT

Constructing and operating the *ISS* have proven to be enormous logistical tasks requiring the coordination of ground control stations and space transportation systems. The *ISS* comprises individual components that were docked together in orbit. Each component was constructed on the ground and then launched into space. Figure 5.1 shows a diagram of the *ISS* including all existing and planned major components as of February 2016. Note that it was expected that the Multipurpose Laboratory Module (MLM) and accessories shown attached to the Zvezda Service Module would be installed in 2017.

The U.S. space shuttle fleet carried most of the large *ISS* pieces into orbit; however, some were launched by Russia's powerful Proton rockets. The station includes various docking ports at which visiting spacecraft can attach. Table 5.1 lists the ISS missions flown from 1998 through May 2011 to transport major elements of the station. Since that time spacecraft have continued to travel back and forth carrying crew members and cargo. As of February 2016, (http://www.nasa.gov/mission_pages/station/main/onthestation/facts_and_figures.html), NASA said the *ISS* weighed approximately 462 tons (419 metric t) (when on Earth) and was about the length of a U.S. football field.

The *ISS* orbits Earth in LEO at around 220 to 250 miles (354 to 402 km) above the planet's surface. In "Higher Altitude Improves Station's Fuel Economy" (February 14, 2011, http://www.nasa.gov/mission_pages/station/expeditions/expedition26/iss_altitude.html), NASA explains that even at these altitudes there are still atmospheric molecules that rub against the station, slowing its speed and gradually dragging it lower. Thus, the *ISS* needs occasional reboosts to compensate.

Picking the safest orbital altitude for the station is complicated by several factors. One of them is the amount of solar and cosmic radiation that the crew members and delicate electronics can safely withstand. In addition, the *ISS* has to orbit at altitudes at which visiting spacecraft can easily reach and dock with it. Another consideration is space debris. As noted in Chapter 3, the U.S. military has cataloged more than 17,000 objects of various sizes circling Earth. These objects could strike and damage, perhaps disable, the station. It sometimes has to take evasive maneuvers to avoid them. In "Two More Collision Avoidance Maneuvers for the International Space Station" (October 2015, *Orbital Debris Quarterly News*, vol. 19, no. 4), NASA indicates that as of October 2015 the *ISS* had taken 25 such maneuvers throughout its history.

According to NASA several tons of propellants are required each year for *ISS* reboosts and maneuvers. The station has its own store of propellants and can also be moved by visiting spacecraft if they are equipped with powerful thrusters. In either case the transportation of propellant from Earth is a significant logistical challenge for the project. Food is another consumable that must be delivered. Numerous other kinds of supplies, equipment, and tools are also transported in addition to crew members and visitors.

FIGURE 5.1

Components of completed *International Space Station*

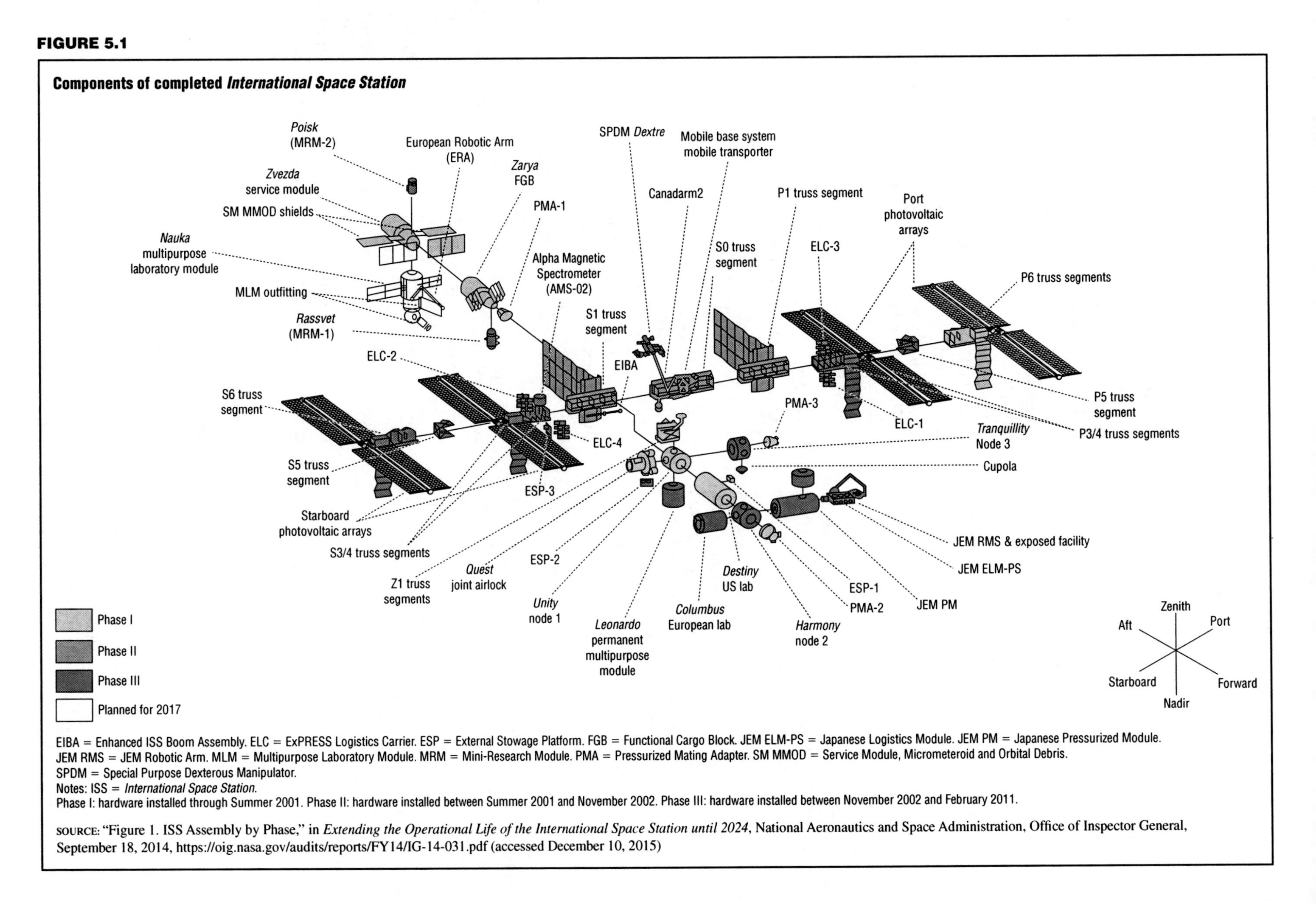

EIBA = Enhanced ISS Boom Assembly. ELC = ExPRESS Logistics Carrier. ESP = External Stowage Platform. FGB = Functional Cargo Block. JEM ELM-PS = Japanese Logistics Module. JEM PM = Japanese Pressurized Module. JEM RMS = JEM Robotic Arm. MLM = Multipurpose Laboratory Module. MRM = Mini-Research Module. PMA = Pressurized Mating Adapter. SM MMOD = Service Module, Micrometeroid and Orbital Debris. SPDM = Special Purpose Dexterous Manipulator.

Notes: ISS = *International Space Station.*

Phase I: hardware installed through Summer 2001. Phase II: hardware installed between Summer 2001 and November 2002. Phase III: hardware installed between November 2002 and February 2011.

SOURCE: "Figure 1. ISS Assembly by Phase," in *Extending the Operational Life of the International Space Station until 2024*, National Aeronautics and Space Administration, Office of Inspector General, September 18, 2014, https://oig.nasa.gov/audits/reports/FY14/IG-14-031.pdf (accessed December 10, 2015)

TABLE 5.1

Principal stages in *International Space Station* construction

Launch date (Eastern time zone)	ISS stage*	ISS element added	Launch vehicle
November 1998	1 A/R	Functional Cargo Block (FGB): Zarya	Proton
December 1998	2A	Node 1 (Unity), Pressurized Mating Adapter (PMA) 1 and 2	Shuttle/STS-88
July 2000	1R	Service Module (SM): Zvezda	Proton
October 2000	3A	Zenith 1 (Z1) truss, PMA 3	Shuttle/STS-92
November 2000	4A	Port 6 (P6) Truss	Shuttle/STS-97
February 2001	5A	U.S. Lab: Destiny	Shuttle/STS-98
March 2001	5A.1	External Stowage Platform (ESP) 1	Shuttle/STS-102
April 2001	6A	Space Station Remote Manipulator System (SSRMS): Canadarm 2	Shuttle/STS-100
July 2001	7A	U.S. Airlock: Quest	Shuttle/STS-104
September 2001	4R	Russian Docking Compartment (DC): Pirs and airlock	Progress
April 2002	8A	Starboard Zero (S0) truss segment and Mobile Transporter (MT)	Shuttle/STS-110
June 2002	UF-2	Mobile Base System (MBS)	Shuttle/STS-111
October 2002	9A	Starboard One (S1) truss segment	Shuttle/STS-112
November 2002	11A	P1 truss segment	Shuttle/STS-113
July 2005	LF-1	ESP 2	Shuttle/STS-114
September 2006	12A	P3/P4 Truss	Shuttle/STS-115
December 2006	12A.1	P5 Truss, retracting P6 arrays	Shuttle/STS-116
June 2007	13A	S3/S4 Truss	Shuttle/STS-117
August 2007	13A.1	S5 Truss and ESP 3	Shuttle/STS-118
October 2007	10A	Node-2: Harmony	Shuttle/STS-120
February 2008	1E	European Space Agency (ESA) Columbus Module	Shuttle/STS-122
March 2008	1J/A	Japanese Experiment Module Experiment Logistics Module Pressurized Section (JEM-ELM-PS) and Canadian Special Purpose Dexterous Manipulator: Dextre	Shuttle/STS-123
May 2008	1J	JEM Pressurized Module (PM)	Shuttle/STS-124
March 2009	15A	S6 Truss	Shuttle/STS-119
July 2009	2J/A	JEM Exposed Facility (JEM-EF)	Shuttle/STS-127
November 2009	5R	Russian Mini-Research Module 2 (MRM 2): Poisk	Progress
November 2009	ULF-3	ExPRESS Logistics Carriers (ELC) 1, 2	Shuttle/STS-129
February 2010	20A	Node 3: Tranquility and Cupola	Shuttle/STS-130
May 2010	ULF-4	Russian MRM 1: Rassvet	Shuttle/STS-132
February 2011	ULF-5	Permanent Multipurpose Module (PMM): Leonardo and ELC-4	Shuttle/STS-133
May 2011	ULF-6	Alpha Magnetic Spectrometer (AMS) and ELC-3	Shuttle/STS-134

*ISS Stage Letter Conventions:
A = U.S. Assembly.
E = European Assembly.
ISS = *International Space Station.*
J = Japanese Assembly.
LF = Logistics.
R = Russian Assembly.
UF = Utilization.
ULF = Utilization/Logistics.
STS = Space Transportation System.

SOURCE: Adapted from "Principal Stages in Construction," in *Reference Guide to the International Space Station: Utilization Edition, July 2015*, National Aeronautics and Space Administration, July 2015, http://www.nasa.gov/sites/default/files/atoms/files/np-2015-05-022-jsc_iss_utilization_guide_2015-508c.pdf (accessed December 1, 2015), and "International Space Station ISS Assembly Progress," in *Human Exploration and Operations*, National Aeronautics and Space Administration, October 24, 2012, http://www.nasa.gov/directorates/heo/reports/iss_assembly_progress.html (accessed December 1, 2015)

As of February 2016, the *ISS* had been visited by the following spacecraft:

- U.S. space shuttles—as shown in Table 2.6 in Chapter 2 and described in that chapter, between 1998 and 2011 space shuttles visited the station 37 times carrying crew members, components, and supplies. They only docked at the U.S. components of the station. Thrusters aboard the shuttles were used to reboost and maneuver the *ISS*.
- Russian Soyuz (see Figure 5.2)—this crewed vehicle is the "workhorse" of the Russian space fleet. Since 2000 dozens of Soyuz craft have transported *ISS* crew members and other visitors to the station and back to Earth. After the U.S. space shuttle program ended in 2011, Soyuz spacecraft became the only means for *ISS* crew transport. Each vehicle carries up to three people. As described in Chapter 3, a Soyuz returning to Earth lands on land with the aid of parachutes.
- Russian Progress (see Figure 5.3)—since 2000 dozens of these unmanned spacecraft have delivered consumables (food and water), spare parts, fuel, and other cargo to the station. Progress capsules are launched by Soyuz rockets. After being unloaded by the *ISS* crew, the capsules are packed with unneeded equipment, wastewater, and trash for their return trips. They are destroyed during fiery reentry to Earth's atmosphere. Progress spacecraft have powerful thrusters that are used to maneuver the *ISS* while they are docked to the station. As of February 2016, two Progress had been lost. In 2011 one of the spacecrafts

FIGURE 5.2

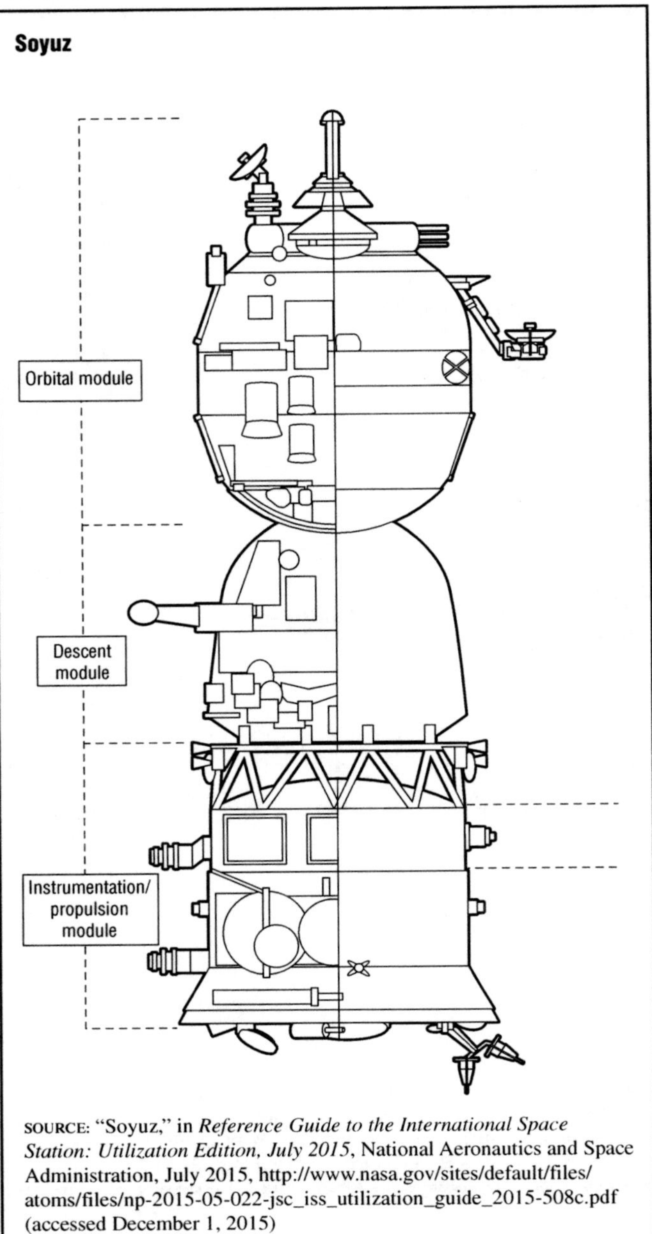

SOURCE: "Soyuz," in *Reference Guide to the International Space Station: Utilization Edition, July 2015*, National Aeronautics and Space Administration, July 2015, http://www.nasa.gov/sites/default/files/atoms/files/np-2015-05-022-jsc_iss_utilization_guide_2015-508c.pdf (accessed December 1, 2015)

crashed soon after liftoff. In 2015 a Soyuz/Progress spun out of control before it reached the proper orbit. It and its cargo were destroyed upon reentry to Earth's atmosphere.

- European automated transfer vehicles (ATVs)—five of these unmanned spacecraft made cargo deliveries to the station between 2008 and 2014. Their thrusters were used to make *ISS* maneuvers. The ATVs and their launch dates were *Jules Verne* (2008), *Johannes Kepler* (2011), *Edoardo Amaldi* (2012), *Albert Einstein* (2013), and *Georges Lemaître* (2014). In 2012 the ESA decided to cease producing any more of the spacecraft. The ATVs were destroyed during atmospheric reentry to Earth.
- Japanese H-II transfer vehicles (HTVs)—as of February 2016, five of these unmanned spacecraft had made cargo deliveries to the *ISS*. HTV-1 visited the station in 2009 and HTV-2 in 2011. According to JAXA (November 11, 2010, http://www.jaxa.jp/press/2010/11/20101111_kounotori_e.html), the HTV-2 was nicknamed "Kounotori" (which translates as "white stork") based on suggestions from the public. Three more HTVs, dubbed "Kuonotori 3" through "Kounotori 5," visited the *ISS* between 2012 and 2015. Future HTV missions are expected to take place through the late 2010s. The spacecraft are destroyed during atmospheric reentry to Earth.
- U.S. Dragon (see Figure 4.4 in Chapter 4)—in 2012 a Dragon made history by becoming the first privately made spacecraft to dock with the *ISS*. As described in Chapter 4, the unmanned spacecraft are built by the American company Space Exploration Technologies Corp. (SpaceX). Dragon capsules are launched by SpaceX's Falcon 9 rockets. The reusable Dragons are protected by heat shields and splash down in the ocean on their return to Earth. As of February 2016, five Dragons had visited the station carrying supplies and equipment. On June 28, 2015, a Falcon 9/Dragon exploded approximately two minutes after lifting off from the launchpad. The spacecraft and its cargo were destroyed. As of February 2016, NASA had not set a date for Dragon missions to resume.
- U.S. Cygnus (see Figure 4.6 in Chapter 4)—the first Cygnus (also a privately made spacecraft) visited the *ISS* in 2013. The unmanned capsule is a product of the American company Orbital ATK (formerly Orbital Sciences Corporation) as explained in Chapter 4. The Cygnus spacecraft are launched by Orbital ATK's Antares rockets. The capsules are destroyed by reentry into Earth's atmosphere. In January 2014 a second Cygnus flight to the station took place. On October 28, 2014, however, an Antares/Cygnus failed only seconds after lifting off from the launchpad and was blown up by controllers. The spacecraft and its cargo were destroyed. In December 2015 the Cygnus returned to flight with a successful trip to the *ISS*.

Shuttle Loss Brings Tragedy and Delays

On February 1, 2003, the space shuttle *Columbia* broke apart as it entered Earth's atmosphere over the western United States. The shuttle had been on a research mission and did not visit the *ISS*. The catastrophe killed the seven crew members and shook the U.S. space program to its core. An investigation revealed that the shuttle's thermal protection tiles were likely damaged by a foam strike shortly after launch. During reentry, hot gases seeped past the tiles into the orbiter structure, and the resulting turbulence tore it apart.

FIGURE 5.3

Progress

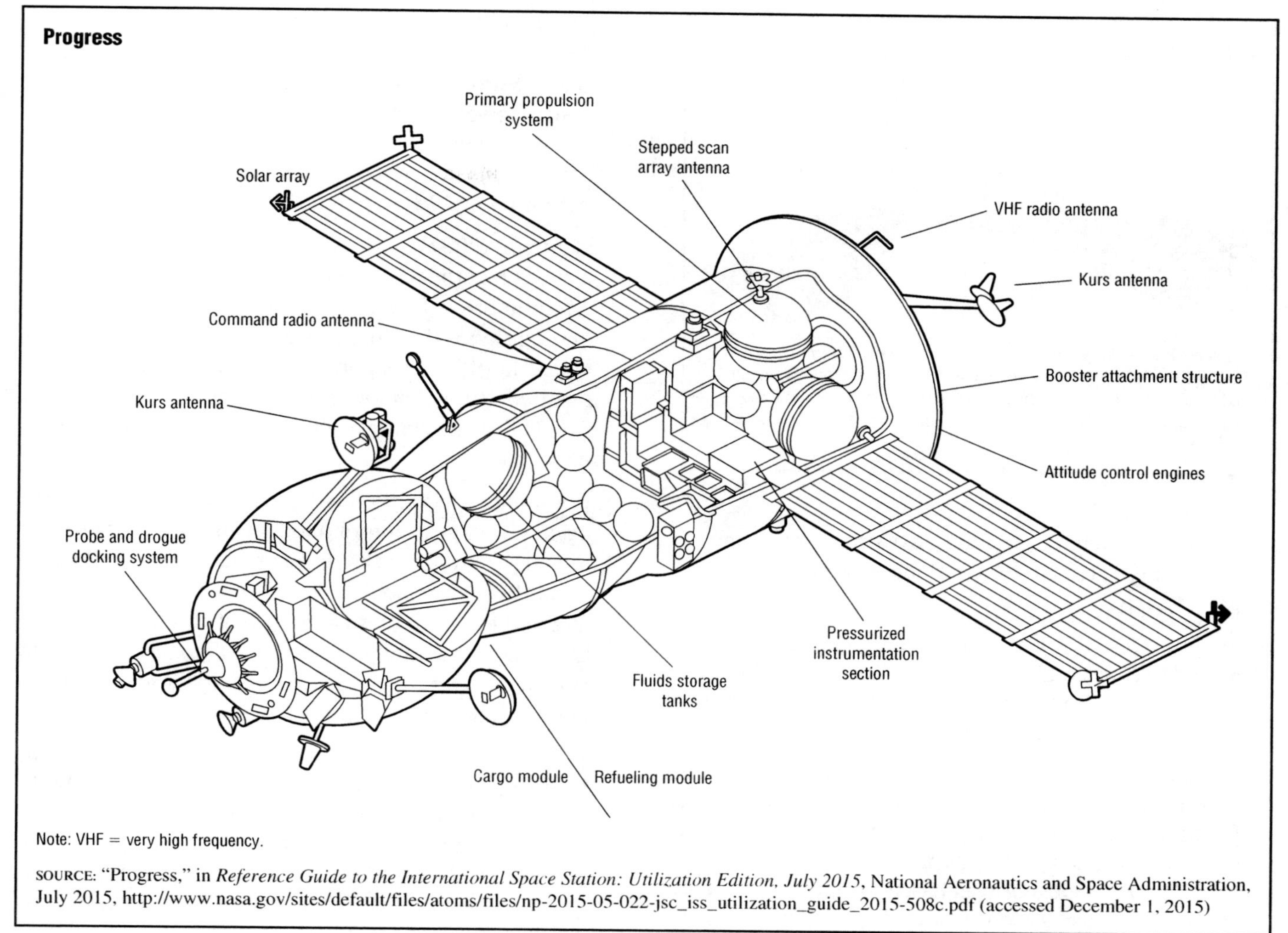

Note: VHF = very high frequency.

SOURCE: "Progress," in *Reference Guide to the International Space Station: Utilization Edition, July 2015*, National Aeronautics and Space Administration, July 2015, http://www.nasa.gov/sites/default/files/atoms/files/np-2015-05-022-jsc_iss_utilization_guide_2015-508c.pdf (accessed December 1, 2015)

The entire shuttle fleet was grounded for more than two years. The shuttle flights that had been scheduled to deliver heavy components to the *ISS* during this time were postponed. There were no other spacecraft capable of transporting them to the station. The *ISS* assembly essentially came to a halt.

The shutdown of the shuttle program had a number of operational and cost effects on the ISS program. In its 2003 report *Space Station: Impact of the Grounding of the Shuttle Fleet*, the GAO estimates that between 1985 and 2002 the United States had spent $32 billion on ISS research, development, training, and construction. However, significant additional costs were expected due to the shuttle shutdown. The GAO notes that the modules and other equipment that were waiting to be shipped to the *ISS* had to be unpacked, undergo maintenance, be retested, and then be repacked before flight. Also, batteries had to be recharged due to prolonged storage. All these problems resulted in unexpected costs in NASA's ISS program.

Grounding the shuttle also introduced significant delays to planned research projects, equipment repairs, and construction activities. Postponement of the latter added unexpected costs to the ISS project.

Major *ISS* Components

As shown in Figure 5.1, the space station includes multiple components that are linked together. Certain of the components make the *ISS* habitable for humans because they are fully pressurized. This means that they are supplied with air at the optimal pressure for breathing. The launch years and brief descriptions of the major pressurized components are provided below.

- Zarya (1998)—this U.S. component is shown on the left in Figure 5.4. NASA paid Roscosmos to build the module and launch it into orbit using a Russian Proton K rocket. According to NASA (October 18, 2013, http://www.nasa.gov/mission_pages/station/structure/elements/fgb.html#.UwDpM55dWrY), Zarya translates as "sunrise." The module's technical name is the Functional Cargo Block, which in Russian has the acronym FGB. It was self-propelled and kept the station in orbit until the Zvezda service module

FIGURE 5.4

Zarya/Unity modules

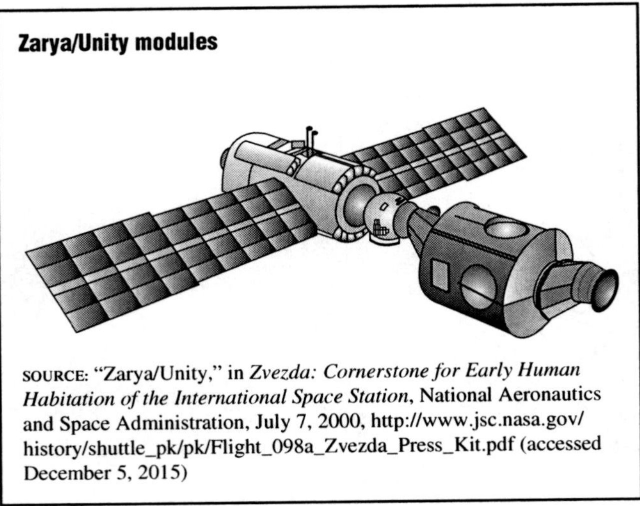

SOURCE: "Zarya/Unity," in *Zvezda: Cornerstone for Early Human Habitation of the International Space Station*, National Aeronautics and Space Administration, July 7, 2000, http://www.jsc.nasa.gov/history/shuttle_pk/pk/Flight_098a_Zvezda_Press_Kit.pdf (accessed December 5, 2015)

arrived in 2000. Since then Zarya has served mostly as a storage area. It does include external fuel tanks.

- Unity (1998)—this U.S. module is shown on the right in Figure 5.4. It was the first *ISS* component delivered by a U.S. space shuttle, the *Endeavour*. Unity provides a node (link) between modules. It is basically a passageway with some internal storage space.
- Zvezda (2000)—the first Russian module forms the core of that nation's area on the station. Zvezda has functions for station control, navigation, communications, and life support systems (including crew quarters). Zvezda's design was based on the core module of the *Mir* space station. It includes docking ports for visiting spacecraft.
- Destiny (2001)—this American lab module is the primary research laboratory for U.S. payloads. (See Figure 5.5.) It has many racks that can support a variety of electrical and fluid systems during the performance of experiments. In 2011 the lab gained a new "occupant" called Robonaut 2 (August 27, 2014, http://www.nasa.gov/mission_pages/station/main/robonaut.html#.Uw9cFp5dWra), a human-shaped robot designed to help astronauts perform tasks around the station.
- Quest (2001)—the U.S. airlock shown in Figure 5.6 is attached to the Unity module. According to NASA (October 18, 2013, http://www.nasa.gov/mission_pages/station/structure/elements/quest.html#.UwIjvJ5dWrY), it is the "primary path" by which crew members in American spacesuits enter and leave the station. (See Figure 5.7.) However, it can also accommodate crew members in Russian spacesuits. (See Figure 5.8.)
- Pirs (2001)—as of February 2016, this Russian docking compartment was attached to Zvezda. It provides crew members in Russian spacesuits with station egress and ingress. Pirs was intended to be a temporary component. It has long been planned to replace it with a more sophisticated airlock on the MLM, a pressurized component that has been nicknamed "Nauka." Figure 5.1 shows the MLM, its airlock, and a European robotic arm in their intended locations. However, on-ground construction of the entire MLM assembly has been fraught with delays. These components are not expected to be installed at the station before 2017.
- Harmony (2007)—this American module is an Italian-made compartment including living and working space. According to NASA (October 18, 2013, http://www.nasa.gov/mission_pages/station/structure/elements/node2.html#.UwIgcp5dWrY), the module's name resulted from a student naming contest. It was the first U.S. component of the *ISS* named by non-NASA personnel.
- Columbus (2008)—this ESA laboratory is the agency's largest contribution to the space station. It includes various racks and apparatus for conducting scientific experiments. The ESA (2016, http://www.esa.int/Our_Activities/Human_Spaceflight/Columbus/Columbus_laboratory) notes that experiments are conducted in the life sciences, materials science, fluid physics, and other disciplines.
- Kibo (2008)—the name of this Japanese Experiment Module (JEM) translates as "hope." According to NASA (October 18, 2013, http://www.nasa.gov/mission_pages/station/structure/elements/jem.html#.UwJDsZ5dWrY), the laboratory facilitates experiments in "space medicine, biology, Earth observations, material production, biotechnology and communications research."
- JEM Experiment Logistics Module (ELM; 2009)—the Japanese ELM is attached to Kibo and provides storage space for equipment and supplies.
- Poisk (2009)—the Russian Mini-Research Module (MRM) 2 known as Poisk is attached to Zvezda. The word *Poisk* translates as "search" or "explore." According to NASA (January 25, 2010, http://www.nasa.gov/externalflash/ISSRG/pdfs/MRM2.pdf), the component is very structurally similar to Russia's Pirs docking compartment in that it provides crew members in Russian spacesuits with station egress and ingress.
- Tranquility (2010)—this American module was built in Italy by the ESA. It provides storage and houses various utility functions. NASA (October 18, 2013, http://www.nasa.gov/mission_pages/station/structure/elements/tranquility.html#.UwJGxp5dWrY) notes that a treadmill is also available in Tranquility. As shown in Figure 5.1, the module includes a cupola, which is a windowed viewing area. (See Figure 5.9.) This provides crew members with a vantage point for observing Earth and the operations going on outside the station.

FIGURE 5.5

U.S. laboratory Destiny

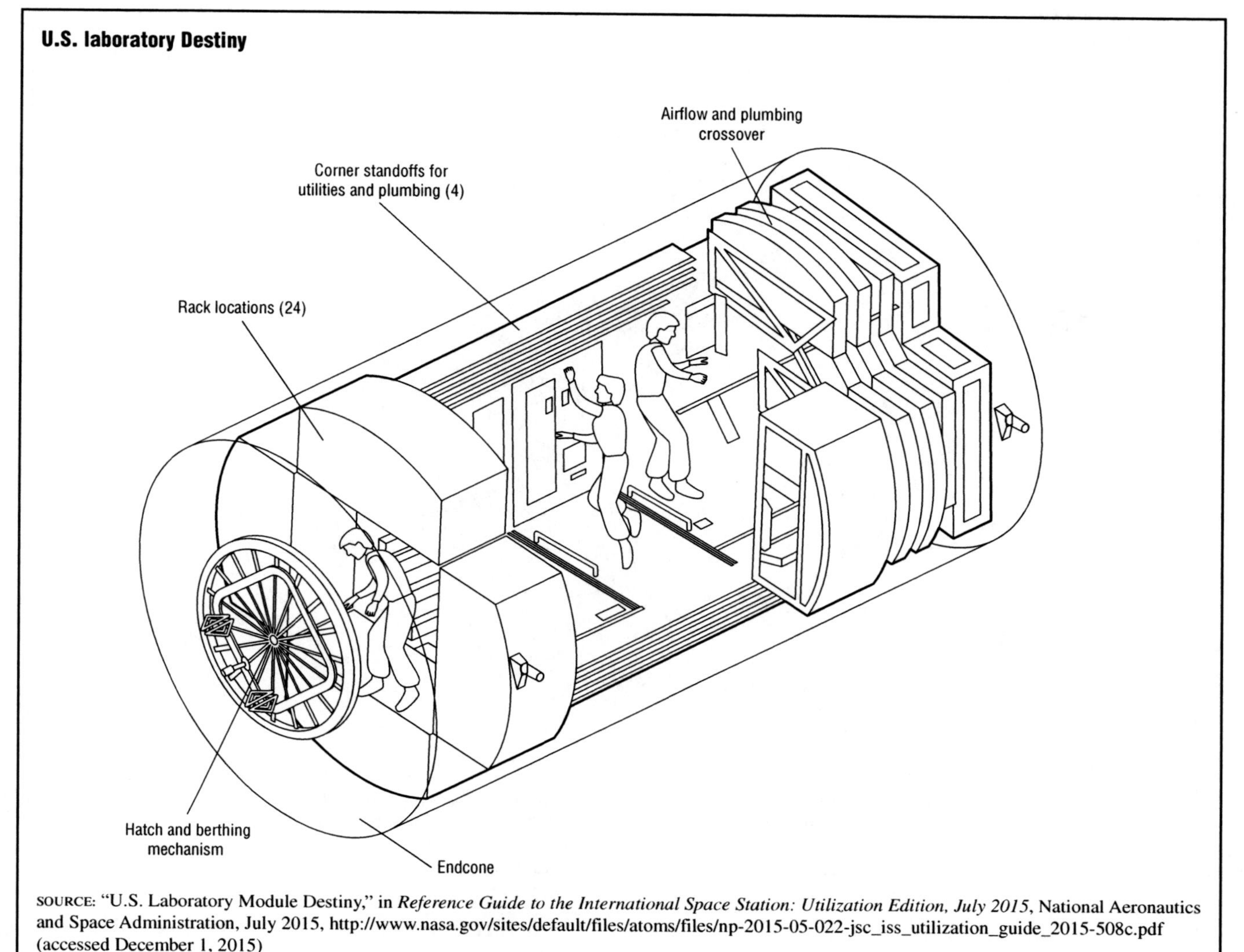

SOURCE: "U.S. Laboratory Module Destiny," in *Reference Guide to the International Space Station: Utilization Edition, July 2015*, National Aeronautics and Space Administration, July 2015, http://www.nasa.gov/sites/default/files/atoms/files/np-2015-05-022-jsc_iss_utilization_guide_2015-508c.pdf (accessed December 1, 2015)

- Rassvet (2010)—this Russian compartment is also known as MRM 1. The word *rassvet* translates as "dawn." It is attached to Zarya. According to NASA (January 21, 2010, http://www.nasa.gov/externalflash/ISSRG/pdfs/MRM1.pdf), the MRM 1 is primarily used for storage and payload operations.
- Leonardo or Permanent Multipurpose Module (PMM; 2011)—this is a U.S. component. NASA explains in *Space Shuttle Mission STS-133 Press Kit: The Final Flight of Discovery* (February 2011, http://www.nasa.gov/pdf/491387main_STS-133%20Press%20Kit.pdf) that the Italian-made PMM (named after Leonardo da Vinci) was originally a cargo transport container called a multipurpose logistics module (MPLM). Leonardo and another MPLM named Raffaello (after the Italian painter Raphael) traveled back and forth to the station aboard space shuttles during *ISS* construction. Leonardo was modified for permanent installation on the station and primarily serves as a storage area. It was transported to the *ISS* by the space shuttle *Discovery* during its final mission in 2011.

LIFEBOAT. One of the most important pressurized items aboard the *ISS* is its lifeboat. An extra Soyuz is kept docked to the station at all times to provide the crew a means of escape in case of a catastrophic event. The lifeboat is replaced on a regular basis as new Soyuz bring crew members and visitors to the *ISS*. When six crew members are aboard the station there are two Soyuz docked there. As of February 2016, no crew members had ever used a Soyuz to escape from the station due to an emergency.

Docked Soyuz spacecraft have provided refuge when the station was in danger of being struck by space debris. According to NASA, in "International Space Station Performs Two Debris Avoidance Maneuvers and a Shelter-in-Place" (July 2015, *Orbital Debris Quarterly News*, vol. 19, no. 3), as of July 2015, station crew members had sheltered-in-place in the lifeboats four times. The most recent close call occurred on July 16, 2015, when a satellite

FIGURE 5.6

Airlock Quest

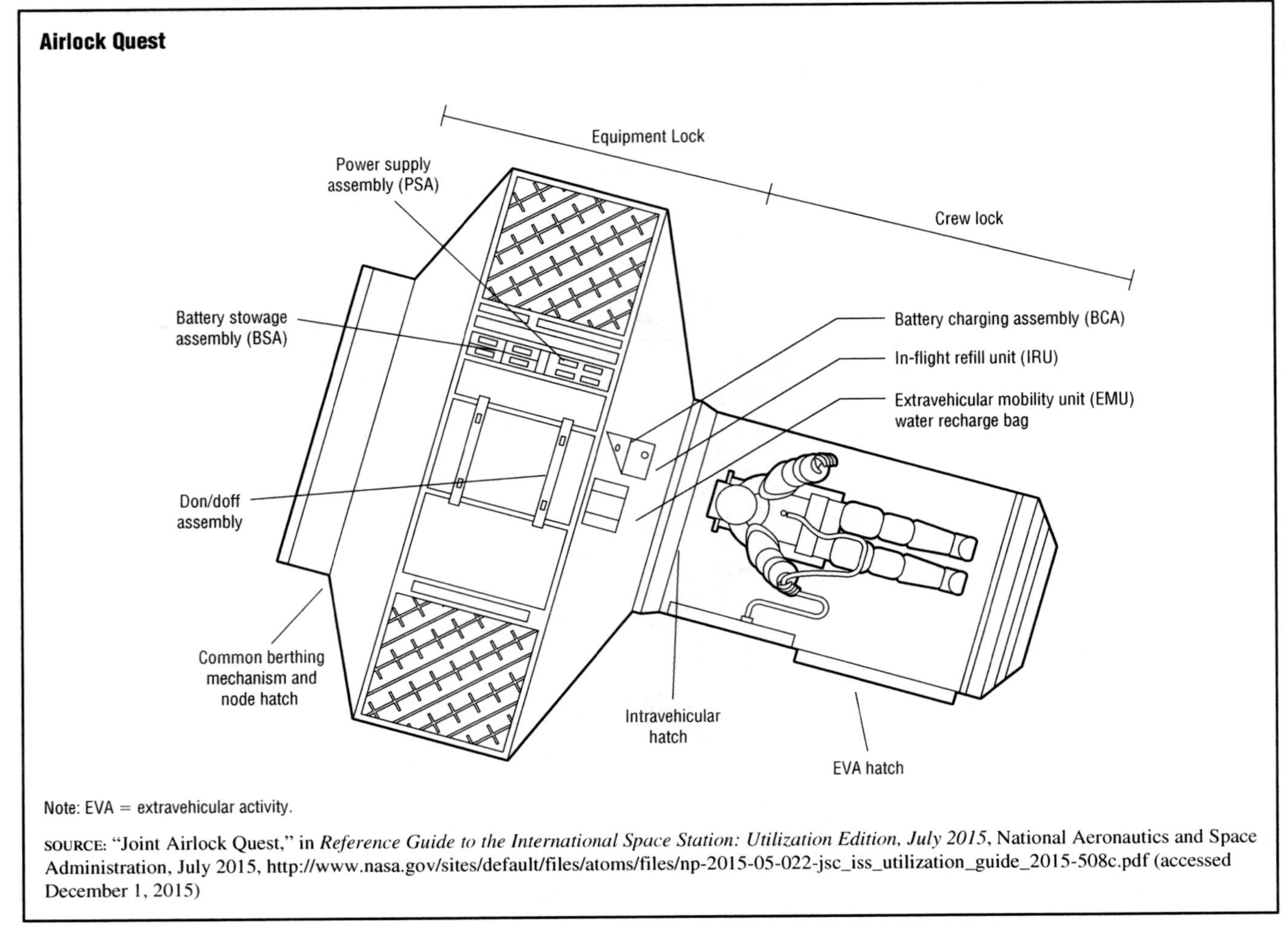

Note: EVA = extravehicular activity.

SOURCE: "Joint Airlock Quest," in *Reference Guide to the International Space Station: Utilization Edition, July 2015*, National Aeronautics and Space Administration, July 2015, http://www.nasa.gov/sites/default/files/atoms/files/np-2015-05-022-jsc_iss_utilization_guide_2015-508c.pdf (accessed December 1, 2015)

fragment was detected hurtling toward the *ISS*. There was insufficient time to maneuver the station to a safer position. The three-member crew took refuge in a Soyuz lifeboat until the debris had safely passed by the *ISS*.

NONPRESSURIZED COMPONENTS. The *ISS* also includes components that are not pressurized. The most notable is the truss, which is a long girderlike structure that is perpendicular to the modules. At each end of the truss sit solar array wings (solar panels) that help power the station. The truss was constructed in segments from 2000 through 2009. In addition, several of the pressurized modules include external platforms or exposed facilities on which scientific experiments are conducted. This research is described in more detail later in this chapter.

As of February 2016, the space station included three external robotic arms or manipulators. The Canadarm2 was installed in 2001. It was built by Canadian companies and is an enhanced version of the robotic arm called Canadarm1 that was used on the U.S. space shuttles. The Canadarm2's dexterous grappling abilities proved useful in attaching components to the *ISS*. It can also help visiting spacecraft dock and can grasp and move crew members during space walks (extravehicular activities or EVAs).

Another Canadian robotic system is called Dextre or the Special Purpose Dexterous Manipulator. It was installed in 2008 and can move along the truss. According to the CSA (January 7, 2016, http://www.asc-csa.gc.ca/eng/iss/dextre/default.asp), this "robotic handyman" performs external repairs and maintenance tasks.

The Japanese module Kibo is also equipped with a robotic arm called the JEM remote manipulator system that was installed in 2008. It facilitates scientific experiments conducted on the Kibo's external platforms. The Russian side of the *ISS* did not include a robotic arm as of February 2016. The Russians do have two cranes that are attached to the Pirs module. As shown in Figure 5.1, a European-built robotic arm is scheduled to be installed as part of the MLM assembly in 2017.

RESIDENT CREWS AND VISITORS

Since 1998 the *ISS* has been visited by more than 130 people. The vast majority have been professional astronauts or cosmonauts employed by national space agencies. In addition, a handful of paying tourists have visited the station.

FIGURE 5.7

Extravehicular Mobility Unit (spacesuit) used by U.S. astronauts

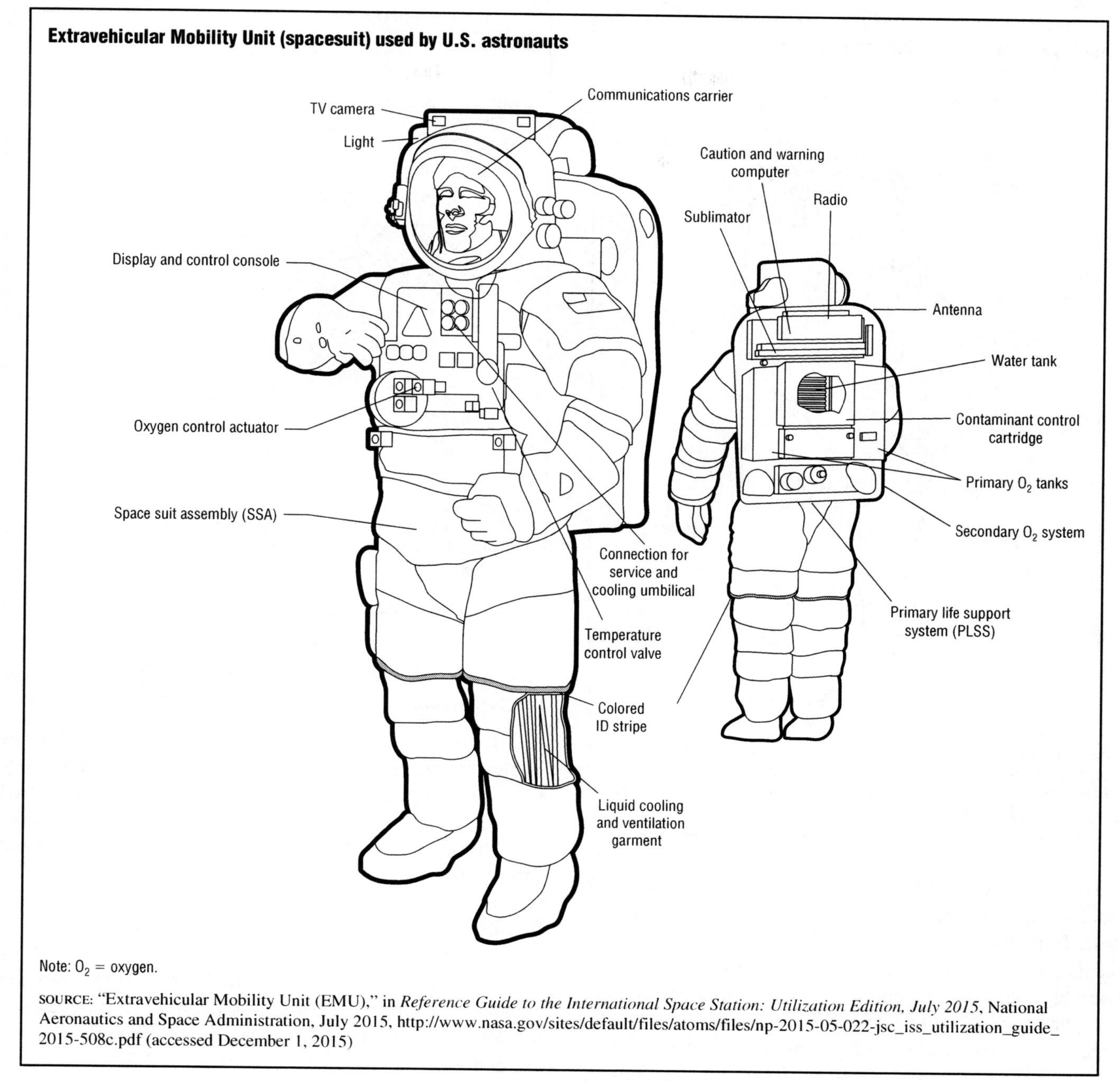

Note: O_2 = oxygen.

SOURCE: "Extravehicular Mobility Unit (EMU)," in *Reference Guide to the International Space Station: Utilization Edition, July 2015*, National Aeronautics and Space Administration, July 2015, http://www.nasa.gov/sites/default/files/atoms/files/np-2015-05-022-jsc_iss_utilization_guide_2015-508c.pdf (accessed December 1, 2015)

Expedition Crew Members

On November 2, 2000, the first resident crew to actually inhabit the *ISS* arrived aboard a Soyuz spacecraft after a two-day trip from Earth. The first crew was called Expedition 1. It included one American astronaut and two Russian cosmonauts. (See Table 5.2.) They lived aboard the station for 136 days. On March 10, 2001, the Expedition 2 crew arrived at the station aboard the space shuttle *Discovery*. The Expedition 1 crew returned to Earth aboard the space shuttle on March 19, 2001, leaving their Soyuz spacecraft at the station to serve as the first lifeboat. Note that the Expedition 1 crew departed the station nine days after the Expedition 2 crew arrived. An overlap period provides time for departing crew members to brief incoming crew members and hand over activities to them.

The Expedition 2 crew included one Russian cosmonaut and two American astronauts. It had been decided to swap out the Expedition crews every four to six months and to rotate back and forth between crews that were predominantly Russian and crews that were predominantly American. As shown in Table 5.2, this pattern was maintained until Expedition 7, which was the first mission following the loss in February 2003 of the space shuttle *Columbia*. The ISS partners decided to downsize future station crews to only two people until the shuttle resumed operations. This made it easier for the Russians

FIGURE 5.8

Orlan spacesuit used by Russian cosmonauts

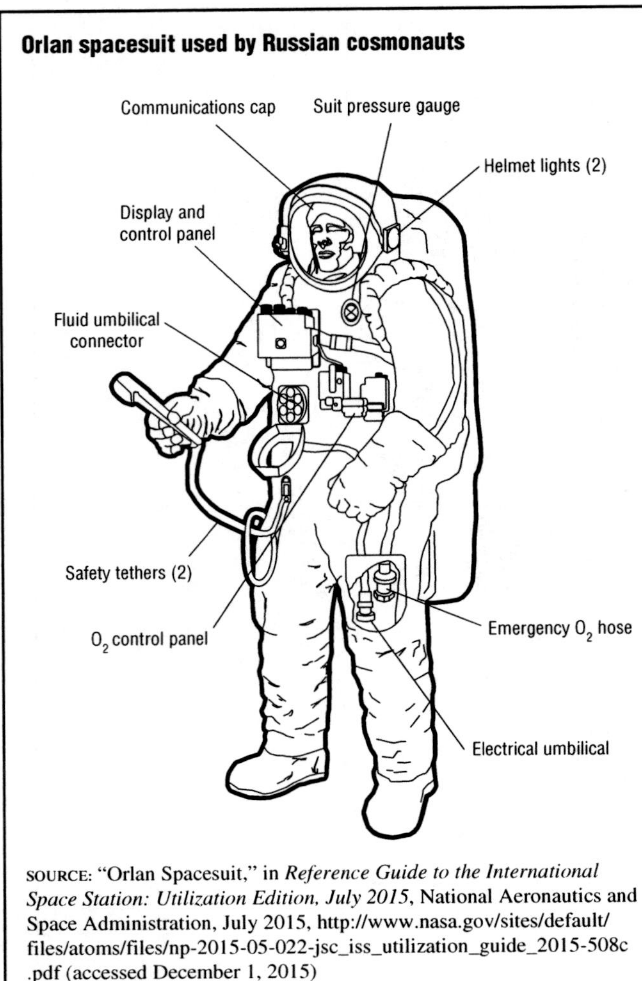

SOURCE: "Orlan Spacesuit," in *Reference Guide to the International Space Station: Utilization Edition, July 2015*, National Aeronautics and Space Administration, July 2015, http://www.nasa.gov/sites/default/files/atoms/files/np-2015-05-022-jsc_iss_utilization_guide_2015-508c.pdf (accessed December 1, 2015)

to assume all responsibility for resupplying the crew with food, water, and other necessities using Progress spacecraft. Expedition 7 included only two crew members—one astronaut and one cosmonaut. Unable to proceed with assembly, they were kept busy maintaining the station and performing limited scientific research. One of the largest obstacles to ISS science was the presence of only two crew members. The number of new and continuing experiments had to be reduced so the crews could devote more time to station maintenance and operation. The two-member crews continued through Expedition 12.

In July 2005 the space shuttle program resumed operations with its first Return to Flight mission, carrying equipment and supplies, but no new crew members, to the *ISS*. The second Return to Flight mission was delayed until July 2006 due to continuing concerns about debris shedding during launch. Two members of the Expedition 13 crew flew to the station aboard a Russian Soyuz spacecraft in March 2006. They were joined in July 2006 by the ESA astronaut Thomas Reiter (1958–) of Germany, who arrived aboard a space shuttle. Reiter's arrival gave the station a three-person crew for the first

FIGURE 5.9

Cupola

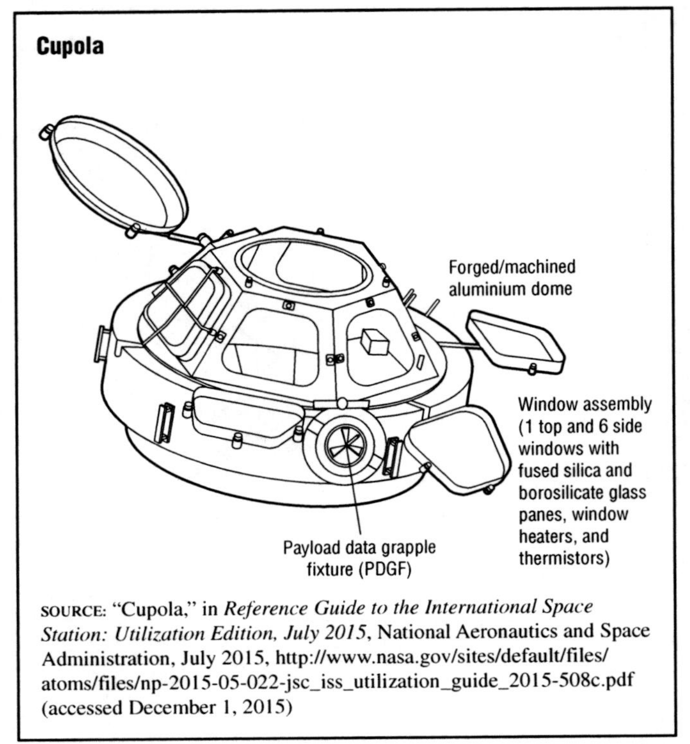

SOURCE: "Cupola," in *Reference Guide to the International Space Station: Utilization Edition, July 2015*, National Aeronautics and Space Administration, July 2015, http://www.nasa.gov/sites/default/files/atoms/files/np-2015-05-022-jsc_iss_utilization_guide_2015-508c.pdf (accessed December 1, 2015)

time since 2002. He was also the first *ISS* crew member from an agency other than NASA or Roscosmos.

Beginning with Expedition 14 in 2006, NASA rotated its flight engineers on the *ISS* at irregular intervals, meaning that some flight engineers only stayed aboard the station for a few weeks before being replaced. In addition, as shown in Table 5.2, some flight engineers were part of overlapping Expeditions. In 2007 the *ISS* temporarily had a six-person crew (Expedition 16) for the first time. In 2009 Robert Thirsk (1953–) became the first CSA astronaut on an *ISS* resident crew. He was part of Expeditions 20 and 21. Also in 2009, Koichi Wakata (1963–) became the first JAXA astronaut to serve as a resident crew member. He was part of Expeditions 18, 19, and 20, and later, Expeditions 38 and 39.

In 2015 NASA and Roscosmos decided to each send a crew member for around a one-year stay on the *ISS*. As described later in this chapter, the intent was to learn about the long-term effects of weightlessness on human health. American astronaut Scott Kelly (1964–) and Russian cosmonaut Mikhail Kornienko (1960–) launched on March 27, 2015, as part of the Expedition 43 crew. Both returned to Earth on March 1, 2016, after 340 days in space.

As shown in Table 5.2, 103 individuals had served as resident crew members aboard the station as of December 2015. (Note that some individuals were members of multiple expeditions.) The agency breakdown for the 103 individuals was: NASA (48), Roscosmos (39), ESA (9), JAXA (5), and CSA (2).

TABLE 5.2

***International Space Station* expedition crews as of December 2015**

Arrival date	Departure date	Crew member	Expedition number(s)	Agency	Crew title(s)
10/31/00	3/18/01	William Shepherd	Expedition 1	NASA	Commander
10/31/00	3/18/01	Yuri Gidzenko	Expedition 1	Roscosmos	Soyuz Commander
10/31/00	3/18/01	Sergei Krikalev	Expedition 1	Roscosmos	Flight Engineer
3/10/01	8/20/01	Yury Usachev	Expedition 2	Roscosmos	Commander
3/10/01	8/20/01	Susan Helms	Expedition 2	NASA	Flight Engineer
3/10/01	8/20/01	James Voss	Expedition 2	NASA	Flight Engineer
8/12/01	12/15/01	Frank Culbertson	Expedition 3	NASA	Commander
8/12/01	12/15/01	Vladimir Dezhurov	Expedition 3	Roscosmos	Soyuz Commander
8/12/01	12/15/01	Mikhail Tyurin	Expedition 3	Roscosmos	Flight Engineer
12/7/01	6/15/02	Yury Onufrienko	Expedition 4	Roscosmos	Commander
12/7/01	6/15/02	Dan Bursch	Expedition 4	NASA	Flight Engineer
12/7/01	6/15/02	Carl Walz	Expedition 4	NASA	Flight Engineer
6/7/02	12/2/02	Valery Korzun	Expedition 5	Roscosmos	Commander
6/7/02	12/7/02	Peggy Whitson	Expedition 5	NASA	Science Officer
6/7/02	12/7/02	Sergei Treschev	Expedition 5	Roscosmos	Flight Engineer
11/25/02	5/3/03	Ken Bowersox	Expedition 6	NASA	Commander
11/25/02	5/3/03	Nikolai Budarin	Expedition 6	Roscosmos	Flight Engineer
11/25/02	5/3/03	Don Pettit	Expedition 6	NASA	Science Officer
4/28/03	10/27/03	Yuri Malenchenko	Expedition 7	Roscosmos	Commander
4/28/03	10/27/03	Ed Lu	Expedition 7	NASA	Science Officer
10/20/03	4/29/04	Michael Foale	Expedition 8	NASA	Commander/Science Officer
10/20/03	4/29/04	Alexander Kaleri	Expedition 8	Roscosmos	Flight Engineer
4/21/04	10/23/04	Gennady Padalka	Expedition 9	Roscosmos	Commander
4/21/04	10/23/04	Mike Fincke	Expedition 9	NASA	Flight Engineer/Science Officer
10/15/04	4/24/05	Leroy Chiao	Expedition 10	NASA	Commander/Science Officer
10/15/04	4/24/05	Salizhan Sharipov	Expedition 10	Roscosmos	Soyuz Commander/Flight Engineer
4/16/05	10/10/05	Sergei Krikalev	Expedition 11	Roscosmos	Commander
4/16/05	10/10/05	John Phillips	Expedition 11	NASA	Flight Engineer/Science Officer
10/3/05	4/8/06	William McArthur	Expedition 12	NASA	Commander/Science Officer
10/3/05	4/8/06	Valery Tokarev	Expedition 12	Roscosmos	Flight Engineer
3/31/06	9/28/06	Pavel Vinogradov	Expedition 13	Roscosmos	Commander
3/31/06	9/28/06	Jeffrey Williams	Expedition 13	NASA	Flight Engineer
7/6/06	12/19/06	Thomas Reiter	Expedition 13/14	ESA	Flight Engineer
9/20/06	4/21/07	Michael Lopez-Alegria	Expedition 14	NASA	Commander
9/20/06	4/21/07	Mikhail Tyurin	Expedition 14	Roscosmos	Flight Engineer
12/11/06	6/19/07	Sunita Williams	Expedition 14/15	NASA	Flight Engineer
4/9/07	10/21/07	Fyodor Yurchikhin	Expedition 15	Roscosmos	Commander
4/9/07	10/21/07	Oleg Kotov	Expedition 15	Roscosmos	Flight Engineer
6/10/07	11/7/07	Clayton Anderson	Expedition 15/16	NASA	Flight Engineer
10/11/07	4/21/08	Peggy Whitson	Expedition 16	NASA	Commander
10/11/07	4/21/08	Yuri Malenchenko	Expedition 16	Roscosmos	Flight Engineer
10/25/07	2/18/08	Daniel Tani	Expedition 16	NASA	Flight Engineer
2/9/08	3/24/08	Léopold Eyharts	Expedition 16	ESA	Flight Engineer
3/13/09	6/12/09	Garrett E. Reisman	Expedition 16/17	NASA	Flight Engineer
4/10/08	10/23/08	Sergei Volkov	Expedition 17	Roscosmos	Commander
4/10/08	10/23/08	Oleg Kononenko	Expedition 17	Roscosmos	Flight Engineer
6/2/08	11/28/08	Gregory E. Chamitoff	Expedition 17/18	NASA	Flight Engineer
10/12/08	4/8/09	E. Michael Fincke	Expedition 18	NASA	Commander
10/12/08	4/8/09	Yury Lonchakov	Expedition 18	Roscosmos	Flight Engineer
11/16/08	3/25/09	Sandra H. Magnus	Expedition 18	NASA	Flight Engineer
3/17/09	7/29/09	Koichi Wakata	Expedition 18/19/20	JAXA	Flight Engineer
3/28/09	10/11/09	Gennady Padalka	Expedition 19/20	Roscosmos	Commander
3/28/09	10/11/09	Michael Barratt	Expedition 19/20	NASA	Flight Engineer
7/17/09	9/9/09	Tim Kopra	Expedition 20	NASA	Flight Engineer
5/29/09	12/1/09	Roman Romanenko	Expedition 20/21	Roscosmos	Flight Engineer
5/29/09	12/1/09	Frank De Winne	Expedition 20/21	ESA	Flight Engineer (20); Commander (21)
5/29/09	12/1/09	Robert Thirsk	Expedition 20/21	CSA	Flight Engineer
8/30/09	11/25/09	Nicole Stott	Expedition 20/21	NASA	Flight Engineer
10/2/09	3/18/10	Jeffrey Williams	Expedition 21/22	NASA	Flight Engineer (21); Commander (22)
10/2/09	3/18/10	Maxim Suraev	Expedition 21/22	Roscosmos	Flight Engineer
12/22/09	6/1/10	Oleg Kotov	Expedition 22/23	Roscosmos	Flight Engineer (22); Commander (23)
12/22/09	6/1/10	Soichi Noguchi	Expedition 22/23	JAXA	Flight Engineer
12/22/09	6/1/10	T.J. Creamer	Expedition 22/23	NASA	Flight Engineer
4/4/10	9/25/10	Alexander Skvortsov	Expedition 23/24	Roscosmos	Flight Engineer (23); Commander (24)
4/4/10	9/25/10	Tracy Caldwell Dyson	Expedition 23/24	NASA	Flight Engineer
4/4/10	9/25/10	Mikhail Kornienko	Expedition 23/24	Roscosmos	Flight Engineer
6/17/10	11/25/10	Doug Wheelock	Expedition 24/25	NASA	Flight Engineer (24); Commander (25)
6/17/10	11/25/10	Shannon Walker	Expedition 24/25	NASA	Flight Engineer
6/17/10	11/25/10	Fyodor Yurchikhin	Expedition 24/25	Roscosmos	Flight Engineer
10/9/10	3/16/11	Oleg Skripochka	Expedition 25	Roscosmos	Flight Engineer
10/9/10	3/16/11	Alexander Kaleri	Expedition 25	Roscosmos	Flight Engineer
10/9/10	3/16/11	Scott Kelly	Expedition 25/26	NASA	Flight Engineer (25); Commander (26)

TABLE 5.2

***International Space Station* expedition crews as of December 2015 [CONTINUED]**

Arrival date	Departure date	Crew member	Expedition number(s)	Agency	Crew title(s)
12/17/10	5/23/11	Dmitry Kondratyev	Expedition 26/27	Roscosmos	Flight Engineer (26); Commander (27)
12/17/10	5/23/11	Paolo Nespoli	Expedition 26/27	ESA	Flight Engineer
12/17/10	5/23/11	Catherine Coleman	Expedition 26/27	NASA	Flight Engineer
4/6/11	9/15/11	Andrey Borisenko	Expedition 28	Roscosmos	Commander
4/6/11	9/15/11	Ron Garan	Expedition 28	NASA	Flight Engineer
4/6/11	9/15/11	Alexander Samokutyaev	Expedition 28	Roscosmos	Flight Engineer
6/9/11	11/21/11	Mike Fossum	Expedition 28/29	NASA	Flight Engineer (28); Commander (29)
6/9/11	11/21/11	Sergei Volkov	Expedition 28/29	Roscosmos	Flight Engineer
6/9/11	11/21/11	Satoshi Furukawa	Expedition 28/29	JAXA	Flight Engineer
11/16/11	4/27/12	Dan Burbank	Expedition 29/30	NASA	Flight Engineer (29); Commander (30)
11/16/11	4/27/12	Anton Shkaplerov	Expedition 29/30	Roscosmos	Flight Engineer
11/16/11	4/27/12	Anatoly Ivanishin	Expedition 29/30	Roscosmos	Flight Engineer
12/23/11	7/1/12	Oleg Kononenko	Expedition 30/31	Roscosmos	Flight Engineer (30); Commander (31)
12/23/11	7/1/12	Don Pettit	Expedition 30/31	NASA	Flight Engineer
12/23/11	7/1/12	Andre Kuipers	Expedition 30/31	ESA	Flight Engineer
5/17/12	9/16/12	Gennady Padalka	Expedition 31/32	Roscosmos	Flight Engineer (31); Commander (32)
5/17/12	9/16/12	Joe Acaba	Expedition 31/32	NASA	Flight Engineer
5/17/12	9/16/12	Sergei Revin	Expedition 31/32	Roscosmos	Flight Engineer
7/17/12	11/18/12	Sunita Williams	Expedition 32/33	NASA	Flight Engineer (32); Commander (33)
7/17/12	11/18/12	Akihiko Hoshide	Expedition 32/33	JAXA	Flight Engineer
7/17/12	11/18/12	Yuri Malenchenko	Expedition 32/33	Roscosmos	Flight Engineer
10/25/12	3/15/13	Kevin Ford	Expedition 33/34	NASA	Flight Engineer (33); Commander (34)
10/25/12	3/15/13	Evgeny Tarelkin	Expedition 33/34	Roscosmos	Flight Engineer
10/25/12	3/15/13	Oleg Novitskiy	Expedition 33/34	Roscosmos	Flight Engineer
12/21/12	5/13/13	Chris Hadfield	Expedition 34/35	CSA	Flight Engineer (34); Commander (35)
12/21/12	5/13/13	Roman Romanenko	Expedition 34/35	Roscosmos	Flight Engineer
12/21/12	5/13/13	Tom Marshburn	Expedition 34/35	NASA	Flight Engineer
3/28/13	9/10/13	Pavel Vinogradov	Expedition 35/36	Roscosmos	Flight Engineer (35); Commander (36)
3/28/13	9/10/13	Alexander Misurkin	Expedition 35/36	Roscosmos	Flight Engineer
3/28/13	9/10/13	Chris Cassidy	Expedition 35/36	NASA	Flight Engineer
5/28/13	11/10/13	Fyodor Yurchikhin	Expedition 36/37	Roscosmos	Flight Engineer (36); Commander (37)
5/28/13	11/10/13	Luca Parmitano	Expedition 36/37	ESA	Flight Engineer
5/28/13	11/10/13	Karen Nyberg	Expedition 36/37	NASA	Flight Engineer
9/25/13	3/10/14	Oleg Kotov	Expedition 37/38	Roscosmos	Flight Engineer (37); Commander (38)
9/25/13	3/10/14	Michael Hopkins	Expedition 37/38	NASA	Flight Engineer
9/25/13	3/10/14	Sergey Ryazanskiy	Expedition 37/38	Roscosmos	Flight Engineer
11/7/13	5/13/14	Koichi Wakata	Expedition 38/39	JAXA	Flight Engineer (38); Commander (39)
11/7/13	5/13/14	Mikhail Tyurin	Expedition 38/39	Roscosmos	Flight Engineer
11/7/13	5/13/14	Rick Mastracchio	Expedition 38/39	NASA	Flight Engineer
3/25/14	9/10/14	Aleksandr Skvortsov	Expedition 39/40	Roscosmos	Flight Engineer
3/25/14	9/10/14	Oleg Artemyev	Expedition 39/40	Roscosmos	Flight Engineer
3/25/14	9/10/14	Steven R. Swanson	Expedition 39/40	NASA	Flight Engineer (39); Commander (40)
5/28/14	11/9/14	Gregory R. Wiseman	Expedition 40/41	NASA	Flight Engineer
5/28/14	11/9/14	Maksim Surayev	Expedition 40/41	Roscosmos	Flight Engineer (40); Commander (41)
5/28/14	11/9/14	Alexander Gerst	Expedition 40/41	ESA	Flight Engineer
9/25/14	3/11/15	Aleksandr Samokutyayev	Expedition 41/42	Roscosmos	Flight Engineer
9/25/14	3/11/15	Yelena Serova	Expedition 41/42	Roscosmos	Flight Engineer
9/25/14	3/11/15	Barry E. Wilmore	Expedition 41/42	NASA	Flight Engineer (41); Commander (42)
11/23/14	6/11/15	Anton Shkaplerov	Expedition 42/43	Roscosmos	Flight Engineer
11/23/14	6/11/15	Samantha Cristoforetti	Expedition 42/43	ESA	Flight Engineer
11/23/14	6/11/15	Terry W. Virts	Expedition 42/43	NASA	Flight Engineer (42); Commander (43)
3/27/15	9/11/15	Genady Padalka	Expedition 43/44	Roscosmos	Flight Engineer (43); Commander (44)
3/27/15	3/1/16	Scott Kelly	Expedition 43/44/45/46	NASA	Flight Engineer (43/44); Commander (45/46)
3/27/15	3/1/16	Mikhail Kornienko	Expedition 43/44/45/46	Roscosmos	Flight Engineer
7/22/15	12/11/15	Kjell Lindgren	Expedition 44/45	NASA	Flight Engineer
7/22/15	12/11/15	Oleg Kononenko	Expedition 44/45	Roscosmos	Flight Engineer
7/22/15	12/11/15	Kimiya Yui	Expedition 44/45	JAXA	Flight Engineer
9/2/15	Ongoing	Sergey Volkov	Expedition 45/46	Roscosmos	Flight Engineer
12/15/15	Ongoing	Yuri Malenchenko	Expedition 46	Roscosmos	Flight Engineer
12/15/15	Ongoing	Timothy Kopra	Expedition 46	NASA	Flight Engineer
12/15/15	Ongoing	Timothy Peake	Expedition 46	ESA	Flight Engineer

Note: CSA = Canadian Space Agency. ESA = European Space Agency. JAXA = Japanese Aerospace Exploration Agency. NASA = National Aeronautics and Space Agency.

SOURCE: Adapted from *International Space Station: Crews and Expeditions*, National Aeronautics and Space Administration, December 2015, http://www.nasa.gov/mission_pages/station/expeditions/index.html (accessed December 11, 2015)

Visiting Astronauts

Overall, 37 space shuttle flights traveled to the station over the life of the space shuttle program. These flights sometimes carried personnel who visited the *ISS* temporarily to help with assembly. They were not considered *ISS* resident crew members. For example, when the space shuttle *Endeavour* visited the *ISS* in April 2001, the crew included the ESA astronaut Umberto Guidoni (1954–). He was the first European to board the space station. Later that year ESA astronaut Claudie Haigneré (1957–) became the first European woman to visit the *ISS*.

TABLE 5.3

Space "tourists" that have visited the *International Space Station* as of January 2014

Name	Country	Date of visit	Transport spacecraft/mission
Dennis Tito	USA	April 2001	Soyuz TM-32
Mark Shuttleworth	South Africa	April 2002	Soyuz TM-34
Gregory Olsen	USA	October 2005	Soyuz TMA-7
Anousheh Ansari	Iran/USA	September 2006	Soyuz TMA-9
Charles Simonyi	Hungary/USA	April 2007 and March 2009	Soyuz TMA-10 and Soyuz TMA-14
Sheikh Muszaphar Shukor	Malaysia	October 2007	Soyuz TMA-11
Richard Garriott	USA	October 2008	Soyuz TMA-13
Guy Laliberté	Canada	September 2009	Soyuz TMA-16

SOURCE: Created by Kim Masters Evans for Gale, © 2016

Astronauts temporarily visiting the space station were on specific missions. For example, in "ESA Astronaut Launched to ISS to Deliver and Retrieve Hardware" (August 29, 2009, http://www.esa.int/Our_Activities/Human_Spaceflight/Alisse_Mission/ESA_astronaut_launched_to_ISS_to_deliver_and_retrieve_hardware), the ESA reports that in 2009 Christer Fuglesang (1957–) conducted mission Alissé to help prepare the *ISS* for installation of the Italian-made Tranquility module among other tasks. Flights have also been made to the station by "taxi" cosmonauts who brought fresh Soyuz spacecraft to serve as *ISS* lifeboats.

Tourists

As of February 2016, eight space tourists had visited the *ISS*. (See Table 5.3.) All traveled aboard Soyuz spacecraft.

An April 2001 taxi mission to the station to replace the Soyuz lifeboat included the U.S. millionaire Dennis Tito (1940–). A year earlier, Tito had paid $20 million for a visit to the *Mir* space station. When the Russians decided to deorbit *Mir*, they rescheduled Tito for a trip to the *ISS*. They hoped he would be the first of many space tourists to pay to fly on *ISS* taxis. The Russian Space and Aviation Agency (Rosaviakosmos) desperately needed the money.

NASA and the European partners in the ISS were not happy with the decision. When Tito and the two taxi cosmonauts showed up at the Johnson Space Center for training, NASA would not let Tito enter the facility. The cosmonauts responded by refusing to undergo training. The standoff resulted in a flurry of negotiations between NASA and Rosaviakosmos. The Americans eventually agreed to allow Tito to train in the United States, but they continued to argue that he posed a safety risk to the station and severely restricted his access to the American segments of the *ISS*. NASA repeatedly asked Rosaviakosmos to postpone Tito's flight, but the Russians refused. During his *ISS* visit Tito spent his time in the Russian segment taking pictures of Earth and listening to opera. The crew stayed at the station for four days before returning to Earth aboard the old Soyuz lifeboat.

In April 2002 another taxi crew visited the *ISS*. It included another space tourist: the South African Internet entrepreneur Mark Shuttleworth (1973–). He paid approximately $20 million to visit the space station.

After two years of negotiation, the ISS partners had worked out an agreement that specified who could visit the station. In November 2001 they signed the Principles Regarding Processes and Criteria for Selection, Assignment, Training, and Certification of *ISS* (Expedition and Visiting) Crewmembers (Spaceref.com, January 31, 2002). The agreement listed strict requirements regarding the personal character and communication skills of any visitor. It disqualified anyone found to have a drinking or drug problem, those with poor employment or military records, convicted criminals, people who had engaged in "notoriously disgraceful conduct," and anyone known to be affiliated with organizations that wished to "adversely affect the confidence of the public" in the space program. Visitors also had to speak English.

The Soyuz that carried the Expedition 14 crew to the station in September 2006 carried the first woman tourist to the *ISS*. The Iranian American entrepreneur Anousheh Ansari (1966–) reportedly paid $20 million to Russian authorities to take the flight. Ansari is best known for her sponsorship (along with her brother-in-law) of the Ansari X Prize that awarded $10 million in 2004 to the developers of *SpaceShipOne*, the first commercially funded private spacecraft to carry a human passenger into space.

In 2007 the *ISS* was visited by two space tourists. One was Sheikh Muszaphar Shukor (1972–), a surgeon from Malaysia and the first practicing Muslim to visit the station. The second was Charles Simonyi (1948–), an American born in Hungary who made a second trip to the *ISS* two years later. In 2008 the station was visited by the American space tourist Richard Garriott (1961–), a video game developer, whose father Owen K. Garriott (1930–) was a NASA astronaut aboard *Skylab* in 1973. Thus, the younger Garriott became the first second-generation space traveler. In 2009 Canadian Guy Laliberté (1959–) visited the *ISS* as a tourist. His visit was arranged by private space tourism company Space Adventures, which is based in the United States.

In "Singer Sarah Brightman Calls Off Flight to Space Station" (Reuters.com, May 13, 2015), Irene Klotz notes that the British singer Sarah Brightman (1960–) began training in Russia in 2015 to visit the space station on a $52 million trip arranged by Space Adventures. However, she changed her plans for "family reasons." According to Klotz, a Japanese entrepreneur named Satoshi Takamatsu (1963–) was expected to be the next tourist visitor to the *ISS*. As of February 2016, Takamatsu had not visited the space station.

EXTRAVEHICULAR ACTIVITIES (SPACE WALKS)

Constructing and maintaining the *ISS* has required a great deal of hands-on work by astronauts and cosmonauts. According to NASA, in "15 Years of Continuous Human Presence aboard the International Space Station" (October 30, 2015, http://www.nasa.gov/mission_pages/station/research/news/infographic_15_years), as of October 2015 more than 180 EVAs had been conducted at the station. These space walks have provided the project with some dramatic moments because EVAs are inherently dangerous.

In July 2013 the Italian astronaut Luca Parmitano (1976–) had to cut a space walk short after water droplets gathered inside his helmet, blocking his vision and making it difficult for him to breathe. He was wearing an American spacesuit. (See Figure 5.7.) He reentered the station safely, and NASA launched an investigation into the problem. In "Snorkels in Space: American Astronauts Prep for Urgent Spacewalk Repairs" (NBCNews.com, December 2013), Jeff Black notes that the agency incorporated new safety features to prevent such an event from happening again. These include an absorbent pad at the back of the helmets to soak up any water leaks. In addition, astronauts will have access to a tubular "snorkel" inside their helmets that can provide an emergency supply of oxygen.

Despite these modifications, in January 2016 an EVA was cut short when a water droplet gathered inside the helmet of NASA astronaut Tim Kopra (1963–). In "Spacewalk Ends Early after Water Detected in Helmet" (January 15, 2016, https://blogs.nasa.gov/spacestation/2016/01/15/spacewalk-ends-early-after-water-detected-in-helmet/), the agency states, "The crew was never in any danger and returned to the airlock in an orderly fashion." The astronauts planned to examine the helmet's absorption pad in an effort to determine why it did not operate as expected.

EVAs are often performed at the station to repair or replace external equipment. Ammonia leaks have been a nagging problem. NASA explains in "Cooling System Keeps Space Station Safe, Productive" (December 11, 2013, http://www.nasa.gov/content/cooling-system-keeps-space-station-safe-productive/#.UwYKfZ5dWrY) that externally the station includes two separate cooling systems that provide redundancy (i.e., each one serves as a backup system to the other). They rely on ammonia as a heat-transfer fluid. Ammonia is toxic to humans; thus, the station's internal cooling systems use water instead. Since 2010, ammonia leaks and equipment malfunctions in the external cooling systems have required several space walks for repairs to be completed.

Space walks cannot be initiated on a moment's notice. In fact, EVAs require at least a few hours of preparation time because spacesuits are operated at a low atmospheric pressure. To avoid decompression sickness, which can be fatal, humans cannot undergo sudden pressure changes. In *In-Suit Light Exercise (ISLE) Prebreathe Protocol Peer Review Assessment* (February 2011, http://ntrs.nasa.gov/archive/nasa/casi.ntrs.nasa.gov/20110007150_2011007443.pdf), Timothy K. Brady and James D. Polk note that astronauts or cosmonauts wearing American spacesuits must spend at least two hours acclimating to the reduced atmospheric pressure in their spacesuits before leaving the station.

ISS SCIENCE

The *ISS* was intended to be a world-class laboratory for conducting experiments under microgravity conditions. However, construction delays and limited crew sizes have thwarted the ambitious science programs that were originally envisioned for the station. Nevertheless, a number of scientific achievements have been obtained. The ISS Program Science Forum (IPSF) includes representatives from the CSA, ESA, the Italian Space Agency, JAXA, NASA, and Roscosmos. As of February 2016, the latest annual report from the IPSF was published in March 2015 and titled *International Space Station Utilization Statistics, Expeditions 0–40, December 1998–September 2014* (http://www.nasa.gov/sites/default/files/atoms/files/iss_utilization_statistics_0-40_mcbapproved_july2015.pdf).

According to the IPSF, 1,765 investigations were conducted aboard the *ISS* from December 1998 to September 2014. The organization defines an investigation as "a set of activities and measurements (observations) designed to test a scientific hypothesis, related set of hypotheses, or set of technology validation objectives." The ISS investigations were conducted by 2,484 investigators representing 83 countries. A breakdown by lead agency is as follows:

- CSA—27 investigations
- ESA—247 investigations
- JAXA—485 investigations

- NASA and the Italian Space Agency—607 investigations
- Roscosmos—399 investigations

Laboratories and Internal Equipment

As of February 2016, the station included three science laboratories, each under the direction of a different nation:

- Destiny—U.S. laboratory deployed in 2001
- Columbus—European laboratory deployed in 2008
- Kibo—Japanese laboratory deployed in 2008

The locations of the laboratories within the *ISS* are shown in Figure 5.1. These three facilities are connected by the Harmony node (node 2). In addition, Russian-sponsored scientific experiments are conducted on equipment within the Russian modules of the station, primarily the Poisk (MRM 2). As noted earlier, the Russian MLM (which is called Nauka) is expected to be deployed in 2017. It is shown in Figure 5.1 attached to the Zvezda service module.

NASA explains in "International Space Station: All Laboratories Are Go . . . for Research" (September 2008, http://www.nasa.gov/pdf/276628main_ISS_Brochure.pdf) that the station includes multiple racks. These structures facilitate experiment performance. Racks hold equipment, such as microscopes, sensors, cameras, and other devices, and supply power, cooling, heating, vacuum, gas and fuel, and other utility needs. By their very nature, racks are crucial to conducting science on the station.

Glove boxes are also important to scientific investigations in space. They are basically enclosed work spaces that prevent materials (particularly liquids) from floating away. One of the difficulties of performing typical chemistry experiments in space is the microgravity condition. Liquids will not stay inside beakers or test tubes because they form into droplets of various sizes and float away. This could be extremely dangerous for the crew and the station's electronic components. To overcome this obstacle, NASA and ESA engineers developed glove boxes for use aboard the station. The NASA version is called the microgravity science glove box (MSG). The MSG includes a pair of built-in gloves that crew members can use to handle tools and equipment within the box.

In September 2014 a 3-D printer was installed on the *ISS*. A 3-D printer performs what is called "additive manufacturing." Basically, it "extrudes streams of heated plastic, metal or other material, building layer on top of layer to create three-dimensional objects." This ability was added to allow astronauts on the station to manufacture certain parts, tools, and other equipment that they need, rather than waiting for them to be delivered from Earth. In "Space Station Astronauts to Test 3-D Printing in Microgravity" (ScientificAmerican.com, August 26, 2013), Larry Greenemeier notes that the demonstration printer is about the size of a microwave oven and manufactures items using thermoplastic resins.

The *ISS* also includes various other devices important to scientific research, including freezers, incubators, growth chambers, and centrifuges that are located inside the station.

External Platform Investigations

One of the goals of space science is to determine the effect of space exposure on various materials. Several of the station's pressurized modules have external platforms on which experiments are performed. In addition, there are four separate external platforms called ExPRESS Logistics Centers (ELCs). According to NASA, in "Goddard Team Develops New Carriers for ISS" (November 13, 2009, http://www.nasa.gov/centers/goddard/news/topstory/2009/iss_carriers.html), "ExPRESS" stands for "Expedite the Processing of Experiments to the Space Station." Figure 5.1 shows the locations of the ELCs.

Materials on the International Space Station Experiment (MISSE) is a series of long-term investigations in which scientists assess the durability of various materials to the harsh space environment in LEO. Hazards include solar radiation, erosion by atomic oxygen, heating, contamination from spacecraft, and impacts by micrometeors and orbital debris. MISSE containers holding material samples are attached to the outside of the station. In "Time in Space Exposes Materials to the Test of Time" (November 24, 2014, http://phys.org/news/2014-11-space-exposes-materials.html), Mike Giannone indicates that more than 4,000 samples were tested through 2014 when the two-year MISSE-8 investigation ended. In 2015 the private company AlphaSpace (http://www.alphaspace.com/nasa-coop.html) signed an agreement with NASA to install a permanent testing platform called MISSE-FF on the *ISS* in 2017.

Focus on Medical Complications

Chapter 2 describes NASA's grand plan for space exploration, the Flexible Path option. Potential destinations include Earth's moon, Mars, and asteroids and other objects in the solar system. (See Table 5.4.) It is anticipated that humans will increasingly travel greater distances from Earth and spend more time in space.

Although science endeavors aboard the space station cover multiple disciplines, one of the most important to NASA's future plans for human space travel is the investigation of medical complications associated with spending long periods in space. In *NASA's Efforts to Manage Health and Human Performance Risks for Space Exploration* (October 29, 2015, https://oig.nasa.gov/audits/reports/FY16/IG-16-003.pdf), NASA's Office of Inspector

TABLE 5.4

Mission types evaluated by NASA for human health and performance risks

Design reference mission	Mission duration	Distance from earth (in miles)	Earth return	Gravity environment	Radiation environment
Low earth orbit (ISS)	6 months	237	less than or equal to 1 day	Microgravity	Low earth orbit
Low earth orbit (ISS)	1 year	237	less than or equal to 1 day	Microgravity	Low earth orbit
Deep Space Sortie	1 month	greater than 237,000	less than 5 days	Microgravity	Deep space
Lunar visit or habitation	1 year	237,000	5 days	1/6 G	Lunar
Deep space journey or near-earth asteroid	1 year	237,000–33,900,000	weeks to months	Microgravity	Deep space
Planetary	3 years	33,900,000*	months	Fractional	Deep space

*Planetary distance is based on the distance of Mars from Earth.
Note: ISS = International Space Station.

SOURCE: "Table 1. Design Reference Missions for Flexible Path," in *NASA's Efforts to Manage Health and Human Performance Risks for Space Exploration*, National Aeronautics and Space Administration, Office of Inspector General, October 29, 2015, https://oig.nasa.gov/audits/reports/FY16/IG-16-003.pdf (accessed December 10, 2015)

General (OIG) notes that the agency has identified dozens of specific human health and performance risks associated with space travel. (See Table 5.5.) Much of the information that scientists have about these risks has come from studying *ISS* crew members. In particular, research has focused on how human bones and eyesight are affected by prolonged exposure to microgravity.

BONE LOSS. Scientists have known for some time that human bones in the legs and feet undergo deterioration during prolonged stays in space. This was first discovered in Soviet and Russian cosmonauts who spent many months aboard space stations. Scientists believe the effect is due to the lack of mechanical loading in microgravity. Mechanical loading refers to the weight of the upper body pressing down on the lower body as a person's body is pulled toward the ground by gravity on Earth. The type of bone loss and muscle deterioration experienced by space travelers is similar to that resulting from prolonged bed rest. Using legs and feet keeps them healthy. In a spaceship people do not experience the full force of gravity through the downward load of the upper body. Also, they rarely use muscles in their legs and feet to move around. They rely much more on muscles in their arms and upper body to maneuver through hatches and accomplish tasks.

During the Russian *Mir* program, cosmonauts reported that the skin on the soles of their feet became very soft. They also lost muscle tone in their legs and feet due to lack of use. These factors made walking extremely difficult when they returned to Earth. Scientists incorporated exercise regimens on the *ISS* to help prevent these problems. For example, stationary bicycles help crew members maintain foot muscle strength. However, the exercises have had little effect on bone loss.

Historical data show that humans experience a rate of bone loss in space of approximately 1% to 2% per month. This means a bone loss of 12% to 24% per year. Scientists know that the bone loss problem must be resolved before humans can make interplanetary journeys or inhabit space or moon stations for long periods.

In "Subregional Assessment of Bone Loss in the Axial Skeleton in Long-Term Space Flight" (January 9, 2014, http://www.nasa.gov/mission_pages/station/research/experiments/118.html), NASA reports on the results of a four-year study of the long-term effects of microgravity on 16 *ISS* crew members. NASA notes that astronauts in Expeditions 2 through 8 lost an average of 11% of their total hip bone mass during a typical six-month mission. Bone scans performed one year after the astronauts had returned to Earth indicate that much of the bone had regrown; however, bone structure, bone density, and hip strength were not fully recovered.

As bones lose mass the calcium and other minerals can build up in the kidneys and cause kidney stones. These small solid objects cause great pain as they travel through the urinary tract. In "Strong Bones and Fewer Renal Stones for Astronauts" (February 23, 2012, http://www.nasa.gov/mission_pages/station/research/news/Strong_Bones.html), NASA reports that *ISS* crew members use specially designed exercise equipment and take nutritional supplements to help minimize bone loss and kidney stone formation.

EYESIGHT PROBLEMS. Another medical problem related to prolonged exposure to microgravity is papilledema (swelling of the optic nerve due to abnormally high intracranial pressure within the skull). This problem only recently became known. In "NASA Astronauts' Eyesight Damaged by Long Space Flights" (ABCNews.com, March 13, 2012), Ned Potter and Gina Sunseri note that in 2011 the brain scans of some astronauts who had spent more than a month on the *ISS* showed "intracranial hypertension—high fluid pressure in the skull." (See Figure 5.10.) This condition has caused vision changes that appear in some cases to be permanent because they lingered even after the astronauts returned to Earth. Even

TABLE 5.5

Human health and performance risks associated with space flight missions

Risk	Risk summary
Acute and chronic carbon dioxide (CO_2)	Given carbon dioxide levels in spacecrafts are 6–20 times higher than on Earth, exposure may impact crew health and performance when complex decisions are necessary.
Adverse behaviorial conditions (Bmed)[a]	Isolated, confined environments can lead to adverse behavioral conditions and mental disorders
Altered immune response (immune)[a]	Changes in immune function can result in reactivation of latent herpes-viruses, rashes and hypersensitivity reactions
Bone fracture (fracture)[a]	Adverse changes in bone strength may result in fractures and early onset of osteoporosis
Cardiac rhythm problems (arrhythmia)[a]	Micro-gravity, radiation, stress, and fluid shifts can result in cardiac rhythm disturbances
Celestial dust exposure (dust)[a]	Celestial dust and volatiles (lunar, Mars, and asteroids) can lead to damage to the respiratory, cardiopulmonary, ocular and dermal systems
Decompression sickness (DCS)[a]	Given that tissue inert gas partial pressure is often greater than ambient pressure during phases of a mission (primarily EVA), there is a possibility of decompression sickness (DCS)
Electrical shock (shock)[a]	Electrical systems on spacecraft have the potential to release an electrical shock from both direct contact and plasma
Exploration atmospheres (ExAtm)[a]	Changes in atmosphere pressure when engaging in or finishing EVA can lead to mild hypobaric hypoxia
Hearing loss (hearing)	Given conditions during spaceflight (e.g. noise and physiological changes), crew auditory systems could experience temporary or permanent reductions in hearing sensitivity
Human-system interaction design (HSID)[a]	Training, procedures, human-robotic and human-computer interactions and vehicle, design habitat and workplace design integration affect efficiencies and injuries
Host-microorganism interactions (microhost)[a]	The combination of increased microbe virulence and compromised immunity can increase illnesses
In-flight medical capabilities (ExMC)[a]	Limited medical capabilities can result in in-flight medical events leading to unacceptable health and mission outcomes
Inadequate nutrition and food storage (nutrition, food)[a]	Inadequate nutrition and food storage can compromise crew health (e.g. bone mass and strength, endurance, cardio-, gastro-, and endocrine function)
Inadequate team performance (team)[a]	Performance and behavioral health decrements due to inadequate cooperation, coordination, communication, and psychosocial adaption with team during spaceflight
Injury, compromised performance due to EVA operations (EVA)[a]	Inadequate EVA Suit Systems can result in injuries which compromise EVA performance and crew health
Intervertebral disc damage (IVD)[b]	The lengthening of the spine in micro-gravity may lead to intervertebral disc damage when re-exposed to gravity
Occupant protection (OP)[a]	The dynamic loads from launching and landing can result in crew injury
Orthostatic intolerance (OI)[a]	Re-exposure to Earth and gravity can result in loss of blood pressure and fainting
Pharmacokinetics (PK/PD)[b]	The body may handle administered medications, the concentration of circulating medication, differently in the spaceflight environment
Reduced aerobic capacity (aerobic)[a]	Exposure to a micro-gravity environment causes cardiovascular fitness to decline
Reduced muscle mass, strength and endurance (muscle)[a]	Muscle mass, strength and endurance can decline during spaceflight
Renal stone formation (renal)[a]	Renal (kidney) stone formation may lead to urinary calculi or urolithiasis, pain, nausea, vomiting, hematuria, infection and/or hydrone-phrosis
Sensorimotor alteration (sensorimotor)[a]	Gravitational transitions can affect eye-hand, balance and locomotor coordination which would impair control of spacecraft
Sleep loss/work overload (sleep)[a]	Performance errors and adverse health outcomes due to fatigue resulting from sleep loss, circadian desynchronization, and work overload
Space adaption back pain (SABP)	Given the physiological changes in spaceflight, crewmembers may experience space adaptation back pain ranging from aches and stiffness to tingling and numbness or radicular pain.
Space radiation exposure (radiation)[a]	Radiation exposure can increase cancer morbidity/mortality and result in degenerative tissue and central nervous system damage and acute radiation sickness
Sunlight exposure (sunlight)	Exposure to solar light during spaceflight missions may lead to crewmembers developing solar retinopathy, photokeratis and/or sunburn
Toxic exposure (toxic)	Various sources of toxic exposures cannot be eliminated during missions, as a result there is a possibility that exposure will cause illness to the crew
Urinary retention	Physiological changes, altered gravity, and limited access to voiding may result in some crew developing urinary retention
Unpredicted medication effects (stablity)[a]	Given the hostile space environment, the ability to treat medical conditions with pharmaceuticals may be significantly reduced
Vision impairments and intracranial pressure (VIIP)[a]	Changes in vision and intracranial pressure could result from fluid shifts due to changes in gravity

[a]Risk considered to be a Human Research Program (HRP) risk.
[b]HMTA and HRP consider issue to be a concern and have not formally accepted it as a risk.
Notes: EVA = extravehicular activity. HMTA = Health and Medical Technical Authority.

SOURCE: Adapted from "Appendix B. Acceptance of Human Health and Performance Risks by Selected DRMs," in *NASA's Efforts to Manage Health and Human Performance Risks for Space Exploration*, National Aeronautics and Space Administration, Office of Inspector General, October 29, 2015, https://oig.nasa.gov/audits/reports/FY16/IG-16-003.pdf (accessed December 10, 2015)

more troubling is the fear that astronauts experiencing this increased pressure will be more likely to suffer brain damage in the event of head injuries. Potter and Sunseri indicate that not all of the 24 astronauts tested showed the increased pressure—60% did, but 40% did not. In addition, some of the astronauts had full or partial recovery, whereas others did not. Scientists are unsure why these differences occurred. As a result, papilledema research has become a top priority for NASA as it continues to plan long-duration space flights for the future.

In *NASA's Efforts to Manage Health and Human Performance Risks for Space Exploration* the agency's OIG indicates that NASA has learned that vision changes have occurred in astronauts who have spent as little as two weeks on the space station. In addition, the OIG states, "Based on data from 300 post-flight questionnaires, approximately 29 percent of short duration and 60 percent of long duration mission astronauts reported a degradation in vision." (Note that the OIG does not specifically define the terms *short duration* and *long duration* in its report.)

THE ONE-YEAR MISSION. As shown in Table 5.2, on March 27, 2015, NASA astronaut Scott Kelly and Russian cosmonaut Mikhail Kornienko launched to the *ISS*.

FIGURE 5.10

Reported changes to vision and eye anatomy due to space flight

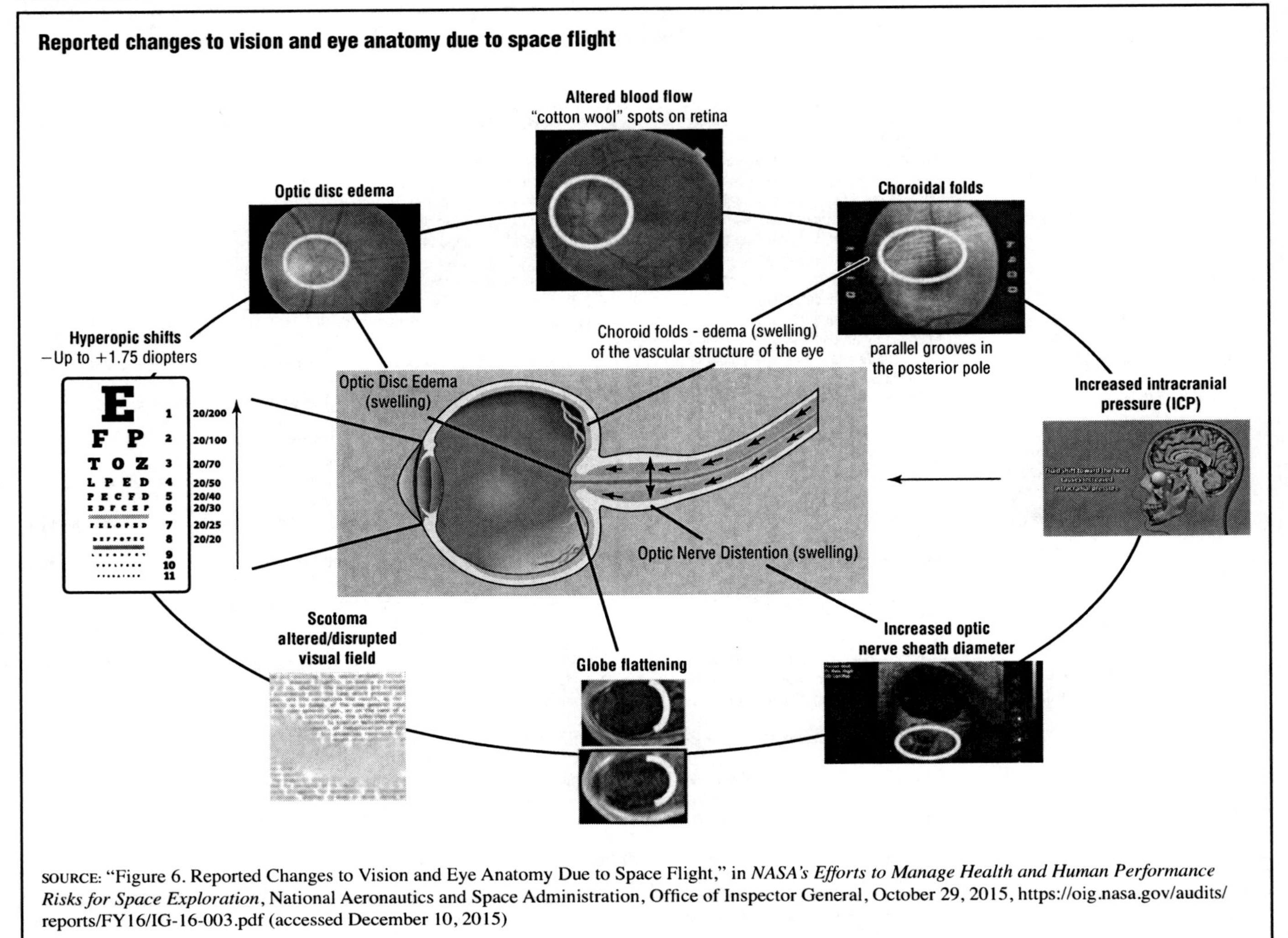

SOURCE: "Figure 6. Reported Changes to Vision and Eye Anatomy Due to Space Flight," in *NASA's Efforts to Manage Health and Human Performance Risks for Space Exploration*, National Aeronautics and Space Administration, Office of Inspector General, October 29, 2015, https://oig.nasa.gov/audits/reports/FY16/IG-16-003.pdf (accessed December 10, 2015)

Each was scheduled for an approximate one-year mission in LEO. Scientists collected detailed medical and psychological data about the two men to investigate the effects of prolonged time in space on human health and well-being. The project had an additional dimension because Scott Kelly's identical twin brother, Mark Kelly, was taking part in the study on the ground. (Mark Kelly is a former NASA astronaut.) Identical twins have the same genetic makeup. Thus, researchers compared the medical data of the Kelly brothers to assess the effects of environment alone on their health. In October 2015 NASA announced that Scott Kelly, who participated in multiple space flights, had achieved the record as the American with the longest total time in space.

Both Scott Kelly and Kornienko returned to Earth on March 1, 2016. Kelly spent 520 total days in space, over two spaceflights, according to NASA's Mark Garcia in "Veteran Station Crew Returns to Earth after Historic Mission" (March 1, 2016, https://blogs.nasa.gov/spacestation/2016/03/01/veteran-station-crew-returns-to-earth-after-historic-mission/).

FUTURE OF THE *ISS*

As explained in Chapter 2, in 2014 President Barack Obama (1961–) pledged continued U.S. support for the *ISS* through at least 2024. The U.S. Commercial Space Launch Competitiveness Act (https://www.faa.gov/about/office_org/headquarters_offices/ast/media/US-Commercial-Space-Launch-Competitiveness-Act-2015.pdf) was passed in November 2015. In it Congress revised the National Aeronautics and Space Administration Authorization Act of 2010 (http://www.nasa.gov/pdf/649377main_PL_111-267.pdf) to read, "It shall be the policy of the United States, in consultation with its international partners in the ISS program, to support full and complete utilization of the ISS through at least 2024."

It remains to be seen whether future presidents and U.S. Congresses will support the space station through 2024. There are a number of challenges associated with extending the space station mission. They include cost, hardware, safety, and political considerations.

Cost Considerations

The space station program has been and will continue to be a very expensive undertaking for the United States. NASA's OIG indicates in *Extending the Operational Life of the International Space Station until 2024* (September 18, 2014, https://oig.nasa.gov/audits/reports/FY14/IG-14-031.pdf) that as of September 2014, the United States had spent nearly $75 billion on *ISS* construction and operation.

NASA spending on the space station was nearly $3 billion in FY 2014. (See Table 2.7 in Chapter 2.) The agency's budget projections through FY 2020 predict that $3 billion to $4 billion per year will be devoted to the ISS program. Nevertheless, NASA's OIG calls these estimates "overly optimistic." Its opinion is that NASA is underestimating future crew transportation costs. As explained in Chapter 4, in 2014 the agency contracted with SpaceX and the Boeing Company to develop spacecraft to transport American astronauts to and from the space station.

In addition, NASA's OIG expresses concern over disappointing commercial spending on ISS science investigations. In the NASA Authorization Act of 2005 (https://www.nasa.gov/pdf/181091main_1-Public%20Law%20109-155.pdf), the U.S. Congress designated the U.S. segment of the *ISS* as a national laboratory. The United States has dozens of such facilities on the ground. Although federally funded, many are operated under contract by private partners. Congress ordered NASA to "increase the utilization of the ISS by other Federal entities and the private sector through partnerships, costsharing agreements, and other arrangements that would supplement NASA funding of the ISS."

In 2011 NASA selected a nonprofit organization, the Center for the Advancement of Science in Space (CASIS; http://www.iss-casis.org/About/AboutCASIS.aspx), to manage the ISS National Laboratory. However, NASA's OIG states, "issues relating to funding and patent licenses and data rights continue to pose challenges to the organization's efforts to attract private entities to ISS research."

Hardware Considerations

NASA's OIG indicates that the space agency "has identified no major obstacles to extending operations to 2024." However, the OIG does raise some concerns. One is the aging infrastructure of the space station. As of 2015 many major *ISS* components had been in orbit for nearly 15 years. For example, the first solar array wing (or solar panel) was installed in 2000. There are eight of the wings attached to the truss segments, as shown in Figure 5.1. They provide solar power to the station. According to NASA's OIG the expected operational life of each solar array wing is 15 years. Thus, the first wing reached the end of its expected life span in 2015. The remaining wings will do likewise between 2021 and 2024. Solar panel output is degrading faster than expected due to radiation exposure and strikes by micrometeorites (bits of dust and tiny rock particles hurtling through space).

The Russian portion of the *ISS* receives its power from the U.S. section. NASA's OIG indicates that the power demand on the solar array wings will drop after the Russians install a planned power module. In "A Rare Look at the Russian Side of the Space Station" (AirSpaceMag.com, September 2015), Anatoly Zak notes that as of September 2015, the power module was "still in [the] early stages of development" and expected to be installed by 2020. Zak points out that another Russian component, the MLM, has suffered severe delays. Originally slated for installation in 2007, its launch has been put off to at least 2017. If similar delays plague the Russian power module, NASA will have to come up with alternative solutions for the *ISS*, including rationing power.

Safety Considerations

The space station faces threats from the thousands of pieces of debris in orbit around Earth. Possible large debris strikes have forced crew members to maneuver the station out of harm's way or take refuge in the Soyuz lifeboats. The number of debris objects in orbit is increasing over time, which raises the risks to the station as its mission is extended. According to NASA's OIG, "NASA officials estimate a 1-in-42 chance in any given 6-month period that orbital debris will puncture the Station and cause a loss of pressurization. Applying this estimate over the life of the ISS equates to 1-in-4 chance of such puncture." In other words, extending the station's mission through 2024 dramatically increases the odds that a life-threatening debris strike could occur.

Political Considerations

The United States and Russia have long been partners in the *ISS* project. Political tensions, however, between the two countries escalated dramatically during the mid-2010s. Chapter 3 describes Russian military engagement in neighboring Ukraine, and the resulting economic sanctions imposed by the U.S. government. Congress has since passed legislation attempting to limit the use of Russian engines on U.S. rockets (such as the Atlas V) that launch American national security satellites. As of February 2016, the Russian space agency had officially committed to continue supporting space station operations through 2024. However, Zak notes that some Russian politicians have advocated for an early end to their nation's participation in the project.

It is widely expected that when the U.S.-Russia collaboration does end, the Russian government will remove its components from the *ISS* and operate its own space station independently.

CHAPTER 6
SPACE SCIENCE ROBOTIC MISSIONS

The Universe is an explosive, energetic and continually changing place.

—National Aeronautics and Space Administration, "Chandra's X-Ray Vision" (February 19, 2004)

Sending humans into space is expensive and risky. It takes great resources to protect and sustain them after they leave the planet. Losing a crewed spacecraft means the loss of life. This is a high price to pay to learn about the universe. This explains why robotic spacecraft are so vital to space science. Since the beginning of the space age, satellites and probes have been sent out to gather data about Earth's surroundings. The earliest ones were rather crude in their technology. People referred to them simply as unmanned spacecraft. Times changed, and technology improved significantly. In the 21st century these mechanized explorers are called robotic spacecraft.

Only a handful of robotic spacecraft are sent to other planets and celestial bodies. Chapters 7 and 8 describe robotic missions to Mars and to the far planets and beyond, respectively. The vast majority of robotic spacecraft orbit Earth or the sun. Spacecraft in Earth orbit serve commercial, military, and scientific purposes. Scientists rely on satellites to collect data about Earth's weather, climate, atmospheric conditions, sea levels, ocean circulation, and gravitational and electromagnetic fields. These satellites are not space explorers but Earth observers that reside in space.

Other satellites in Earth orbit look outward toward the cosmos. These space observatories carry high-powered telescopes that beam images back to Earth. They can detect the radiation of celestial objects that are hidden from human view. They peer into the deep, dark regions of space to explore what is out there. Scientists use the images from such satellites to learn about the origins of stars and planets and unravel some of the mysteries of the universe.

Closer to Earth is the sun. It emits radiation, such as infrared radiation (heat), and produces a flow of energetic particles called the solar wind that constantly blows against Earth. Gusts of solar wind can upset the planet's electromagnetic field. Every so often the sun spits out globs of plasma and intense bursts of radiation. These, too, can have a profound effect on Earth. Investigating the sun-Earth connection is a major goal of modern space science. Robotic spacecraft are put into Earth orbit or sent out into space to collect data about this vital connection.

NASA'S SPACE SCIENCE PROGRAM

The vast majority of robotic spacecraft are operated by the National Aeronautics and Space Administration (NASA), under the agency's Science Mission Directorate (SMD; http://science.nasa.gov/missions). This directorate focuses on areas of research including the origin and evolution of the universe, the nature of life, the solar system, and the relationship between Earth and the sun. It also supports human spaceflight by studying the formation of Mars and Earth's moon; searching for exoplanets (planets outside the solar system; also called extrasolar planets), particularly those that are potentially habitable; and investigating space phenomena that could harm robotic and human explorers.

In *SMD Mission Handbook* (November 2012, http://science.nasa.gov/media/medialibrary/2013/01/22/Mission_Handbook.pdf), the SMD classifies space science missions by science category: earth science, heliophysics (the physics of the sun), planetary science, and astrophysics (the physics of the universe). The SMD also categorizes space science missions administratively, that is, by mission lines and by the NASA program under which they are developed.

Space Science Goals

NASA's science priorities are shaped by the National Research Council (NRC; http://www.nationalacademies.org/nrc/). The NRC and three other bodies—the National

Academy of Sciences, the Institute of Medicine, and the National Academy of Engineering—together make up the National Academies. They rely on independent experts within academia, industry, and other fields to advise the federal government on public policy and decision making related to scientific endeavors. The NRC notes in *The Space Science Decadal Surveys: Lessons Learned and Best Practices* (2015, http://www.nap.edu/catalog/21788/the-space-science-decadal-surveys-lessons-learned-and-best-practices) that it issues reports called decadal surveys in which it recommends scientific objectives for the upcoming decade. NASA uses the surveys devoted to space science to conduct its long-term planning. As of February 2016, the most recent decadal surveys for the SMD's four science categories were:

- *Earth Science and Applications from Space: National Imperatives for the Next Decade and Beyond* (2007, http://www.nap.edu/catalog.php?record_id=11820)
- *New Worlds, New Horizons in Astronomy and Astrophysics* (2010, http://www.nap.edu/catalog.php?record_id=12951)
- *Vision and Voyages for Planetary Science in the Decade 2013–2022* (2011, http://www.nap.edu/catalog.php?record_id=13117)
- *Solar and Space Physics: A Science for a Technological Society* (2013, http://www.nap.edu/catalog.php?record_id=13060)

At the direction of Congress, the NRC conducts performance assessment reviews around the middle of the 10-year period covered by a decadal survey. These are known informally as midterm reviews. They provide lawmakers and the public with updates on how federal agencies are progressing toward meeting their decadal goals. The midterm reviews relevant to NASA's space science missions are discussed later in this chapter.

NASA's Space Science Budget

NASA's overall budget is discussed in Chapter 2. Table 2.7 in Chapter 2 shows that the agency spent $5.1 billion on space science during fiscal year (FY) 2014. This amount was 29% of its total budget of $17.6 billion for the year. A detailed budget allotment for FY 2015 is not available because lawmakers could not agree on specific appropriations. NASA operated that year under an agreement called a Continuing Resolution that basically kept its overall budget at around the same amount as the previous year. For FY 2016 President Barack Obama (1961–) requested a total of $18.5 billion for NASA, of which $5.3 billion would be devoted to space science.

NASA's Space Science Mission Lines

Space science mission lines are characterized primarily by complexity, leadership, and cost. These elements are defined by NASA in *Summary of the Science Plan for NASA's Science Mission Directorate 2007–2016* (January 2007, http://science.hq.nasa.gov/strategy/Science_Plan_07_summary.pdf). NASA has two complexity classes: strategic and competed. Strategic missions are major complex projects that address broad science goals and have long development times and high costs. These missions are managed and led directly by NASA. Competed missions are smaller missions with narrowly focused science goals, short schedules, and relatively low-cost caps. These missions are proposed and led by entities called principal investigators (PIs). A PI can be one or more universities, research organizations, companies, or even NASA centers. Strategic missions may include components that are competed (e.g., specific science instruments onboard a spacecraft).

NASA alerts the scientific community to potential opportunities to propose PI-led missions through announcements of opportunity (AOs).

Since its inception during the 1950s, NASA has operated more than a dozen mission lines. Each mission line differs in terms of cost, goals, or other mission criteria. The agency's two oldest mission lines are the Explorers and Discovery Programs, which are briefly described here.

NASA'S EXPLORERS PROGRAM. The Explorers Program includes robotic missions that conduct relatively low-cost scientific investigations with specific objectives in the fields of astrophysics and heliophysics. NASA (January 19, 2016, http://nssdc.gsfc.nasa.gov/multi/explorer.html) indicates that more than 90 launches have been conducted under the program. The first satellite ever launched into space by the United States was *Explorer 1* in 1958. (See Figure 1.3 in Chapter 1.) It took temperature and radiation measurements in space. As of January 2016, the most recent Explorers spacecraft to be put into orbit was *Interface Region Imaging Spectrograph* (*IRIS*). It was launched in 2013 to collect data about the sun.

According to NASA, in "Astrophysics & Heliophysics Explorers Missions" (February 19, 2016, http://explorers.gsfc.nasa.gov/missions.html), the Explorers Program missions are categorized by complexity and cost as follows:

- Medium-class Explorer—total NASA cost less than $200 million
- Small Explorer—total NASA cost less than $120 million
- University-class Explorer—total NASA cost less than $15 million

There are two other types of Explorer missions. The first is a mission of opportunity (MO). An MO is a mission that is operated by another office within NASA or by the space agency of another country in which an SMD investigation "hitches a ride" with the mission of

FIGURE 6.1

Lagrange points in Earth-Sun space

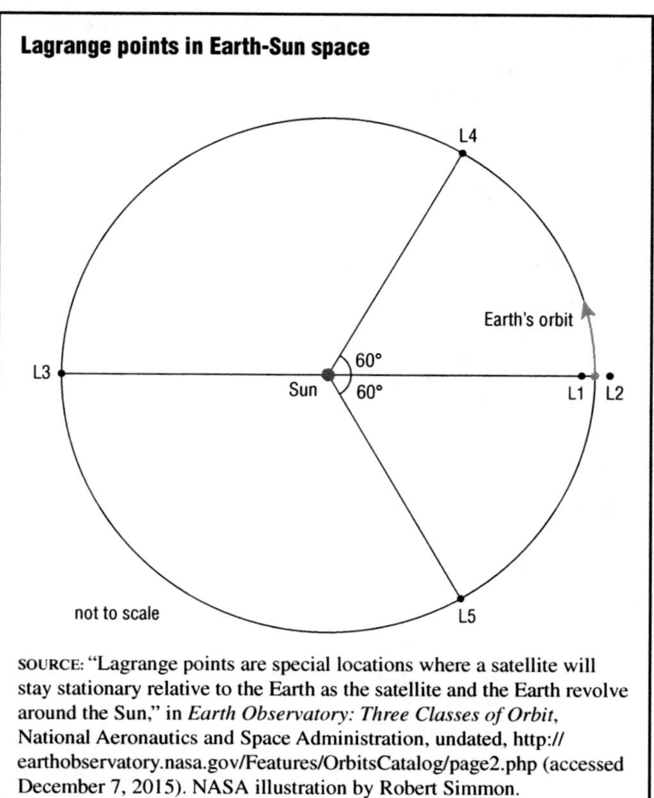

SOURCE: "Lagrange points are special locations where a satellite will stay stationary relative to the Earth as the satellite and the Earth revolve around the Sun," in *Earth Observatory: Three Classes of Orbit*, National Aeronautics and Space Administration, undated, http://earthobservatory.nasa.gov/Features/OrbitsCatalog/page2.php (accessed December 7, 2015). NASA illustration by Robert Simmon.

the other agency or country. The total cost to NASA of an MO mission cannot exceed $55 million. A second type of Explorer mission is called Internationals and features international partners.

Most operational Explorers Program spacecraft are small observatories in low Earth orbit (LEO; 100 to 1,200 miles [160 to 1,900 km] above Earth's surface). Some spacecraft are positioned at Earth-sun Lagrange points L1 or L2. (See Figure 6.1.) These are points at which the gravitational pulls of Earth and the sun are relatively even, meaning that a spacecraft can stay "parked" between the two heavenly bodies. Point L1 is approximately 1 million miles (1.6 million km) from Earth toward the sun; point L2 is approximately 1 million miles from Earth in the other direction (away from the sun).

NASA'S DISCOVERY PROGRAM. The Discovery Program was initiated in 1989. NASA notes in "Proposing a Mission" (2016, http://discovery.nasa.gov/p_mission.cfml) that total NASA costs for an entire mission must be less than $425 million. In addition, the development time from mission start date to launch date must be 36 months or less. Discovery missions are PI-led missions that allow scientists to conduct small space investigations that complement NASA's larger and more expensive interplanetary missions. MO projects are also allowed under the program.

Because of the relatively high costs of Discovery missions (excluding MO projects), relatively few of them have taken place. In "NASA's Discovery Program" (May 10, 2011, http://nssdc.gsfc.nasa.gov/planetary/discovery.html), NASA lists 11 missions for the program; however, only 10 of them were successful. Some of the missions have targeted Mercury, Mars, Earth's moon, asteroids, comets, and exoplanets. Many past and operational missions under the program are discussed later in this chapter. The Mars mission is described in Chapter 7.

NASA's Flagship Missions

Since the 1970s NASA has occasionally conducted science missions that are so ambitious and costly they are dubbed "flagship" missions. In *Summary of the Science Plan for NASA's Science Mission Directorate 2007–2016*, NASA notes that flagship missions are characterized by their great "scope, strategic objectives, and technology challenges." As a result, a flagship mission costs significantly more than a typical mission within a mission line. When the *Science Plan* was written in 2007, a flagship mission was one with a total cost exceeding $1 billion. In previous decades the cost amount distinguishing a flagship mission would have been lower, reflecting the economic conditions of the era.

NASA has launched only a handful of flagship missions since its founding. The first was the Voyager mission of the 1970s. It included two spacecraft—*Voyager 1* and *Voyager 2*—that were designed to pass by Jupiter and Saturn and then make their way out of the solar system. The mission is described in detail in Chapter 8. In "Voyager 1" (February 12, 2016, http://nssdc.gsfc.nasa.gov/nmc/spacecraftDisplay.do?id=1977-084A), NASA's National Space Science Data Center indicates that the original estimated cost to conduct the Voyager mission was around $250 million. This seems low by 21st-century standards but was quite expensive for its time. Successive flagship missions included Galileo (a Jupiter orbiter) during the 1980s and Cassini (a Saturn orbiter) during the 1990s. Both missions are discussed in Chapter 8.

Budgetary restrictions in the 21st century have severely limited NASA's ability to conduct flagship missions. However, one such flagship mission is the Mars Science Laboratory mission (including the *Curiosity* rover), which was launched in 2011. It is described in Chapter 7. As of February 2016, NASA was developing a flagship mission to go to Europa, a moon of Jupiter, in the early 2020s. The mission is described in Chapter 8.

NASA'S EARTH SCIENCE PROGRAM

NASA indicates in *SMD Mission Handbook* that the overall goal of the earth science program is to "advance Earth system science to meet the challenges of climate and environmental change." Long-term objectives for the program were laid out in the NRC's 2007 decadal survey *Earth Science and Applications from Space: National Imperatives for the Next Decade and Beyond.* A midterm assessment of the program's activities is provided by the NRC in *Earth Science and Applications from Space: A Midterm Assessment of NASA's Implementation of the Decadal Survey* (2012, http://www.nap.edu/catalog/13405/earth-science-and-applications-from-space-a-midterm-assessment-of). In general, the NRC is pleased with NASA's progress, noting that the agency "responded favorably and aggressively" to the recommendations laid out in the decadal survey. The NRC complains, however, that the earth science program has been hampered by budget shortfalls.

The earth science program is the SMD's most highly funded program. (See Table 2.7 in Chapter 2.) For fiscal year (FY) 2014 it was appropriated $1.8 billion by Congress. NASA's FY 2016 budget request for the program was $1.9 billion. This represented 37% of the agency's total $5.3 billion request for its science operations.

As of December 2015, 26 operational NASA space missions were devoted to the earth science goal. (See Table 6.1.) These are satellites in Earth orbit that collect data about the planet's atmosphere, winds, rainfall, oceans, clouds, radiation levels, and land surface. Nearly all the missions are operated under the Earth Systematic Missions line or the Earth System Science Pathfinder line. The former are led by NASA, whereas the latter are competed missions led by PIs.

Figure 6.2 depicts the *Soil Moisture Active Passive* (*SMAP*) spacecraft with its Delta II launch vehicle. Delta rockets are supplied by the United Launch Alliance, a private company described in Chapter 4. Launched in January 2015, *SMAP* is equipped with instruments that measure surface soil moisture levels and freeze-thaw states. In January 2016 the weather satellite *JASON-3* was launched by a Falcon 9 rocket supplied by the Space Exploration Technologies Corp. (SpaceX). SpaceX is a private company described in Chapter 4. The JASON-3 mission is led by the National Oceanic and Atmospheric Administration (NOAA; February 12, 2016, http://www.nesdis.noaa.gov/jason-3/).

As of January 2016, NASA (http://science.nasa.gov/earth-science/missions/) planned to launch four more earth science spacecraft during 2016. The *SAGE III-ISS* will be installed aboard the *International Space Station* to study Earth's ozone layer. The *CYGNSS* satellite will measure ocean surface winds during hurricanes and tropical storms. The *GOES-R* and *OCO-3* satellites will carry on the work of their predecessors, which are described in Table 6.1.

NASA'S HELIOPHYSICS PROGRAM

According to NASA, in *SMD Mission Handbook*, the overall goal of the SMD's heliophysics program is to "understand the Sun and its interactions with the Earth and the solar system." In 2013 the NRC published its latest decadal survey for the program, *Solar and Space Physics: A Science for a Technological Society* (http://www.nap.edu/catalog.php?record_id=13060). The NRC recommends objectives for the program to pursue through 2022 related to solar phenomena and "space weather," a term that is explained in the next section.

Solar Phenomena and Space Weather

The sun emits radiation, charged particles, and other substances that can disrupt electrical systems and electronic equipment that are vital to modern societies. The term *space weather* was created by scientists to refer to the overall effects of solar activities on the space around Earth (or geospace). Space weather encompasses solar phenomena and the resulting geospace conditions.

Solar phenomena include the following:

- Solar wind—this is a constant flow of charged particles from the sun. The region of affected outer space is called the heliosphere. The solar wind constantly pushes and shapes Earth's magnetosphere, which is a region around Earth dominated by the planet's magnetic field. (See Figure 6.3.) The magnetosphere helps protect Earth from dangerous electromagnetic radiation moving through space. A high gust of solar wind can energize the magnetosphere, producing beautiful auroras, but wreaking havoc on sensitive electronics.
- Sunspots—these look like dark blemishes on the surface of the sun. They are actually areas of plasma that are slightly cooler than their surroundings. Sunspots can be enormous in size, even bigger than planet Earth, and can appear, change size, and disappear. Each typically lasts from a few hours to a few months. Scientists know that the number of sunspots peaks approximately every 11 years and then drops off dramatically. (See Figure 6.4.) This is called the solar cycle. The period of peak sunspot activity is called the solar maximum, and the period of lowest activity is called the solar minimum. As shown in Figure 6.4, the most recent solar maximum occurred in 2014. The next peak is expected around 2025.
- Solar flares—these are sudden energy releases that burst out from the sun near sunspots. They are most common during the solar maximums. Solar flares can last from minutes to hours. Those that erupt in Earth's direction can shower the planet with energetic particles. Solar flares are associated with auroras and magnetic storms on Earth and can cause radio interference.
- Solar prominences—these are giant clouds of dense plasma suspended in the sun's corona (outermost atmosphere). They are usually calm, but occasionally erupt and snake out from the sun along magnetic field

TABLE 6.1

Earth science program operational missions as of December 2015

Mission name	Mission description	Launch date	NASA mission line
AirMOSS	Using airborne ultra-high frequency synthetic aperture radar to measure root-zone soil moisture in North America	03/01/12	Earth system science pathfinder
Aqua	Obtain a set of precise atmosphere and oceans measurements to understand their role in Earth's climate and its variations	05/04/02	Earth systematic missions
ATTREX	Measures atmospheric water vapor and other gases	12/01/11	Earth system science pathfinder
Aura	Make the most comprehensive measurements ever undertaken of atmospheric trace gases.	07/15/04	Earth systematic missions
CALIPSO	Provide new information about the effects of clouds and aerosols (airborne particles) on changes in the Earth's climate	04/28/06	Earth system science pathfinder
CARVE	Provide experimental insights into Arctic carbon cycling	07/04/05	Earth system science pathfinder
CloudSat	Uses advanced radar to "slice" through clouds to see their vertical structure	04/28/06	Earth system science pathfinder
DISCOVER-AQ	Improve Interpretation of satellite observations to diagnose near-surface conditions relating to air quality	07/01/11	Earth system science pathfinder
Earth Observing-1 (NMP)	An advanced land-imaging mission that demonstrates new instruments and spacecraft systems	11/21/00	Earth systematic missions
GOES N - P	Primary element of U.S. weather monitoring and forecast operations and are a key component of NOAA's National Weather Service operations and modernization program.	03/04/10	GOES/POES
GPM	Provides next-generation observations of rain and snow worldwide every three hours.	02/01/14	Earth systematic missions
GRACE	Accurately map variations in the Earth's gravity field	03/17/02	Earth system science pathfinder
HS3	Investigate processes that underlie hurricane intensity change in the Atlantic Ocean basin	08/01/12	Earth system science pathfinder
LAGEOS 1 & 2	Provides an orbiting benchmark for geodynamical studies of the Earth	05/04/76	Earth science program
Landsat 7	Observing and measuring Earth's continental and coastal landscapes	04/15/99	Earth systematic missions
LDCM/Landsat 8	Observing and measuring Earth's continental and coastal landscapes	02/11/13	Earth systematic missions
NOAA-N	Imaging and measurement of the Earth's atmosphere, its surface, and cloud cover	05/20/05	GOES/POES
OCO-2	Acquire precise measurements of atmospheric carbon dioxide	07/01/14	Earth system science pathfinder
Operation IceBridge	Characterize annual changes in thickness of sea ice, glaciers, and ice sheets. Collects data used to predict response of earth's polar ice to climate change and resulting sea-level rise.	10/15/09	Earth science program
OSTM	Measure sea surface height to an accuracy of <3.3 cm every ten days.	06/20/08	Earth systematic missions
QuikSCAT	Record sea-surface wind speed and direction data under all weather and cloud conditions over Earth's oceans.	06/19/99	Earth systematic missions
SeaWinds (ADEOS II)	Measure near-surface wind velocity (both speed and direction) under all weather and cloud conditions over Earth's oceans. Flies on the Japanese *ADEOS-II* Spacecraft	12/14/02	Earth systematic missions
SMAP	Combines low-frequency microwave radiometer and radar to measure surface soil moisture and freeze-thaw state	01/01/15	Earth systematic missions
SORCE	Provides state-of-the-art measurements of incoming x-ray, ultraviolet, visible, near-infrared, and total solar radiation.	01/25/03	Earth systematic missions
Suomi NPP	Long term monitoring of climate trends and of global biological productivity.	10/25/11	Earth systematic missions
Terra	Provides global data on the state of the atmosphere, land, and oceans	12/18/99	Earth systematic missions

ADEOS = Advanced Earth Observing Satellite.
AirMOSS = Airborne Microwave Observatory of Subcanopy and Subsurface.
ATTREX = Airborne Tropical Tropopause Experiment.
CALIPSO = Cloud-Aerosol Lidar and Infrared Pathfinder Satellite Observation.
CARVE = Carbon in Arctic Reservoirs Vulnerability Experiment.
DISCOVER = Deriving Information on Surface Conditions from Column and Vertically Resolved Observations Relevant to Air Quality.
GOES = Geostationary Satellite Server.
GPM = Global Precipitation Measurement.
GRACE = Gravity Recovery and Climate Experiment.
HS3 = Hurricane and Severe Storm Sentinel.
LAGEOS = Laser Geodynamics Satellites.
LDCM = Landsat Data Continuity Mission/Landsat 8.
NMP = New Millennium Program.
NOAA = National Oceanic and Atmospheric Administration.
NPP = NPOESS (National Polar-orbiting Operational Environmental Satellite System) Preparatory Project.
OCO = Orbiting Carbon Observatory.
OSTM = Ocean Surface Topography Mission.
POES = Polar-Orbiting Operational Environmental Satellite.
SMAP = Soil Moisture Active Passive.
SORCE = Solar Radiation and Climate Experiment.

SOURCE: Adapted from "Missions," in *NASA Science: Missions*, National Aeronautics and Space Administration, November 12, 2015, http://science.nasa.gov/missions/ (accessed December 7, 2015), and "Science and Technology Programs," in *NASA Science: Science and Technology Programs*, National Aeronautics and Space Administration, 2015, http://science1.nasa.gov/about-us/smd-programs/ (accessed December 7, 2015)

lines. Prominences can break away from the sun and hurtle through space carrying large amounts of solar material.

- Coronal mass ejections—these occur when billions of tons of particles are slung away from the sun by broken magnetic field lines. These ejections happen frequently and are most common during the solar maximums, when they can occur several times a day. Coronal mass ejections are known informally as solar storms. When they blast toward Earth, they can harm satellites and disrupt telecommunications and electric power generation around the world.

NOAA monitors space weather and issues warnings about geomagnetic and solar radiation storms (increases

FIGURE 6.2

***Soil Moisture Active Passive* observatory and launch vehicle**

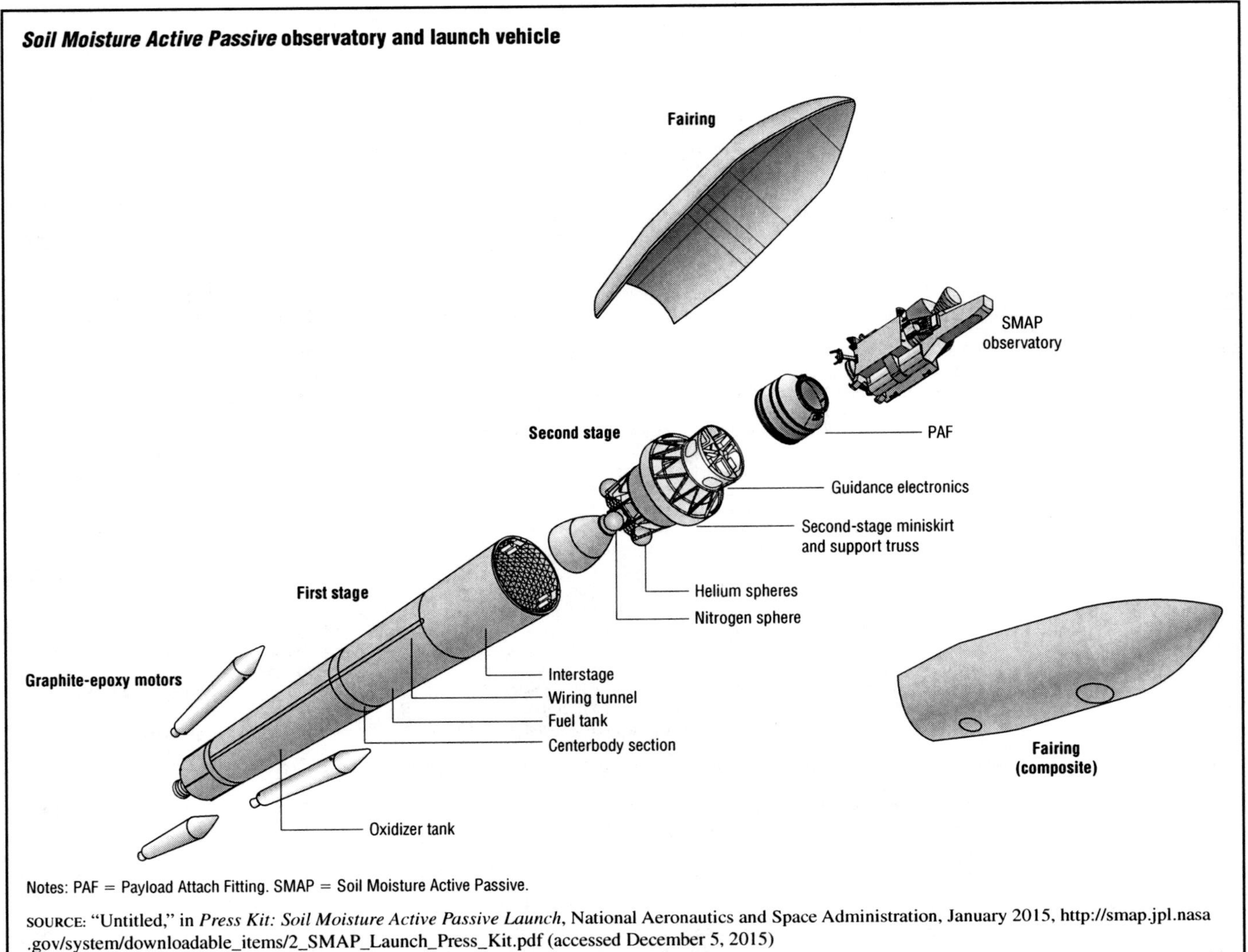

Notes: PAF = Payload Attach Fitting. SMAP = Soil Moisture Active Passive.

SOURCE: "Untitled," in *Press Kit: Soil Moisture Active Passive Launch*, National Aeronautics and Space Administration, January 2015, http://smap.jpl.nasa.gov/system/downloadable_items/2_SMAP_Launch_Press_Kit.pdf (accessed December 5, 2015)

in the number of energy particles in geospace). NOAA's Space Weather Prediction Center (http://www.swpc.noaa.gov) is located in Boulder, Colorado, and is operated by the National Weather Service. It serves as the national and worldwide warning center for space weather disturbances.

NASA's Heliophysics Missions

As of December 2015, 20 space science missions were operating within NASA's heliophysics program. (See Table 6.2.) These missions fall under various mission lines, but the largest number are operated under the Explorers Program. Most of these spacecraft are stationed in Earth orbit and study the effects of solar activity on the magnetosphere or upper atmosphere.

As shown in Table 2.7 in Chapter 2, in FY 2014 the heliophysics program was appropriated $641 million. NASA's FY 2016 budget request for the program was $651 million. This represented 12% of the agency's total science budget of $5.3 billion for that year.

In 2015 NASA conducted two heliophysics missions. The *Magnetospheric Multiscale* (*MMS*) launched in March 2015. According to the agency (December 14, 2015, http://science.nasa.gov/missions/mms/), the mission includes four identical satellites that are investigating "how the Sun's and Earth's magnetic fields connect and disconnect, explosively transferring energy from one to the other in a process that is important at the Sun, other planets, and everywhere in the universe, known as magnetic reconnection." Another operational heliophysics mission relies on the *Deep Space Climate Observatory* (*DSCVR*), which launched in February 2015. Table 6.3 lists facts about the mission and the spacecraft, which orbits at the L1 point in Earth-sun space. (See Figure 6.1.) It is taking measurements of the solar wind.

As of February 2016, NASA (http://science.nasa.gov/heliophysics/missions/) had one other heliophysics mission slated to launch in 2016. The agency (December 14, 2015, http://science.nasa.gov/missions/space-environment-testbeds/) indicates that the Space Environment Testbeds

FIGURE 6.3

Earth and the magnetosphere

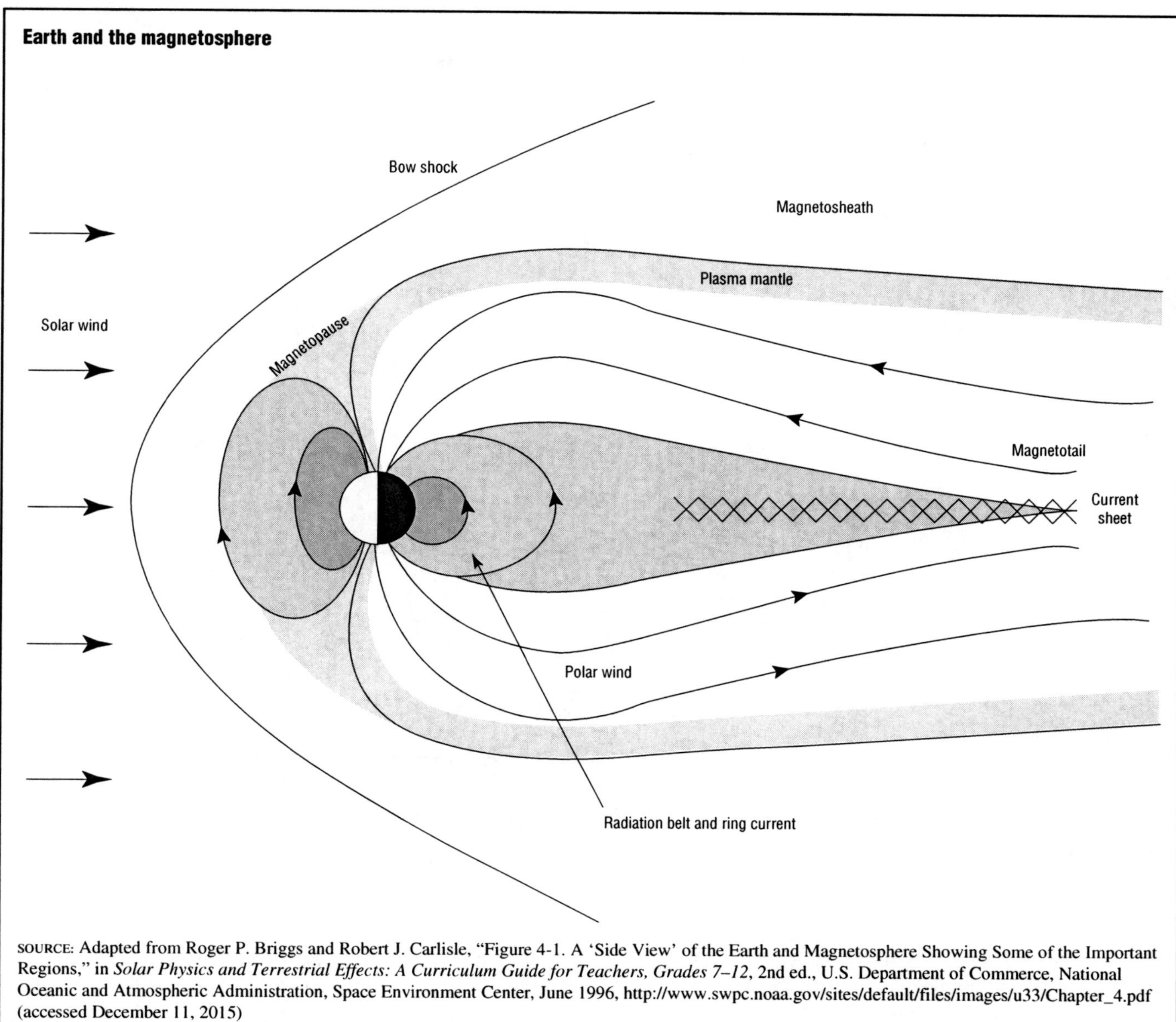

SOURCE: Adapted from Roger P. Briggs and Robert J. Carlisle, "Figure 4-1. A 'Side View' of the Earth and Magnetosphere Showing Some of the Important Regions," in *Solar Physics and Terrestrial Effects: A Curriculum Guide for Teachers, Grades 7–12*, 2nd ed., U.S. Department of Commerce, National Oceanic and Atmospheric Administration, Space Environment Center, June 1996, http://www.swpc.noaa.gov/sites/default/files/images/u33/Chapter_4.pdf (accessed December 11, 2015)

mission will test hardware components to see how they withstand varying solar conditions in the space environment.

NASA'S PLANETARY SCIENCE PROGRAM

In *SMD Mission Handbook*, NASA notes that the overall goal of the SMD's planetary science program is to "ascertain the content, origin, and evolution of the solar system, and the potential for life elsewhere." The program's exploratory targets include planets, dwarf planets, comets, and asteroids.

In 2006 the International Astronomical Union (IAU) designated a new category of planetary bodies called dwarf planets. By official definition, a dwarf planet is planetlike in that it orbits the sun and has sufficient mass and self-gravity to be nearly round in shape. However, the IAU explains in the press release "IAU 2006 General Assembly: Result of the IAU Resolution Votes" (August 24, 2006, http://www.iau.org/news/pressreleases/detail/iau0603/) that unlike a planet, a dwarf planet "has not cleared the neighbourhood around its orbit." In other words, a dwarf planet, unlike a planet, does not clear smaller objects out of its orbit.

Comets orbit the sun, but typically in large elliptical patterns that take them far away from the sun at apogee. Also, comets have shells (called comas) and tails that are actually gases or dust blown off the comets by the sun's heat and solar wind. The sun is believed to have hundreds, maybe even thousands of comets orbiting it. An asteroid is a small planetary body that orbits the sun.

FIGURE 6.4

The sunspot cycle, January 1900–April 2015

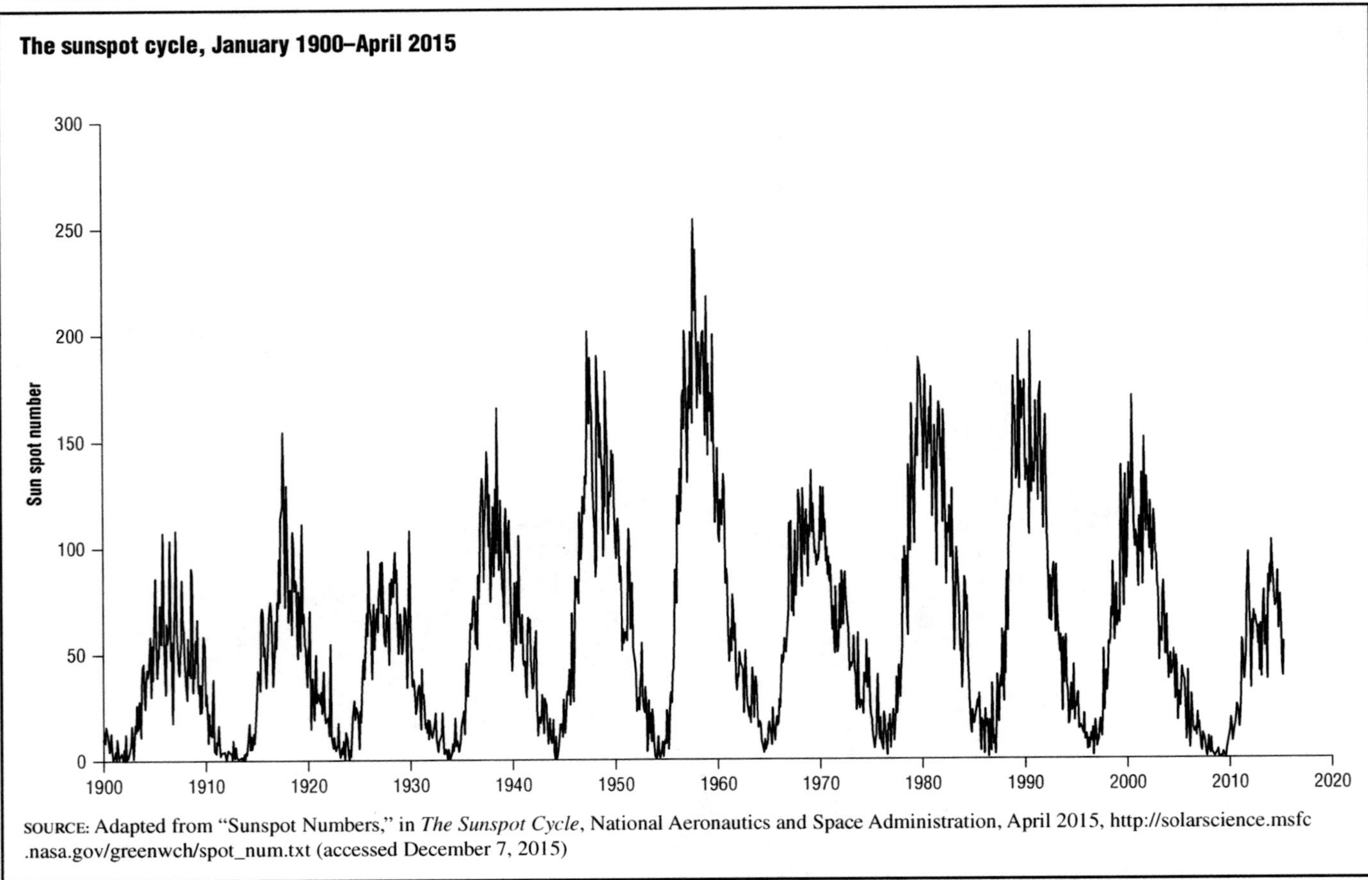

SOURCE: Adapted from "Sunspot Numbers," in *The Sunspot Cycle*, National Aeronautics and Space Administration, April 2015, http://solarscience.msfc.nasa.gov/greenwch/spot_num.txt (accessed December 7, 2015)

Unlike comets, asteroid orbits are typically circular. Asteroids are sometimes called planetoids or minor planets. Scientists believe there are millions of asteroids circling the sun.

The spacecraft involved in NASA's planetary science program are designed for long voyages, and hence are relatively expensive to build and operate. In fact, all of NASA's flagship missions were initiated for planetary science purposes.

The NRC laid out long-term objectives for the program in the 2011 decadal survey *Vision and Voyages for Planetary Science in the Decade 2013–2022* (http://www.nap.edu/catalog.php?record_id=13117). The primary recommendation is to conduct a Mars sample-return mission. If funding permits, the NRC also recommends a mission to Jupiter's icy moon Europa. Past and proposed missions to Mars and the far planets and beyond are described in detail in Chapters 7 and 8, respectively.

As shown in Table 2.7 in Chapter 2, NASA's planetary science program as a whole was appropriated $1.3 billion in FY 2014. Its FY 2016 budget request for the planetary science program was nearly $1.4 billion. This represented 26% of the agency's total science budget of $5.3 billion for that year.

NASA's Past Planetary Science Missions

NASA's planetary science missions have collected significant amounts of data and garnered public relations benefits for the agency due to their high-profile nature.

NEAR EARTH ASTEROID RENDEZVOUS. The Near Earth Asteroid Rendezvous (NEAR) mission was the first mission launched under the Discovery Program. In 2001 the *NEAR* spacecraft softly landed on the asteroid 433 Eros. The asteroid has an elliptical orbit that carries it outside the Martian orbit and then in close to Earth orbit, and is 21 miles (34 km) long. *NEAR* orbited 433 Eros for nearly a year taking high-resolution photographs. On February 12, 2001, it became the first spacecraft in history to land on an asteroid. At the time, 433 Eros was 196 million miles (315 million km) from Earth.

Before it landed, the spacecraft was renamed *NEAR Shoemaker* in honor of the geologist Eugene M. Shoemaker (1928–1997). According to NASA (December 14, 2015, http://science.nasa.gov/missions/near/), the spacecraft conducted scientific measurements on the surface of the asteroid for more than two weeks before going dead. Its last communication with Earth occurred on February 28, 2001. The first Discovery mission was considered an overwhelming success.

TABLE 6.2

Heliophysics program operational missions as of December 2015

Mission name	Mission description	Launch date	NASA mission line
ACE	Observes particles of solar, interplanetary, interstellar, and galactic origins, spanning the energy range from solar wind ions to galactic cosmic ray nuclei	08/25/97	Explorers/Earth systematic missions
AIM	A mission to determine the causes of the highest altitude clouds in the Earth's atmosphere	04/25/07	Explorers
CINDI/CNOFS	A mission to understand the dynamics of the Earth's ionosphere	04/16/08	Explorers (Mission of Opportunity)
Cluster-II	Provide a detailed 3-dimensional map of the magnetosphere	07/16/00	Under operation of the European Space Agency
DSCOVR	Joint mission with NOAA and the U.S. Air Force to take space weather measurements from the Sun-Earth Lagrange point L1.	02/01/15	Heliophysics research
Geotail	Study the tail of the Earth's magnetosphere. Joint mission with Japanese Institute of Space and Astronautical Science.	07/24/92	Designed and built by Japan's Institute of Space and Astronautical Science
Hinode (Solar-B)	Partnering with Japan's Institute of Space and Astronautical Science. Measure the strength and direction of the Sun's magnetic field in the Sun's low atmosphere, also called the photosphere.	09/23/06	Solar terrestrial probes
IBEX	Designed to detect the edge of the Solar System	10/19/08	Explorers
IRIS	Traces the flow of energy and plasma through the chromosphere and transition region into the corona using spectrometry and imaging	06/27/13	Explorers
MMS	Investigate how the Sun's and Earth's magnetic fields connect and disconnect	03/01/15	Solar terrestrial probes
RHESSI	Studies solar flares in X-rays and gamma-rays.	02/05/02	Explorers
SOHO	Solar observatory studying the structure, chemical composition, and dynamics of the solar interior	12/02/95	Heliophysics research with European Space Agency
Solar Dynamics Observatory (SDO)	Study how solar activity is created and how space weather results from that activity.	02/11/10	Living With A Star
STEREO	Understand the origin the Sun's coronal mass ejections (CMEs) and their consequences for Earth.	10/25/06	Solar terrestrial probes
THEMIS	A study of the onset of magnetic storms within the tail of the Earth's magnetosphere.	02/17/07	Explorers
TIMED	Explores the energy transfer into and out of the Mesosphere and Lower Thermosphere/Ionosphere (MLTI) region of the Earth's atmosphere.	12/07/01	Solar terrestrial probes
TWINS A & B	Provide stereo imaging of the Earth's magnetosphere, the region surrounding the planet controlled by its magnetic field and containing the Van Allen radiation belts and other energetic charged particles.	03/13/08	Explorers (Mission of Opportunity)
Van Allen Probes	Studying the Earth's radiation belts on various scales of space and time	08/01/12	Living With A Star
Voyager	Explored all the giant outer planets of our solar system and then put on a trajectory to leave the solar system	09/05/77	Outer planets flagship
Wind	Studies the solar wind and its impact on the near-Earth environment.	11/01/94	Heliophysics research with European Space Agency

ACE = Advanced Composition Explorer.
AIM = Aeronomy of Ice in the Mesosphere.
CINDI/CNOFS = Coupled Ion-neutral Dynamics Investigations/Communication and Navigation Outage Forecast System.
DSCOVR = Deep Space Climate Observatory.
IBEX = The Interstellar Boundary Explorer.
IRIS = Interface Region Imaging Spectrograph.
MMS = Magnetospheric Multiscale.
RHESSI = Reuven Ramaty High Energy Solar Spectroscope Imager.
SOHO = Solar and Heliospheric Observatory.
STEREO = Solar Terrestrial Relations Observatory.
THEMIS = Thermal Emission Imaging System.
TIMED = Thermosphere, Ionosphere, Mesosphere Energetics and Dynamics.
TWINS = Two Wide-Angle Imaging Neutral-Atom Spectrometers.

SOURCE: Adapted from "Missions," in *NASA Science: Missions*, National Aeronautics and Space Administration, November 12, 2015, http://science.nasa.gov/missions/ (accessed December 7, 2015), and "Science and Technology Programs," in *NASA Science: Science and Technology Programs*, National Aeronautics and Space Administration, 2015, http://science1.nasa.gov/about-us/smd-programs/ (accessed December 7, 2015)

GENESIS. Genesis was a revolutionary Discovery Program mission designed to collect charged particles emitted from the sun and return them safely to Earth. The spacecraft launched in 2001 and assumed a tight orbit around the Earth-sun L1 point. (See Figure 6.1.) It spent 2.5 years there collecting samples and then headed back to Earth. After reentering Earth's atmosphere, the *Genesis* spacecraft was supposed to deploy its parachutes for a slow descent toward the surface. A specially equipped helicopter was to snag the spacecraft midair and carry it to land.

On September 8, 2004, *Genesis* began its descent to a Utah landing site. Its parachutes, however, did not open, and the spacecraft plummeted at high speed into the ground. The capsule containing the samples split open during the crash, exposing the sample medium to the outside atmosphere. Nevertheless, approximately 4 milligrams of samples were saved and have since undergone analysis. NASA (June 23, 2011, http://www.nasa.gov/mission_pages/genesis/main/index.html) notes that scientists have gained valuable knowledge about Earth's moon and the composition of the sun and planets from these samples.

Genesis was the first spacecraft to return extraterrestrial materials to Earth since the Apollo missions and the first spacecraft to return extraterrestrial materials from beyond the moon.

DEEP IMPACT. In January 2005 the spacecraft *Deep Impact* was launched into space. The mission was part of

TABLE 6.3

DSCOVR mission facts

Launch date and location:	
February 1, 2015, from Cape Canaveral, FL	
Mission	
Joint mission between NOAA, NASA, and U.S. Air Force	
Spacecraft	
570 kilograms at launch, 54 inches by 72 inches	
Launch vehicle	
SpaceX Falcon 9 v 1.1 provided by U.S. Air Force	
Primary instruments	
Solar Wind Plasma Sensor and Magnetometer (PlasMag), National Institute of Standards and Technology Advanced Radiometer (NISTAR), Earth Polychromatic Imaging Camera (EPIC), Electron Spectrometer (ES), Pulse Height Analyzer (PHA)	
Orbit	
L1 (Lagrangian point 1) orbit, about 1 million miles from Earth	
Primary measurements	
Solar wind observations, including velocity distribution and magnetic field	
Observational mitigation	
Replaces NASA's ACE satellite, which was launched in 1997	
Benefits	
Supports a 15–60 minute lead time on geomagnetic storm warnings; these storms, often caused by solar wind, can disrupt transportation, power grids, telecommunications and GPS.	

ACE = Advanced Composition Explorer.
DSCOVR = Deep Space Climate Observatory.
NASA = National Aeronautics and Space Administration.
NOAA = National Oceanic and Atmospheric Administration.

SOURCE: Adapted from "DSCOVR Quick Facts," in *Press Kit—DSCOVR: Deep Space Climate Observatory*, National Oceanic and Atmospheric Administration, February 2, 2015, http://www.nesdis.noaa.gov/DSCOVR/press.html (accessed December 5, 2015)

the Discovery Program. On July 4, 2005, the spacecraft encountered the comet Tempel 1 and launched a probe that crashed into the comet, which orbits the sun between Mars and Jupiter. The probe relayed data to the flyby portion of the spaceship. NASA scientists are using the data to determine the physical and chemical makeup of Tempel 1. The *Deep Impact* spacecraft has been given a new mission called EPOXI, which is described later in this chapter.

STARDUST. The comet Wild 2 (pronounced "Vilt" 2) was the focus of the Stardust mission, which was managed under the Discovery Program. Launched in 1999, the spacecraft carried a particle collector about the size of a tennis racket that collected interstellar dust and comet dust particles. As the spacecraft neared Earth, it released the sample-return capsule and continued on its way. On January 15, 2006, the capsule entered Earth's atmosphere. Parachutes were deployed, and the capsule landed safely in the Utah desert. The spacecraft was put into orbit around the sun. In 2011 it was reused for the Stardust-NExT mission, which was a flyby of comet Tempel 1. On March 24, 2011, the spacecraft made its last transmission to Earth before running out of fuel.

LUNAR CRATER OBSERVATION AND SENSING SATELLITE. The *Lunar Crater Observation and Sensing Satellite* (*LCROSS*) spacecraft was launched in 2009 on a dual mission with the *Lunar Reconnaissance Orbiter*, which is described later in this chapter. The two spacecraft were operated under NASA's Exploration Systems mission line (May 22, 2011, http://www.nasa.gov/exploration/about/esmd_mission.html) to support the agency's human spaceflight program.

LCROSS was designed to collect data that would help scientists determine if water ice exists near the moon's south pole. On October 9, 2009, the spacecraft launched a Centaur rocket that impacted the lunar surface in Cabeus crater, an area that lies in permanent shadow. Approximately four minutes after the impact the remainder of the spacecraft (called the shepherding spacecraft) was flown into the moon at the same point. Before it impacted, it collected sensory data on the particles in the plume that had been flung up by the Centaur impact. In November 2009 NASA announced that preliminary data collected by *LCROSS* indicated the existence of water molecules within the crater.

EPOXI. The EPOXI mission was a Discovery Program MO. It used the spacecraft *Deep Impact*, which was launched in 2005 as described earlier. In *EPOXI Comet Encounter Fact Sheet* (November 2, 2010, http://www.nasa.gov/pdf/494786main_EPOXI-FlybyFactSheet.pdf), NASA explains that the EPOXI mission name was derived from the names of two of its components—the Extrasolar Planet Observations and Characterization (EPOCh) and the Deep Impact Extended Investigation (DIXI)—which involve exoplanet observation and a comet flyby, respectively. In November 2011 the spacecraft conducted a flyby of comet 103P/Hartley 2. According to NASA, in "SOHO Watches a Comet Fading Away" (July 28, 2011, http://www.nasa.gov/mission_pages/soho/hartley2-fades.html), the spacecraft came within 450 miles (724 km) of the small potato-shaped comet, which takes approximately 6.5 years to orbit the sun.

GRAVITY RECOVERY AND INTERIOR LABORATORY. The Gravity Recovery and Interior Laboratory (GRAIL) mission was operated under the Discovery Program. It included two identical spacecraft that were launched in 2011 and orbited Earth's moon for more than a year measuring and mapping its gravitational field. Each spacecraft was about the size of a washing machine. They were called *Ebb* and *Flow*, names that were suggested by a fourth-grade class in Montana. Each spacecraft was equipped with an imager called a MoonKAM that provided still and video images for educational and public viewing. In December 2012 the two spacecraft were purposely crashed into the moon near its north pole.

LADEE. The *LADEE* spacecraft launched in 2013 and was put into orbit around Earth's moon. The mission was part of NASA's Robotic Lunar Exploration mission line.

(LADEE is the abbreviation for Lunar Atmosphere and Dust Environment Explorer.)

According to NASA (December 17, 2015, http://www.nasa.gov/mission_pages/ladee/main/#.UwoFRJ5dWrY), the spacecraft's name was pronounced "laddie." *LADEE* collected data about the composition and structure of the lunar atmosphere. In April 2014 the spacecraft was purposely crashed into the surface of the moon.

MESSENGER. The *MESSENGER* spacecraft was launched in 2004 on a voyage to Mercury. (MESSENGER is short for Mercury Surface, Space Environment, Geochemistry, and Ranging.) The mission was operated under the Discovery Program. In March 2011 *MESSENGER* became the first spacecraft ever to orbit Mercury. NASA's *Mariner 10* flew by Mercury during the 1970s, but it captured limited images of the planet's surface.

MESSENGER included a complex suite of instruments that captured tens of thousands of detailed images of the planet and investigated its geology and atmosphere. (See Figure 6.5.) The payload was specially designed to endure the high-temperature environment near the sun. The mission was originally scheduled to end in 2012, but it was extended. In April 2015 the spacecraft was purposely crashed into the planet's surface.

In the press release "NASA Completes MESSENGER Mission with Expected Impact on Mercury's Surface" (April 30, 2015, https://www.nasa.gov/press-release/nasa-completes-messenger-mission-with-expected-impact-on-mercurys-surface), the agency states, "Among its many accomplishments, the MESSENGER mission determined Mercury's surface composition, revealed its geological history, discovered its internal magnetic field is offset from the planet's center, and verified its polar deposits are dominantly water ice."

FIGURE 6.5

MESSENGER

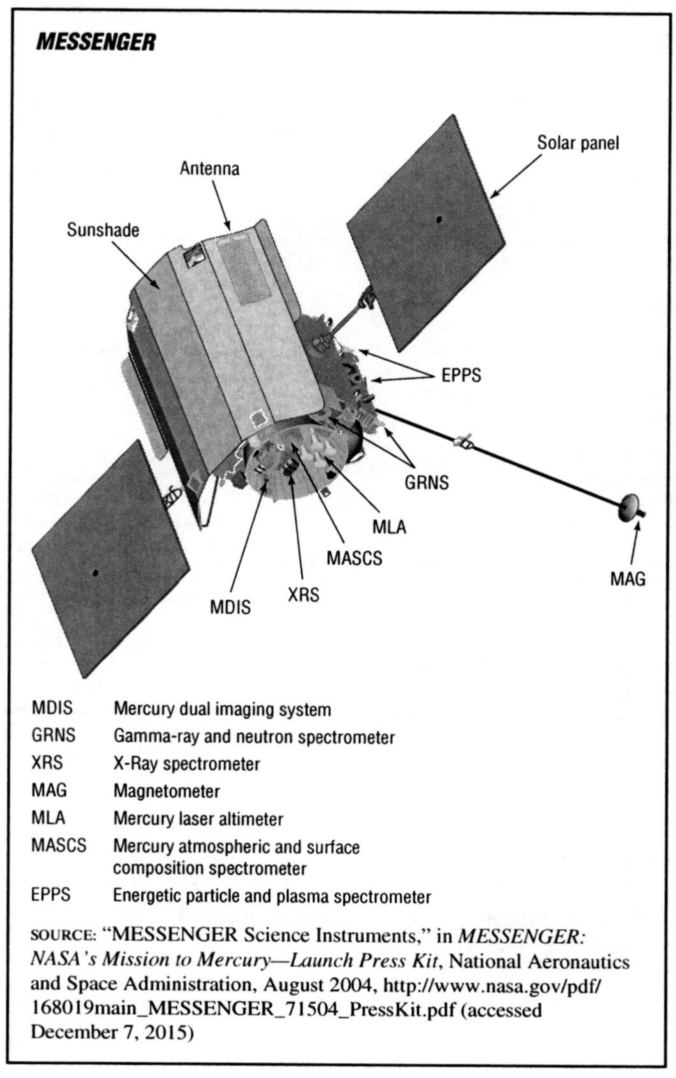

MDIS	Mercury dual imaging system
GRNS	Gamma-ray and neutron spectrometer
XRS	X-Ray spectrometer
MAG	Magnetometer
MLA	Mercury laser altimeter
MASCS	Mercury atmospheric and surface composition spectrometer
EPPS	Energetic particle and plasma spectrometer

SOURCE: "MESSENGER Science Instruments," in *MESSENGER: NASA's Mission to Mercury—Launch Press Kit*, National Aeronautics and Space Administration, August 2004, http://www.nasa.gov/pdf/168019main_MESSENGER_71504_PressKit.pdf (accessed December 7, 2015)

NASA's Operational Planetary Science Missions

Table 6.4 describes the three space missions (excluding those to Mars and the far planets and beyond) that were operational under NASA's planetary science program as of December 2015. The Rosetta mission is led by the European Space Agency (ESA) and is described later in this chapter. NASA supplied three of the instruments flying aboard the spacecraft. The remaining two operational missions are described in the following sections.

DAWN. The Dawn mission is part of NASA's Discovery Program and is designed to study the solar system's asteroid belt, which lies mostly between the orbits of Mars and Jupiter. (See Figure 6.6.) Most of the asteroids orbiting the sun lie within this vast belt. Its largest body, Ceres, is 580 miles (933 km) in diameter. For more than two centuries Ceres was considered to be an asteroid. In 2006 the IAU reclassified it as a dwarf planet. NASA indicates in "Ceres: A Dwarf Planet" (2016, http://starchild.gsfc.nasa.gov/docs/StarChild/solar_system_level2/ceres.html) that Ceres is spherical (nearly round) and takes 4.6 years to make one orbit around the sun. Its surface is believed to contain water ice, carbonates, and clay. The possible presence of water ice makes the dwarf planet of particular interest to scientists because of the association on Earth between water and biological life.

Another occupant of the asteroid belt is Vesta. According to NASA, in "Hubble Images of Asteroids Help Astronomers Prepare for Spacecraft Visit" (June 20, 2007, http://www.nasa.gov/mission_pages/hubble/news/vesta.html), Vesta is 330 miles (530 km) in diameter and takes 3.6 years to make one orbit around the sun. Figure 6.6 shows the relative locations of Ceres and Vesta within the asteroid belt.

In 2011–12 the *Dawn* spacecraft studied Vesta using its sensors and other instruments. It then left for Ceres, which it began orbiting in 2015. In "New Details on Ceres Seen in

TABLE 6.4

Planetary science program operational missions as of December 2015

Mission name	Mission description	Launch date	NASA mission line
Cassini	Detailed study of Saturn, its rings, icy satellites, magnetosphere, and Titan	10/15/97	Outer planets flagship
Dawn	Orbit Vesta and Ceres, two of the largest asteroids in the solar system	09/27/07	Discovery
Juno	A scientific investigation of the planet Jupiter	08/05/11	New Frontiers
Lunar Reconnaissance Orbiter	Find safe lunar landing sites, locate potential resources, characterize the radiation environment, and demonstrate new technology.	06/17/09	Robotic lunar exploration
Mars Exploration Rover - Opportunity	Rover exploring the surface of Mars	07/07/03	Mars exploration
Mars Express (ASPERA-3)	ASPERA-3 is instrument aboard Mars Express, a project of the European Space Agency	06/02/03	Mars exploration
Mars Odyssey	Orbits Mars collecting images and data about Martian geology, climate, and mineralogy	04/07/01	Mars exploration
Mars Reconnaissance Orbiter	Orbits Mars searching for evidence that water persisted on the surface of the planet	08/12/05	Mars exploration
Mars Science Laboratory - Curiosity	Rover exploring the surface of Mars	11/26/11	Mars exploration
MAVEN	Orbiter investigating Mars' upper atmosphere, ionosphere and interactions with the sun and solar wind.	11/18/13	Mars exploration
MESSENGER	A scientific investigation of the planet Mercury	08/03/04	Discovery
New Horizons	A scientific investigation of the planet Pluto and the icy worlds beyond it	01/19/06	New Frontiers
Rosetta	Study the origin of comets, the relationship between cometary and interstellar material and its implications with regard to the origin of the Solar System.	03/02/04	Under operation of the European Space Agency

ASPERA = Analyzer of Space Plasmas and Energetic Atoms.
MAVEN = Mars Atmosphere and Volatile Evolution.
MESSENGER = Mercury Surface, Space Environment, Geochemistry and Ranging.

SOURCE: Adapted from "Missions," in *NASA Science: Missions*, National Aeronautics and Space Administration, November 12, 2015, http://science.nasa.gov/missions/ (accessed December 7, 2015), and "Science and Technology Programs," in *NASA Science: Science and Technology Programs*, National Aeronautics and Space Administration, 2015, http://science1.nasa.gov/about-us/smd-programs/ (accessed December 7, 2015)

FIGURE 6.6

Diagram of inner solar system showing asteroid belt containing Vesta and Ceres

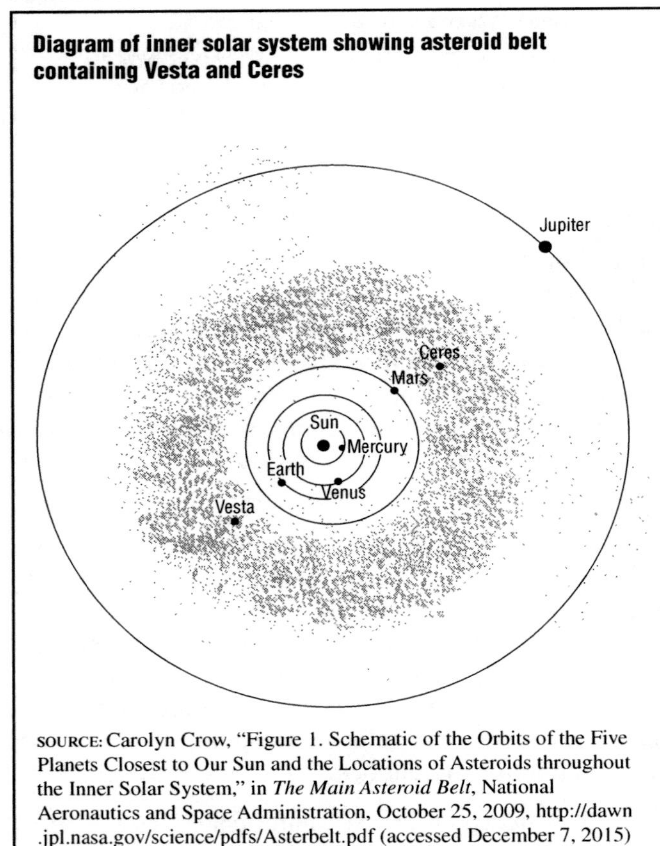

SOURCE: Carolyn Crow, "Figure 1. Schematic of the Orbits of the Five Planets Closest to Our Sun and the Locations of Asteroids throughout the Inner Solar System," in *The Main Asteroid Belt*, National Aeronautics and Space Administration, October 25, 2009, http://dawn.jpl.nasa.gov/science/pdfs/Asterbelt.pdf (accessed December 7, 2015)

Dawn Images" (January 12, 2016, https://www.nasa.gov/feature/jpl/new-details-on-ceres-seen-in-dawn-images), NASA notes that the spacecraft reached its lowest orbit around the world in December 2015, returning detailed images of Ceres's surface features. As of February 2016, *Dawn* continued to orbit the small world.

LUNAR RECONNAISSANCE ORBITER. In 2009 the Lunar Reconnaissance Orbiter (LRO) and LCROSS missions were launched together and sent their respective spacecraft toward the moon. The *LCROSS* spacecraft and its mission were described earlier in this chapter. The *LRO* spacecraft assumed a lunar orbit and began imaging the moon and studying its surface characteristics and radiation environment. (See Figure 6.7.) NASA indicates in "NASA Missions Uncover the Moon's Buried Treasures" (October 21, 2010, http://www.nasa.gov/centers/ames/news/releases/2010/10-89AR.html) that *LCROSS* and *LRO* data confirm that the moon contains water ice and "useful materials." The latter include volatile chemicals, such as ammonia, carbon monoxide, carbon dioxide, and hydrogen gas. As of February 2016, the *LRO* continued to orbit the moon and collect images and other data.

NASA's Future Planetary Science Missions

Planned missions to Mars and the far planets and beyond are discussed in Chapters 7 and 8, respectively.

As of February 2016, NASA (http://science1.nasa.gov/about-us/smd-programs/new-frontiers/) indicated

FIGURE 6.7

Lunar Reconnaissance Orbiter

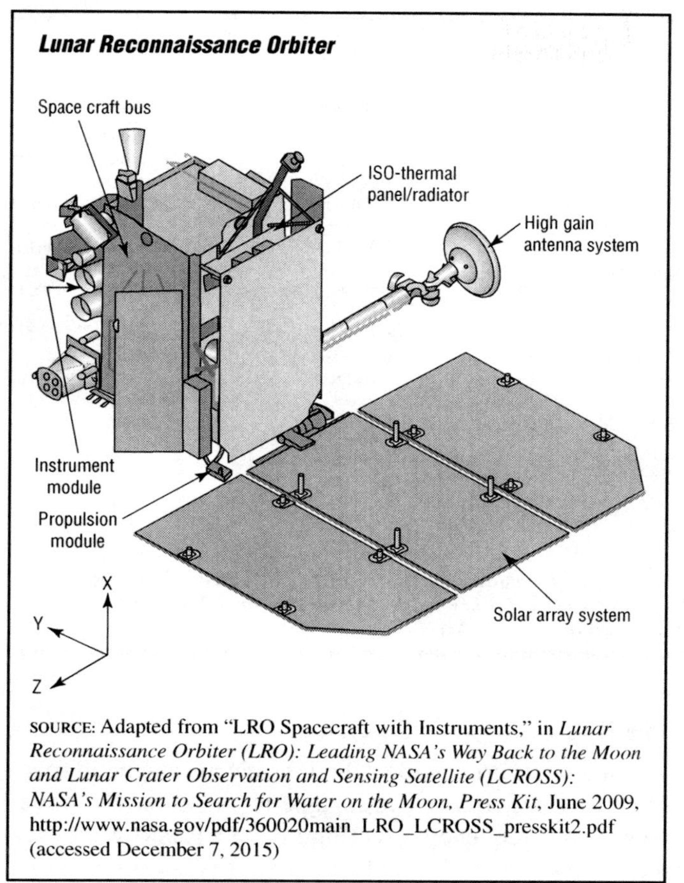

SOURCE: Adapted from "LRO Spacecraft with Instruments," in *Lunar Reconnaissance Orbiter (LRO): Leading NASA's Way Back to the Moon and Lunar Crater Observation and Sensing Satellite (LCROSS): NASA's Mission to Search for Water on the Moon, Press Kit*, June 2009, http://www.nasa.gov/pdf/360020main_LRO_LCROSS_presskit2.pdf (accessed December 7, 2015)

that there was only one planetary science mission scheduled to launch in 2016: the Origins-Spectral Interpretation-Resource Identification-Security-Regolith Explorer (OSIRIS-REx) mission, which will orbit and sample asteroid 101955 (formerly called 1999 RQ36). The Planetary Society is a private organization that advocates space exploration as described in Chapter 4. In "Nine-Year-Old Names Asteroid Target of NASA Mission in Competition Run by the Planetary Society" (May 1, 2013, http://www.planetary.org/press-room/releases/2013/nine-year-old-names-asteroid.html), the organization notes that it held a naming contest for the asteroid in 2013 in conjunction with two universities. The winning name was Bennu, after an ancient Egyptian god. The name was suggested by a third-grade student in North Carolina and was submitted to the IAU for consideration as the official name of the asteroid.

The OSIRIS-REx mission is part of NASA's New Frontiers Program. According to the agency (December 14, 2015, http://science.nasa.gov/missions/osiris-rex/), this will be the first U.S. mission to ever return asteroid samples to Earth. The spacecraft is expected to rendezvous with the asteroid in 2019 and return to Earth in 2023. Although most sun-orbiting asteroids are found in the asteroid belt between Mars and Jupiter, some travel closer to Earth. The asteroid 101955 Bennu is of particular interest because it is classified as a near-Earth object (NEO). According to the IAU (October 7, 2013, http://www.iau.org/public/themes/neo/nea2/) NEOs are asteroids and comets that come within less than 0.3 astronomical units (27.9 million miles [44.9 million km]) of Earth. In "Orbit and Bulk Density of the OSIRIS-REx Target Asteroid (101955) Bennu" (*Icarus*, vol. 235, June 2014), Steven R. Chesley et al. indicate that 101955 has a one in 2,700 chance of impacting Earth during the late 2100s.

According to NASA, in "NASA Selects Investigations for Future Key Planetary Mission" (September 30, 2015, http://www.jpl.nasa.gov/news/news.php?feature=4727), its next mission under the Discovery Program is expected to launch during the early 2020s. In late 2015 NASA selected five proposals from dozens of prospects. The final choice is expected by September 2016. Two of the five proposed missions target Venus: the Venus Emissivity, Radio Science, InSAR, Topography, and Spectroscopy mission and the Deep Atmosphere Venus Investigation of Noble Gases, Chemistry, and Imaging mission. Two other proposed missions would investigate asteroids in the asteroid belt: the Psyche mission and the Lucy mission. The fifth proposed mission is the NEO Camera; it would look for and characterize NEOs, particularly asteroids that might pose an impact hazard to Earth.

NASA's Astrophysics Program

NASA (June 9, 2015, http://science.nasa.gov/astrophysics/) indicates that the overall goal of its astrophysics program is to "discover how the universe works, explore how it began and evolved, and search for life on planets around other stars." In 2010 the NRC issued the decadal survey *New Worlds, New Horizons in Astronomy and Astrophysics* (http://www.nap.edu/catalog.php?record_id=12951), in which the council recommends that NASA construct new space- and ground-based telescopes to investigate phenomena and test theories related to astrophysics.

In FY 2014 the astrophysics program was appropriated $678.3 million. (See Table 2.7 in Chapter 2.) NASA's FY 2016 budget request for the program was $709.1 million. It should be noted that the future *James Webb Space Telescope*, which is also a line item in the SMD's budget, will conduct astrophysics research. It is discussed later in this chapter.

As of December 2015, 10 space missions were operating within the astrophysics program. (See Table 6.5.) These missions fall under various mission lines.

NASA's Space Telescopes

Celestial objects emit electromagnetic waves that give clues about their shape, size, age, and location. Earth's atmosphere filters and blocks out most of this radiation, keeping it from reaching the planet's surface.

TABLE 6.5

Astrophysics program operational missions as of December 2015

Mission name	Mission description	Launch date	NASA mission line
Chandra	Performing detailed studies of black holes, supernovas, and dark matter	07/23/99	Physics of the Cosmos
Fermi	Gamma-ray observatory	06/11/08	Physics of the Cosmos
Hubble Space Telescope (HST)	An ultraviolet, visible and infrared orbiting telescope	04/24/90	Cosmic Origins
INTEGRAL	Seeks to unravel the secrets of the highest-energy - and therefore the most violent - phenomena in the Universe.	10/17/02	Physics of the Cosmos
Kepler	Surveys our region of the Milky Way Galaxy to detect and characterize hundreds of Earth-size and smaller planets in or nearby the habitable zone.	03/06/09	Discovery/exoplanet exploration
LBTI	The LBTI is mounted on the Large Binocular Telescope (LBT) located on Mt. Graham, AZ. It is used to characterize planets and ultimately life beyond our solar system.	Earth-based telescope	Exoplanet exploration
NuSTAR	Studies the universe in high-energy x-rays	06/13/12	Explorers
SOFIA	The largest airborne observatory in the world; it is aboard a Boeing 747-SP aircraft modified to accommodate a 2.5 meter gyro-stabilized telescope.	Began operating in May 2014	Cosmic Origins
Spitzer	Conducts infrared astronomy from space.		Cosmic Origins
Swift	Three-telescope space observatory for studying the position, brightness, and physical properties of gamma ray bursts.	08/25/03 11/20/04	Explorers

INTEGRAL = International Gamma Ray Astrophysics Laboratory.
LBTI = Large Binocular Telescope Interferometer.
NuSTAR = Nuclear Spectroscopic Telescope Array.
SOFIA = Stratospheric Observatory for Infrared Astronomy.

SOURCE: Adapted from "Missions," in *NASA Science: Missions*, National Aeronautics and Space Administration, November 12, 2015, http://science.nasa.gov/missions/ (accessed December 7, 2015), and "Science and Technology Programs," in *NASA Science: Science and Technology Programs*, National Aeronautics and Space Administration, 2015, http://science1.nasa.gov/about-us/smd-programs/ (accessed December 7, 2015)

This is one reason biological life is possible on Earth. The atmospheric shield is good for human health but bad for observing the universe.

To clearly detect all electromagnetic waves traveling through space, an observer must be located above the shielding effects of Earth's atmosphere. The human eye can only detect one type of electromagnetic wave: visible light. This wave makes up only a small fraction of the radiation that is located throughout the cosmos. This is why scientists invented instruments that can detect and measure waves that are invisible to humans. These instruments are put on satellites and launched into space to provide a clearer picture of the universe.

In April 1966 NASA launched its first space telescope, the *Orbital Astronomical Observatory 1*. This launch was the first in a series of many satellites that have been put into Earth orbit to detect energy sources in space. Most of these projects have included NASA and international partners. Some of the largest and most powerful space telescopes in the world are NASA's Great Observatories. These observatories were developed over many years. As of February 2016, NASA had put four Great Observatories into space:

- *Hubble Space Telescope*
- *Chandra X-Ray Observatory*
- *Compton Gamma Ray Observatory*
- *Spitzer Space Telescope*

These observatories were designed to examine the universe across a wide range of the energies that make up the electromagnetic spectrum.

The Electromagnetic Spectrum

Scientists use a scale called the electromagnetic spectrum to categorize electromagnetic radiation by wavelength and frequency. (See Figure 6.8.) Radio waves are the largest type, followed by microwaves, infrared waves, visible waves, ultraviolet waves, X-rays, and gamma rays. Much of the electromagnetic radiation that stars emit is visible. This explains why humans can see stars in the nighttime sky. However, stars emit radiation at other wavelengths across the spectrum, particularly at infrared. There are huge clouds of dust and gas that are located throughout the universe. These clouds, called nebulae, can hide visible light from human sight. Scientists use specially designed telescopes that enable humans to peer through the nebulae and see the stars that lie beyond.

Many of the most mysterious objects and explosions in the cosmos are associated with extremely high temperatures and releases of gamma-ray, X-ray, and ultraviolet radiation. Space-based telescopes that are equipped with special instruments can detect these waves. This is important because X-rays and gamma rays are associated with phenomena such as black holes, supernova remnants, and neutron stars. Table 6.6 defines these terms and describes other celestial objects.

Telescopes as Time Machines

It takes a long time for electromagnetic radiation waves to travel through space. When a telescope captures an image of an object in the universe, the image shows what that object looked like in the past. The farther away a telescope can observe, the further back into time it

FIGURE 6.8

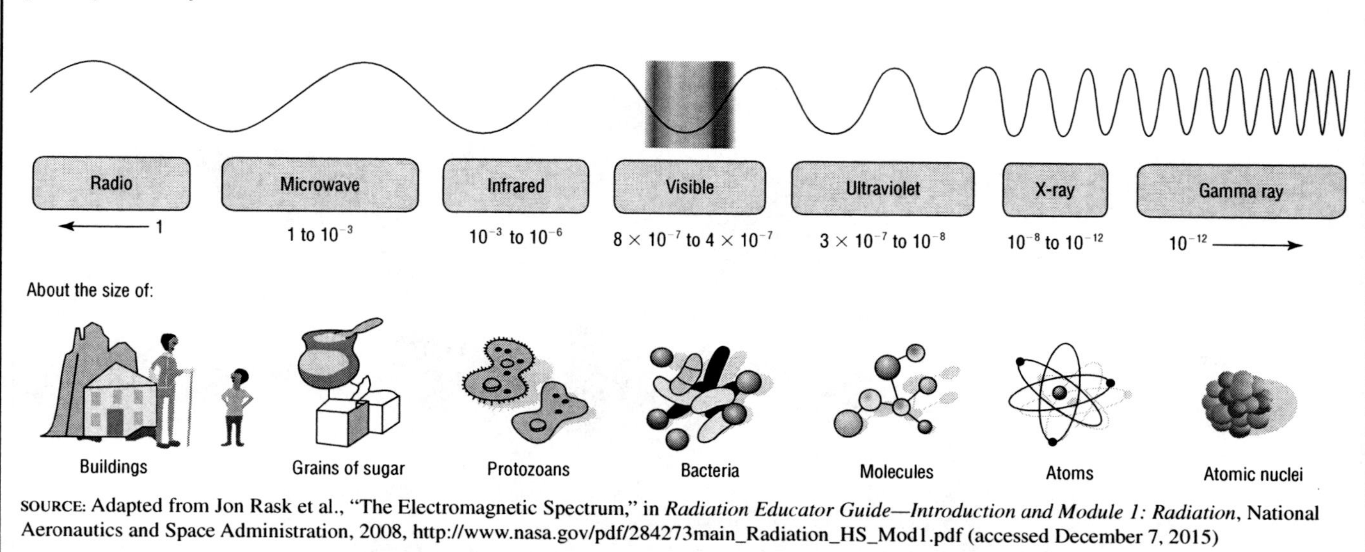

SOURCE: Adapted from Jon Rask et al., "The Electromagnetic Spectrum," in *Radiation Educator Guide—Introduction and Module 1: Radiation*, National Aeronautics and Space Administration, 2008, http://www.nasa.gov/pdf/284273main_Radiation_HS_Mod1.pdf (accessed December 7, 2015)

TABLE 6.6

Definitions

Term	Meaning
Black hole	Invisible celestial object believed created when a massive star collapses. Its gravitational field is so strong that light cannot escape from it.
Brown dwarf	Celestial object much smaller than a typical star that radiates energy, but does not experience nuclear fusion.
Dark matter	Hypothetical matter that provides the gravity needed to hold hot gases within a galaxy cluster.
Event horizon	The boundary of a black hole.
Extrasolar planet	Planet orbiting a star other than Earth's sun.
Galaxy cluster	Huge collection of galaxies bound together by gravity.
Nebulae	Large gas and dust clouds that populate a galaxy.
Neutron star	Hypothetical dense celestial object consisting primarily of closely packed neutrons. Believed to result from the collapse of a much larger star.
Nova	Star that suddenly begins emitting much more light than before and continues to do so for days or months before returning to its former state of illumination.
Pulsar	Celestial body emitting pulses of electromagnetic radiation at short relatively constant intervals.
Quasar	Mysterious celestial object that resembles a star, but releases tremendous amount of energy for its size.
Superflare	Massive explosion from a young star. A superflare is thousands of times stronger than a solar flare emitted by Earth's sun.
Supernova	Catastrophic explosion of a star.
Supernova remnant	Bubble of multimillion degree gas released during a supernova.
White dwarf	Small dying star that has used up all its fuel and is fading.

SOURCE: Created by Kim Masters Evans for Gale, © 2016

looks. NASA's Great Observatories capture images of celestial objects that are extremely distant from Earth. This means that the electromagnetic radiation reaching the observatories left its source a long time ago. This is why NASA refers to its observatories as time machines: They allow humans to look back into time and learn about the origins of the universe.

Hubble Space Telescope

During the 1970s NASA began developing the *Large Space Telescope* (*LST*). Following the moon landings, NASA's budget was severely cut, which forced scientists to downsize the *LST* project several times. In 1975 the ESA joined the project and agreed to fund a percentage of the *LST*'s costs in exchange for a guaranteed amount of telescope time for its scientists. Two years later Congress authorized funding for the construction and assembly of the *LST*.

At the same time that the *LST* was under construction, the space shuttle was also undergoing development. *LST* planners decided to use shuttle crews to deploy and service the telescope and ultimately return it to Earth. This decision would prove to be a fateful one. By 1985 the observatory was finished and given a new name: *Hubble Space Telescope* (*HST*).

The *HST* is the first of NASA's Great Observatories. It is named after the American astronomer Edwin Hubble (1889–1953). It detects three types of electromagnetic radiation: ultraviolet, visible, and near-infrared (the shortest wavelength within the infrared spectrum). It is the only one of the Great Observatories that captures images in visible light.

Figure 6.9 shows the general layout of the *HST*. It can hold eight scientific instruments besides the primary and secondary mirrors. These instruments are powered by sunlight that is captured by the satellite's two solar arrays. Gyroscopes and flywheels are used to point the telescope and keep it stable.

FIGURE 6.9

The configuration of the *Hubble Space Telescope*

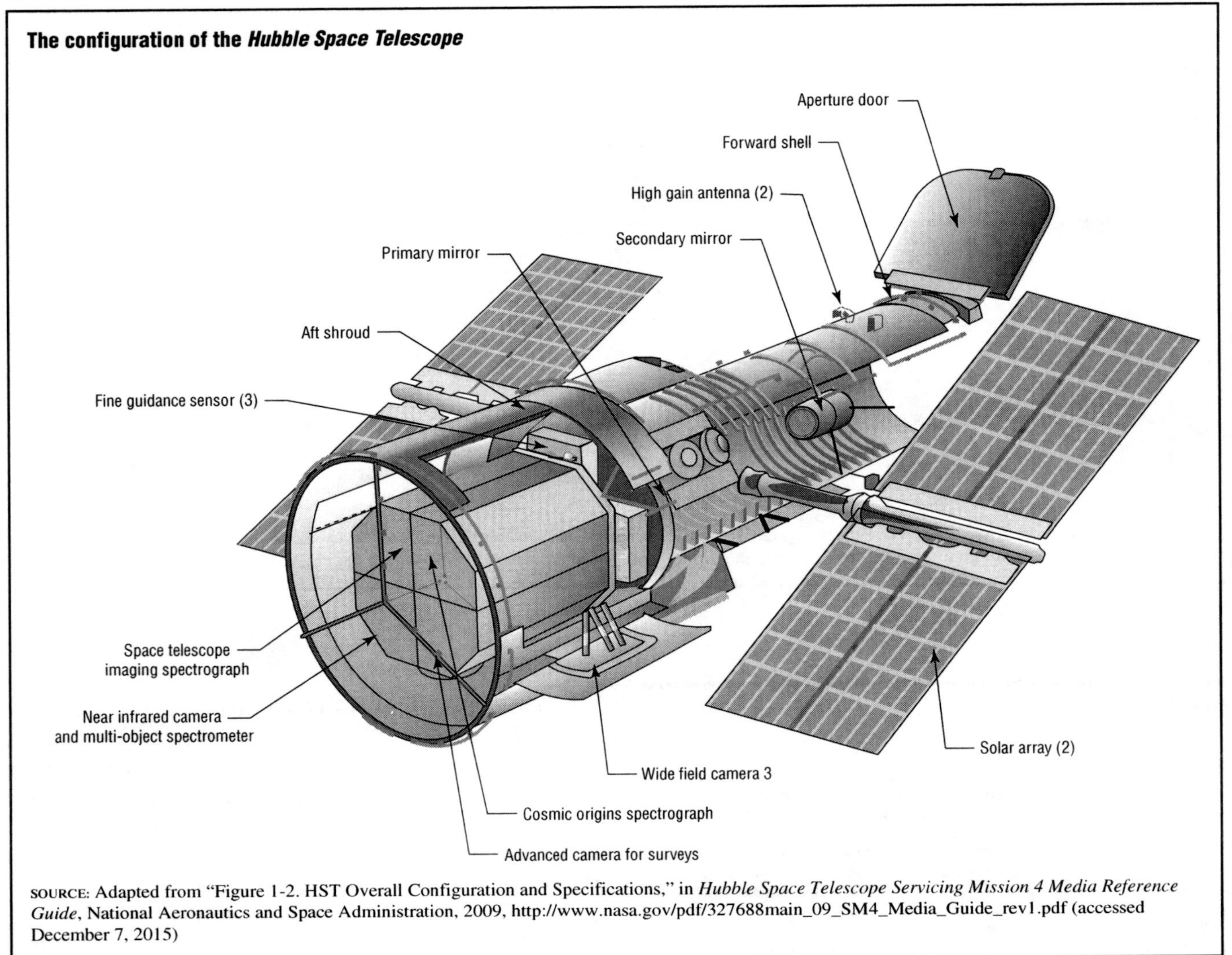

SOURCE: Adapted from "Figure 1-2. HST Overall Configuration and Specifications," in *Hubble Space Telescope Servicing Mission 4 Media Reference Guide*, National Aeronautics and Space Administration, 2009, http://www.nasa.gov/pdf/327688main_09_SM4_Media_Guide_rev1.pdf (accessed December 7, 2015)

NASA's Goddard Space Flight Center in Greenbelt, Maryland, is responsible for oversight of all *HST* operations and for servicing the satellite. The Space Telescope Science Institute in Baltimore, Maryland, selects targets for the observatory based on proposals submitted by astronomers. The astronomical data that are generated by the observatory are also analyzed by the institute. The institute is operated by the Association of Universities for Research in Astronomy, an international consortium of dozens of universities.

The *HST* was originally supposed to go into space in 1986. However, the explosion of the space shuttle *Challenger* that year delayed the *HST*'s launch for four years. On April 24, 1990, the *HST* was carried into orbit by the space shuttle *Discovery*. The following day the astronauts deployed the satellite approximately 380 miles (610 km) from Earth. It assumed a nearly circular orbit.

When scientists first started using the *HST*, they discovered that its images were fuzzy due to a defect in the telescope's optical mirrors. The defect had existed before the satellite was launched into space. Media publicity about the problem caused an uproar and brought harsh criticism of NASA. In December 1993 astronauts aboard the space shuttle *Endeavour* were sent to intercept the *HST* and install corrective devices.

This was the first of five servicing missions that were conducted by shuttle crews between 1993 and 2009. During these missions astronauts performed repairs and maintenance, installed new components to enhance the *HST*'s performance, and reboosted the satellite to a higher altitude. Like all objects in LEO, the *HST* experiences some drag and gradually loses altitude. The satellite does not have its own propulsion system.

As of February 2016, the future deorbit date of the *HST* was uncertain, but estimates varied from the late 2010s to the early 2040s. The deorbit date will depend on many factors, including the stability of its orbit and availability of funding for continued operations.

During its mission the *HST* has captured around 1 million images of celestial objects throughout the universe, including nebulae, galaxies, stars, and exoplanets.

Compton Gamma Ray Observatory

The *Compton Gamma Ray Observatory* (*CGRO*) was the second of NASA's Great Observatories. It was designed to detect high-energy gamma rays, the most powerful type of energy in the electromagnetic spectrum. The observatory was named after the American physicist Arthur Holly Compton (1892–1962).

On April 5, 1991, the *CGRO* was launched into space aboard the space shuttle *Atlantis*. The original plan was for a five-year mission. However, the observatory proved to be much more durable than expected and remained in space for nine years. In early 2000 NASA learned that one of the *CGRO*'s gyroscopes had failed. Although the observatory's instruments were still in working order, NASA decided to purposely deorbit the satellite. Scientists feared that if left to reenter on its own, the *CGRO* could rain pieces down on populated areas. This was of particular concern because the satellite weighed 17 tons (15.4 t) on Earth and included a fair amount of titanium in its components. Titanium does not disintegrate easily during atmospheric reentry.

On June 4, 2000, the *CGRO*'s propulsion system was used to guide the satellite to a safe reentry over a deserted area of the Pacific Ocean.

Chandra X-Ray Observatory

The *Chandra X-Ray Observatory* (*CXRO*) is the third of NASA's Great Observatories and a powerful X-ray telescope. (See Figure 6.10.) It was named after the Indian American scientist Subrahmanyan Chandrasekhar (1910–1995). On July 23, 1999, the *CXRO* was launched into space by the space shuttle *Columbia*. The observatory then moved into a high orbit above Earth. Solar power is used to run the *CXRO*'s equipment and charge its batteries.

Scientists around the world are using *CXRO* images to create an X-ray map of the universe. They hope to use this map to learn more about black holes, supernovas, superflares, quasars, extremely hot gases at the center of galaxy clusters, and the mysterious substance known as dark matter.

As of February 2016, the telescope had been operating for more than 16 years. In *2014 Senior Review of the Chandra X-Ray Observatory* (March 24–March 27, 2014, http://science.nasa.gov/media/medialibrary/2014/05/15/Final_Report_Chandra2014_SeniorReview_Panel.pdf),

FIGURE 6.10

Chandra X-Ray Observatory

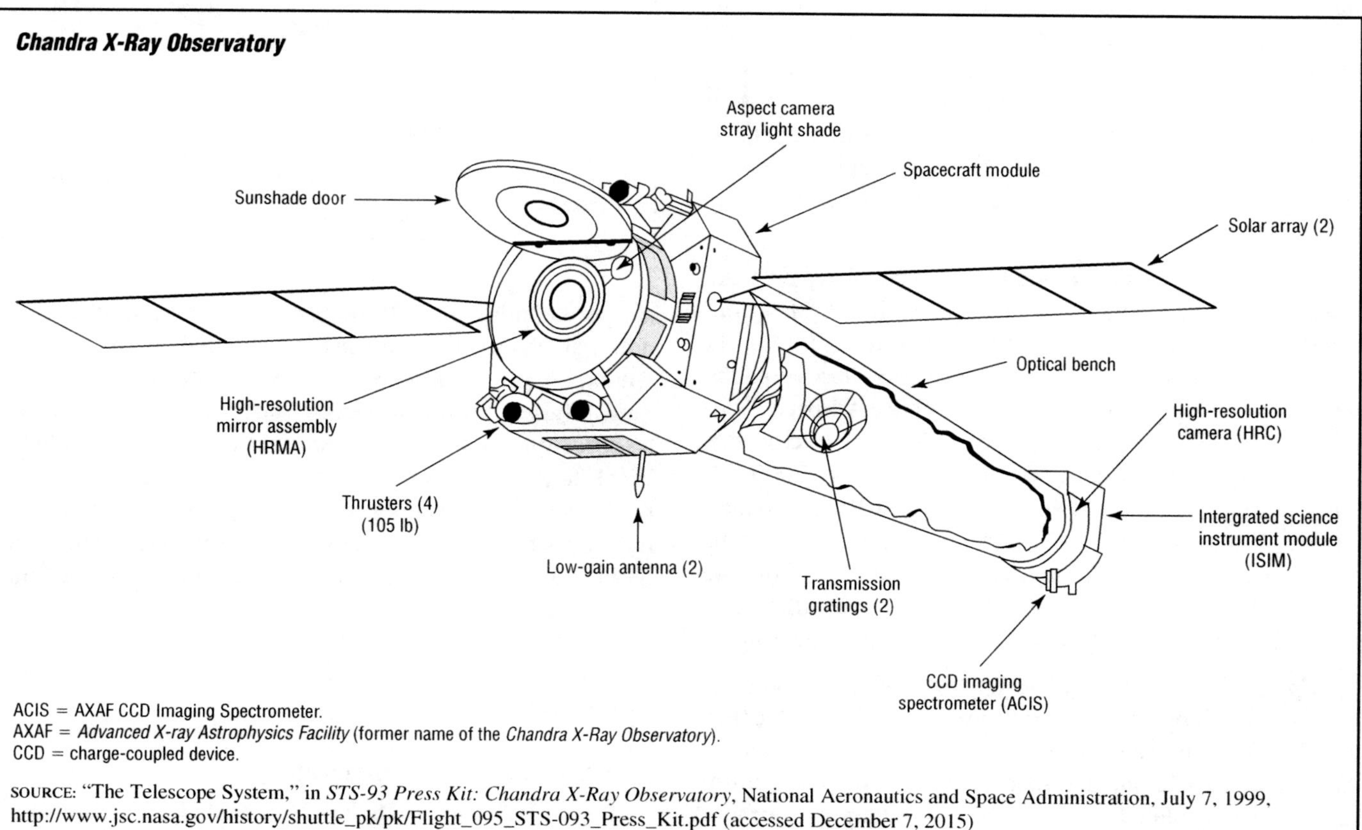

ACIS = AXAF CCD Imaging Spectrometer.
AXAF = *Advanced X-ray Astrophysics Facility* (former name of the *Chandra X-Ray Observatory*).
CCD = charge-coupled device.

SOURCE: "The Telescope System," in *STS-93 Press Kit: Chandra X-Ray Observatory*, National Aeronautics and Space Administration, July 7, 1999, http://www.jsc.nasa.gov/history/shuttle_pk/pk/Flight_095_STS-093_Press_Kit.pdf (accessed December 7, 2015)

an expert panel convened by NASA notes, "There are no technical obstacles to Chandra's continuing scientific productivity." The panel recommends that the *CXRO* continue operating as long as funding permits.

The *CXRO* is managed by the Marshall Space Flight Center for NASA's Office of Space Science. NASA's Jet Propulsion Laboratory, which is a division of the California Institute of Technology in Pasadena, California, provides communications and data links. The Smithsonian Astrophysical Observatory in Cambridge, Massachusetts, controls science and flight operations. NASA's High-Energy Astrophysics Science Archive collects and maintains all astronomical data that are obtained from the *CXRO* and NASA's smaller gamma-ray, X-ray, and extreme ultraviolet observatories.

Spitzer Space Telescope

The *Spitzer Space Telescope* is the fourth of NASA's Great Observatories and detects infrared radiation. It is named after the American astrophysicist Lyman Spitzer Jr. (1914–1997). The project was first conceived during the 1970s. During the 1990s NASA's science programs suffered harsh budget cuts. The telescope was redesigned several times and downsized from a large $2 billion telescope with many capabilities to a modest telescope costing less than half a billion dollars.

On August 25, 2003, it was launched into orbit atop a Delta rocket. It was the only Great Observatory not carried into space by a space shuttle. The telescope was housed inside a protective case called a fairing that fell away when the satellite reached orbit. Figure 6.11 shows details of the observatory's structure. The dark circle at the top of the observatory is actually a door that opens to expose the telescope to space. *Spitzer* is unique among the Great Observatories because it orbits the sun instead of Earth.

FIGURE 6.11

Exterior of the *Spitzer Space Telescope*

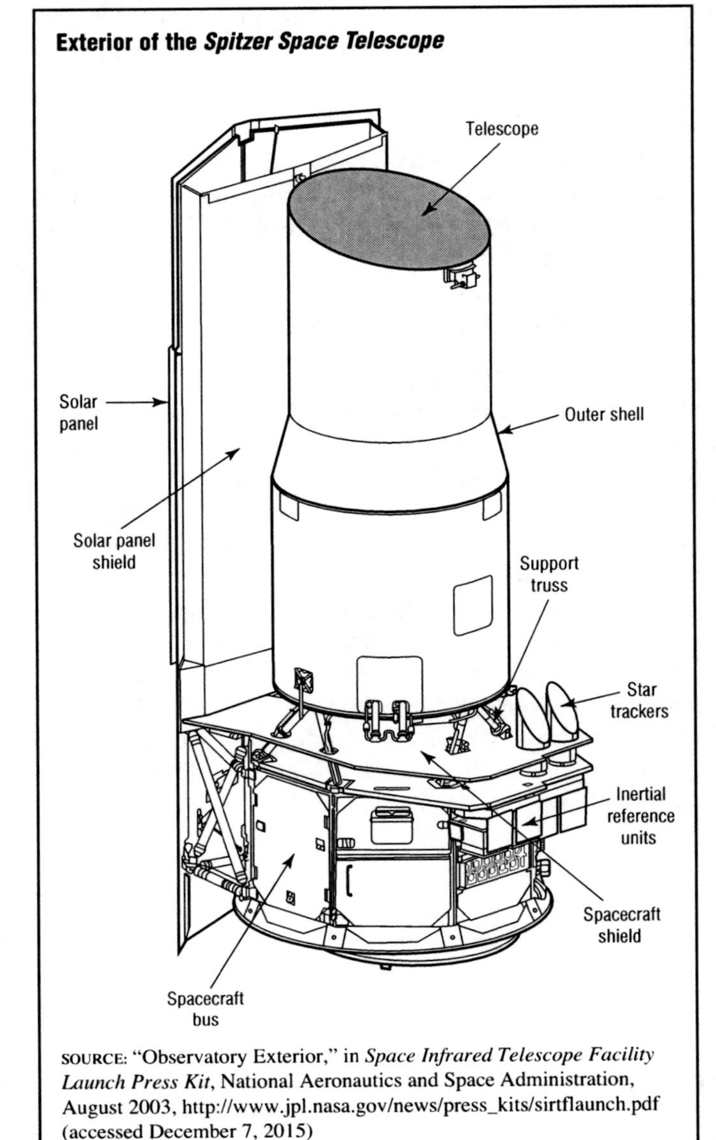

SOURCE: "Observatory Exterior," in *Space Infrared Telescope Facility Launch Press Kit*, National Aeronautics and Space Administration, August 2003, http://www.jpl.nasa.gov/news/press_kits/sirtflaunch.pdf (accessed December 7, 2015)

The telescope has provided images of many celestial features including galaxies and nebulae. It has also contributed to discoveries within the solar system. In October 2009 NASA announced that the observatory discovered the largest known ring around Saturn. In "NASA Space Telescope Discovers Largest Ring around Saturn" (October 6, 2009, http://spitzer.caltech.edu/news/966-ssc2009-19-NASA-Space-Telescope-Discovers-Largest-Ring-Around-Saturn), Whitney Clavin of NASA notes that the enormous ring begins about 3.7 million miles (6 million km) from the planet and is approximately 7.4 million miles (12 million km) wide. Phoebe, one of Saturn's moons, lies within the ring.

Spitzer is managed by the Jet Propulsion Laboratory. Nearby the laboratory is the Spitzer Science Center, which conducts science operations for the telescope. According to Clavin, the spacecraft's cryogenic helium supply was depleted in May 2009, ending its primary mission. However, some of its instruments have continued to operate, allowing astronomers to conduct limited astronomical observations during a so-called warm mission. An expert panel convened by NASA notes in *2014 NASA Astrophysics Senior Review* (March 31–April 3, 2014, http://science.nasa.gov/media/medialibrary/2014/05/15/Final_Report_Astro2014_SeniorReview_Panel.pdf) that the telescope's "observational capabilities are significantly reduced since its prime mission [ended]." The panel recommends that the observatory, which has high operating costs, cease to operate. However, NASA's budget request for FY 2016 (https://www.nasa.gov/sites/default/files/files/NASA_FY2016_Summary_Brief_corrected.pdf) includes funding for *Spitzer*.

NASA's Next Great Observatory

As of February 2016, NASA's next Great Observatory was under development and expected to launch in 2018 or

2019. It is called the *James Webb Space Telescope* (*JWST*) in honor of James Edwin Webb (1906–1992), the NASA director of flight operations during the Apollo Program. The *JWST* will detect infrared radiation from its orbit location at the Earth-sun L2 point. (See Figure 6.1.)

The *JWST* has been fraught with controversy due to significant cost overruns and scheduling delays. Ralph Vartabedian and W. J. Hennigan report in "NASA's Webb Telescope: Revolutionary Design, Runaway Costs" (LATimes.com, February 24, 2012) that the cost of the highly sophisticated telescope climbed from around $500 million when it was first conceived during the late 1990s to $8.8 billion in February 2012. According to Vartabedian and Hennigan, the spiraling costs led some in Congress to vote in 2011 to terminate the project; however, it survived. As shown in Table 2.7 in Chapter 2, the *JWST* was appropriated $658.2 million in FY 2014. NASA's FY 2016 budget request included $620 million for the project. Many analysts blame the *JWST*'s cost overruns for limiting the funds that are available for planetary science missions, particularly to Mars.

NASA's Other Space Observatories

Besides the Great Observatories, NASA operates several smaller space observatories, as listed in Table 6.5. These spacecraft maintain a sun or Earth orbit and operate under a variety of programs. For example, the Nuclear Spectroscopic Telescope Array (NuSTAR) mission is part of the Explorers Program. As shown in Figure 6.12, the observatory includes two coaligned telescopes that can be extended on a deployable mast. NASA (July 30, 2015, http://www.nasa.gov/mission_pages/nustar/overview/index.html#.UtAczZ5dWrZ) notes that the observatory focuses on high-energy X-rays of the electromagnetic spectrum.

The *Kepler* space telescope operates under NASA's Discovery and Exoplanet Exploration Programs. The spacecraft is illustrated in Figure 6.13. It is named after the German astronomer Johannes Kepler (1571–1630), who formulated the laws of planetary motion. Launched on March 7, 2009, *Kepler* surveys a region of the Milky Way galaxy for exoplanets that are similar to Earth.

In January 2010 NASA (http://www.nasa.gov/mission_pages/kepler/news/kepler-5-exoplanets.html) announced that eight exoplanets had been discovered. The first three exoplanets—dubbed Kepler 1b, Kepler 2b, and Kepler 3b—were discovered using ground-based telescopes. The other five exoplanets—Kepler 4b through Kepler 8b—were discovered using the space observatory. Since

FIGURE 6.12

NuSTAR

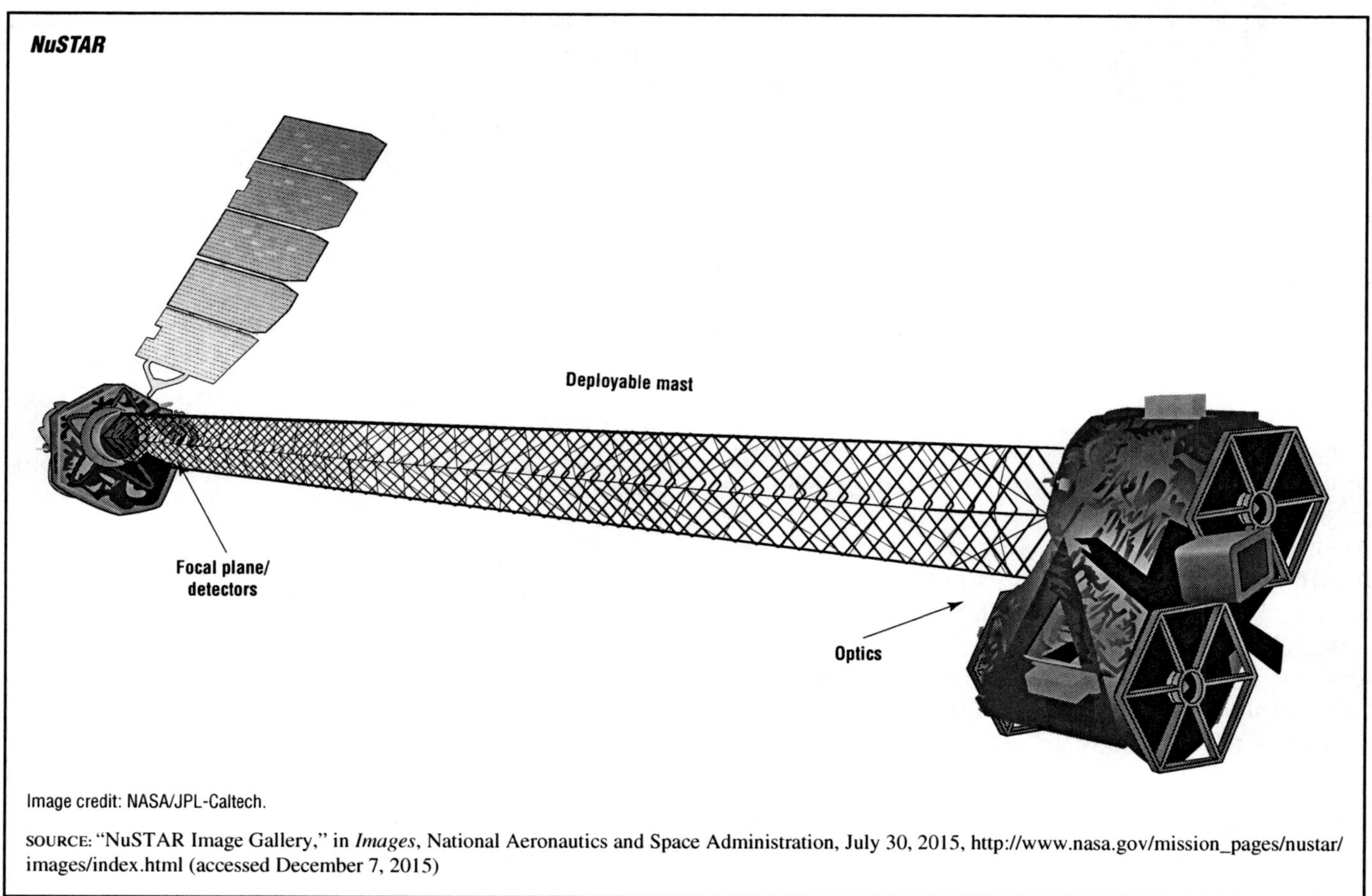

Image credit: NASA/JPL-Caltech.

SOURCE: "NuSTAR Image Gallery," in *Images*, National Aeronautics and Space Administration, July 30, 2015, http://www.nasa.gov/mission_pages/nustar/images/index.html (accessed December 7, 2015)

FIGURE 6.13

***Kepler* spacecraft and photometer**

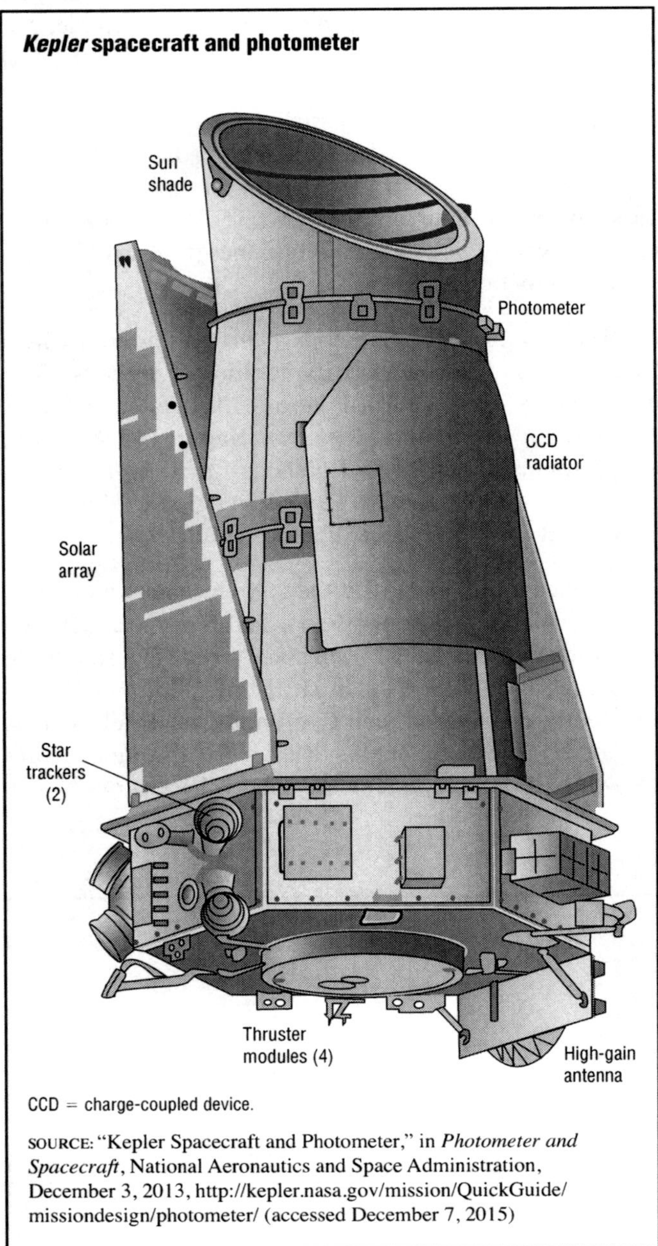

CCD = charge-coupled device.

SOURCE: "Kepler Spacecraft and Photometer," in *Photometer and Spacecraft*, National Aeronautics and Space Administration, December 3, 2013, http://kepler.nasa.gov/mission/QuickGuide/missiondesign/photometer/ (accessed December 7, 2015)

then, *Kepler* has been used to investigate thousands of suspected exoplanets. NASA maintains a list of confirmed exoplanets at http://kepler.nasa.gov/Mission/discoveries/. As of February 2016, there were more than 1,000 confirmed exoplanets discovered by *Kepler*. In addition, the NASA Exoplanet Archive (http://exoplanetarchive.ipac.caltech.edu/) indicated that, as of August 2015, nearly 4,700 suspected exoplanets had been discovered using *Kepler*. Studying exoplanets helps scientists learn about how planets and solar systems formed.

As of February 2016, NASA (http://science.nasa.gov/astrophysics/missions) had one astrophysics mission—Astro-H—scheduled for launch in 2016. The agency is collaborating with the Japan Aerospace Exploration Agency (JAXA) to provide a Soft X-Ray Spectrometer for the JAXA spacecraft, which will focus on X-ray investigations.

FOREIGN SPACE SCIENCE MISSIONS

As noted in Chapter 3, numerous space agencies operate around the world. Besides NASA, some of the most active programs are those operated by France, the ESA, Russia, China, Japan, and India. Although NASA and the China National Space Administration do not collaborate on missions, there is typically considerable cooperation between international agencies in developing and operating spacecraft devoted to scientific missions. For example, one agency may provide instruments that fly on the spacecraft of another agency.

France

The French space agency is called the Centre National d'Études Spatiales (CNES; National Space Study Center). The CNES operated the space-borne telescope *Convection Rotation et Transits planétaires* (*COROT*) until it ceased operating in 2013. *COROT* was launched in 2006 and measured light from distant stars. In 2007 *COROT* discovered its first exoplanet orbiting a yellow star approximately 1,500 light years from Earth.

Since the early 1990s the CNES has collaborated with NASA, JAXA, and the ESA on multiple space science missions, particularly by supplying spacecraft instruments. As of February 2016, the CNES indicated in "CNES Projects Library" (https://cnes.fr/en/fiches_mission_statut) that it was participating in dozens of space science missions in the preliminary or developmental stage, mostly with the ESA.

European Space Agency

The ESA has a long history of conducting space science missions, particularly in concert with international partners. For example, in August 1975 the ESA launched *Celestial Observation Satellite B*, which provided the first complete gamma-ray map of the Milky Way galaxy. Since then, the agency has operated other space observatories. Figure 6.14 shows the *Herschel Space Observatory*, which was launched in 2009. At the time, it was the largest space telescope ever built. It is named after the British astronomer William Frederick Herschel (1738–1822), who is credited with discovering infrared radiation in 1800. The *Herschel Space Observatory* was launched from French Guiana on an Ariane rocket that also carried the ESA's *Planck* satellite (named after the German physicist Max Planck [1858–1947]). The latter was an all-sky surveyor that measured cosmic background radiation until it ceased operating in 2013.

FIGURE 6.14

Herschel Space Observatory

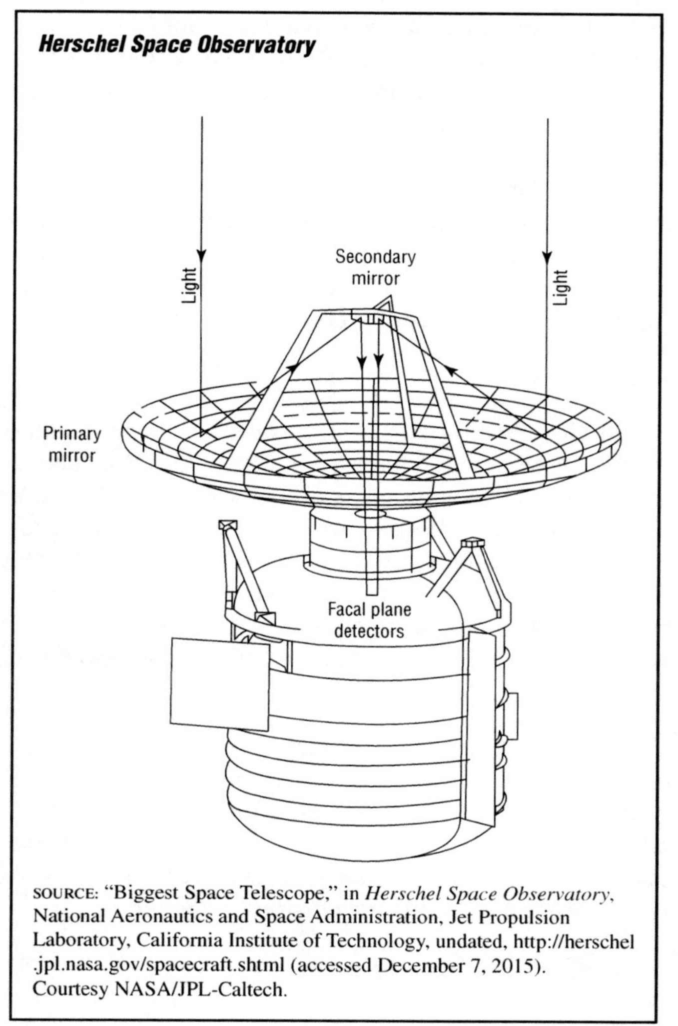

SOURCE: "Biggest Space Telescope," in *Herschel Space Observatory*, National Aeronautics and Space Administration, Jet Propulsion Laboratory, California Institute of Technology, undated, http://herschel.jpl.nasa.gov/spacecraft.shtml (accessed December 7, 2015). Courtesy NASA/JPL-Caltech.

In 2003 the ESA began participating in Double Star, the first Sino-European space mission. This joint effort with the China National Space Administration included two satellites—*Double Star 1* and *Double Star 2*—that investigated physical processes occurring in Earth's magnetic environment. The mission was completed in 2007.

ESA's operational space missions as of December 2015 are listed in Table 6.7. They include several space observatories: the *X-Ray Multi-mirror Mission-Newton* (*XMM-Newton*), the *International Gamma Ray Astrophysics Laboratory* (*INTEGRAL*), and *Gaia*, which was launched in 2013. According to the ESA (February 22, 2016, http://sci.esa.int/gaia/), *Gaia* will "chart a three-dimensional map of our Galaxy, the Milky Way, in the process revealing the composition, formation and evolution of the Galaxy." In addition, the agency's Cluster II satellites collect data about the solar wind and Earth's magnetosphere. The ESA collaborates with NASA in the *HST* project and the Solar and Heliospheric Observatory (SOHO) mission, which studies the sun's interior.

TABLE 6.7

European Space Agency operational space science missions as of December 2015

Mission name	Launch date	Mission description
XMM-Newton	12/10/99	Space observatory
Integral[a]	10/17/02	Space observatory
Mars Express	6/2/03	Mars orbiter
Rosetta	3/2/04	Investigate comet 67P/ Churyumov-Gerasimenko
Herschel	5/14/09	Space observatory
Proba-2	11/2/09	Monitors plasma environment and solar activity
Gaia	12/19/13	Space observatory (astrometry)
LISA Pathfinder	12/3/15	Test key technologies for future LISA-like missions to study gravitational effects
In partnership with NASA		
Hubble Space Telescope	4/24/90	Space observatory
SOHO	12/2/95	Perform solar observations
Cassini[b]	10/15/97	Orbiting Saturn
Cluster II	7/16/00 and 8/9/00	Four spacecraft Investigate solar wind and earth's magnetosphere

LISA = Laser Interferometer Space Antenna.
Proba = Project for OnBoard Autonomy.
SOHO = Solar and Heliospheric Observatory.
XMM = X-ray Multi-Mirror.
[a]Collaboration with Russia and USA.
[b]Collaboration with Italy.

SOURCE: Created by Kim Masters Evans for Gale, © 2016

The ESA has led missions devoted to investigating planets and other bodies in the solar system. Past, present, and future ESA missions to Mars and the far planets are described in Chapters 7 and 8, respectively. In 2003 the ESA launched Small Missions for Advanced Research in Technology (SMART), the first European mission to the moon. The *SMART-1* spacecraft assumed orbit in 2004 and began investigating the lunar surface. In 2006 the spacecraft was purposely crashed into the moon at the Lake of Excellence. *SMART-1* gathered data on the morphology and mineralogical composition of the lunar surface.

The ESA's *Venus Express* was launched in 2005 and went into orbit around Venus the following year. It collected data for eight years before running out of propellant in December 2014.

Rosetta was launched in 2004 on a 10-year journey to rendezvous with the comet 67 P/Churyumov-Gerasimenko. In November 2014 the spacecraft released the probe *Philae* that was supposed to land on the comet and attach itself using a harpoon. However, in "Rosetta and Philae: One Year since Landing on a Comet" (November 12, 2015, http://www.esa.int/Our_Activities/Space_Science/Rosetta/Rosetta_and_Philae_one_year_since_landing_on_a_comet), the ESA indicates that the harpoon did not deploy. Instead, the lander "bounced" away from its intended resting spot. *Philae* did manage to collect some data over several days before its primary battery was exhausted.

NASA (2016, http://rosetta.jpl.nasa.gov/) reports that it contributed three of *Rosetta*'s instruments.

One of the ESA's most anticipated future space science missions is called BepiColombo (named after the Italian scientist and mathematician Giuseppe "Bepi" Colombo [1920–1984]). According to the agency (2016, http://sci.esa.int/bepicolombo/), it includes two spacecraft that will map Mercury and investigate its magnetosphere. As of February 2016, the BepiColombo mission was expected to launch in 2017.

Russia

The Soviet space agency, which operated until the early 1990s, had a very active space science program. Between the 1960s and the 1980s a series of spacecraft in the Venera line investigated Venus. In 1970 *Venera 7* landed on Venus and became the first spacecraft ever to transmit data from the surface of another planet. Five years later *Venera 9* was the first spacecraft to orbit Venus. As explained in Chapter 7, the Soviet Union also launched more than a dozen spacecraft to explore Mars and its moon Phobos. Nevertheless, the vast majority of these missions failed or returned little useful data. In 1996 the Russian Federal Space Agency's (Roscosmos) *Mars 96* orbiter failed after a launch vehicle malfunctioned in Earth orbit.

In general, budget constraints have severely restricted the scope of Roscosmos's endeavors in space science. The agency cooperated with international partners to launch three solar observatories: *Complex Orbital Observations Near-Earth of Activity of the Sun I* (*CORONAS-I*; 1994), *CORONAS-F* (2001), and *CORONAS-Photon* (2009). According to the ESA, in "Permanent Mission in Russia" (July 7, 2006, http://www.esa.int/esaMI/ESA_Permanent_Mission_in_Russia/SEMQJCA6CPE_0.html), Russia was granted approximately a quarter of the research time for the ESA's *INTEGRAL* space observatory in exchange for launching the spacecraft aboard a Russian rocket in 2002. The Russian-built Lunar Exploration Neutron Detector instrument was included on NASA's *Lunar Reconnaissance Orbiter*, which was launched in 2009. In July 2011 Russia launched *Spektr-R* into Earth orbit. The spacecraft carries a radio telescope (believed to be the largest space radio telescope ever put into space) and is part of an international project called RadioAstron. According to the Astro Space Center at the Lebedev Physical Institute of Russian Academy of Sciences (May 2008, http://www.asc.rssi.ru/radioastron/description/intro_eng.htm), the RadioAstron collaboration includes more than 20 countries.

During the early 2010s various media sources indicated that Roscosmos was developing several science projects, including the space observatory *Spektr-RG*, and missions to Earth's moon and Venus. As noted in Chapter 3, the Russian economy suffered in 2015 due to low oil prices and international sanctions. In January 2016 Roscosmos was transformed into a state-run corporation following allegations of corruption and wasted funds at the agency. In "Russia to Rewrite Space Program as Economic Crisis Bites" (Reuters.com, December 29, 2015), Dmitry Solovyov indicates that the new Roscosmos State Corporation may lose billions of dollars in expected funding because of the poor state of the nation's economy. Thus, Russia's ability to conduct future space science endeavors was in question.

China

In October 2007 the China National Space Administration used a Long March 3A rocket to launch a robotic moon orbiter into space. The orbiter was called *Chang'e 1* in honor of the Chinese moon goddess. The spacecraft entered lunar orbit several weeks later and conducted observations for 16 months. In March 2009 it was purposely crashed into the moon. In October 2010 *Chang'e 2* was launched. It orbited the moon and then was moved to the Earth-sun L2 point. (See Figure 6.1.)

In early December 2013 China launched *Chang'e 3*, a lunar lander. It set down on the moon's surface two weeks later. The spacecraft carried a rover called *Yutu*, which translates as "jade rabbit" and is named after the pet of the mythical moon goddess. The rover was designed for a three-month mission on the moon.

It takes the moon just over 28 days to make a complete spin around its axis (as opposed to 24 hours for Earth). Thus, the moon experiences continuous daylight and nightlight periods of approximately 14 days each. The temperature drops dramatically during the lunar nights. The *Chang'e 3* lander and *Yutu* rover were put into sleep (hibernation) mode prior to a lunar night that began in late December 2013. They were awakened afterward for the lunar daylight period.

However, the rover experienced mechanical problems prior to spending its second night period on the moon. Ken Kremer notes in "China's Yutu Moon Rover Starts Lunar Day 4 Awake but Ailing" (UniverseToday.com, March 16, 2014) that *Yutu* was unable to lower its mast, which contains delicate instruments, into a warmed box. Thus, the instruments were exposed to brutally cold conditions during their second and third lunar nights. Kramer quotes China's official government news agency as saying that some of the rover's instruments, including its camera, survived the cold temperatures and continued to operate. Unfortunately, the rover was unable to activate its wheels and move around. Although immovable, it continued to transmit data through March 2015, well beyond its expected three-month lifetime. Meanwhile, the *Chang'e 3* lander was reported by Chinese authorities to be operating normally. As of early 2016, it was still transmitting data.

As noted in Chapter 3, in December 2011 the Chinese government announced its latest five-year plan for space exploration. Besides human spaceflights, China plans to conduct several space science missions, including putting a second lander on Earth's moon. As of February 2016, the Chang'e 5 mission was expected to launch in 2017. The lander will collect samples and return them to Earth.

Japan

Prior to 2003 Japan's space science endeavors were carried out by three separate agencies: the Institute of Space and Astronautical Science, the National Aerospace Laboratory of Japan, and the National Space Development Agency. In July 1998 they launched *Nozomi* (Japanese for "wish" or "hope"), a spacecraft that was intended to travel to Mars. It would have been Japan's first space mission to another planet. However, the spacecraft suffered numerous technical problems and had to be put into a solar orbit.

In 2003 Japan's three space agencies merged to form the Japan Aerospace Exploration Agency (JAXA). *Hayabusa* (Japanese for "peregrine falcon") was launched that year to intercept the asteroid Itokawa. The asteroid orbits the sun between Earth and Mars and is about 2,300 feet by 1,000 feet (700 m by 305 m) in size. It is named after Hideo Itokawa (1912–1999), who is considered to be the founder of Japan's space program. The robotic explorer was designed to land on the asteroid, take a surface sample, and return to Earth by 2007. In November 2005 JAXA lost contact with the spaceship during the touchdown procedure. Although contact was regained after a few days, it was not known if *Hayabusa* was able to collect a surface sample. Thruster problems delayed *Hayabusa*'s return to Earth until June 2010, when it set down in a desolate region of the Australian Outback.

According to NASA, in "Hayabusa Asteroid Itokawa Samples" (November 16, 2015, http://curator.jsc.nasa.gov/hayabusa/index.cfm), the spacecraft's sample containers held less than a milligram of asteroid particles. It is believed that these particles were not actually collected, but rather were forced into the containers when the spacecraft impacted the asteroid surface. Nevertheless, NASA notes that "these are the first direct samples of an asteroid" and as such "have great scientific value." In December 2011 JAXA provided NASA with 10% of the samples in exchange for the latter's technical support during the mission.

In 2007 JAXA launched the lunar mission Kaguya (also called the Selenological and Engineering Explorer mission), which is named after a mythical Japanese moon princess. The mission included three satellites: the primary satellite carrying scientific instruments and two smaller communication relay satellites named *Okina* and *Ouna*. All three satellites eventually crashed into the moon after completing their primary missions. On May 21, 2010, JAXA launched *Akatsuki* (Japanese for "dawn"), which was supposed to begin orbiting Venus later that year. The spacecraft was plagued by technical problems and was not able to achieve orbit until December 2015. As of February 2016, *Akatsuki* remained operational and JAXA (http://www.jaxa.jp/projects/sat/planet_c/index_e.html) was testing the spacecraft's instruments.

Also on May 21, 2010, JAXA (http://www.jspec.jaxa.jp/e/activity/ikaros.html) launched the experimental spacecraft *Interplanetary Kite-craft Accelerated by Radiation of the Sun* (*IKAROS*) to conduct a flyby of Venus. According to JAXA (2016, http://www.jspec.jaxa.jp/e/activity/ikaros.html), the spacecraft is the world's first demonstration of solar sail technology. A solar sail consists of giant ultrathin silvery blades that unfurl after launch to reflect sunlight. The electromagnetic radiation of sunlight exerts force on the objects on which it shines. This force is fairly strong in outer space due to the absence of atmospheric friction, and it could potentially push a solar sail in much the same way that the wind pushes sailing ships on Earth's oceans. A solar sail is a novel technology that could gain far greater use in the future.

On December 3, 2014, JAXA(http://global.jaxa.jp/projects/sat/hayabusa2/orbit.html) launched *Hayabusa 2* to intercept the asteroid Ryugu, which is a NEO with the official designation of 162173 (formerly 1999 JU3). The agency expects the spacecraft to reach the asteroid in 2018 and collect samples from its surface. They will be returned to Earth near the end of 2020.

As of February 2016, JAXA's (http://www.jaxa.jp/projects/index_e.html) operational missions (besides *Akatsuki*, *Hayabusa 2*, and *IKAROS*) included Earth-observing satellites and astronomical observatories. JAXA was also participating in a number of scientific satellite projects with international partners. Future Japanese space projects include missions to Earth's moon, Venus, and Mercury.

INDIA. The Indian Space Research Organisation (ISRO) is an agency of the Indian government. *Chandrayaan 1* was India's first lunar explorer. It was launched in October 2008, went into lunar orbit, and sent a probe to the moon's surface. In late August 2009 controllers lost contact with the spacecraft and the mission was terminated. Among the 11 scientific instruments that were placed aboard the spacecraft was NASA's Moon Mineralogy Mapper (M3). Ker Than reports in "India's First Moon Probe Lost, but Data May Yield Finds" (NationalGeographic.com, August 31, 2009) that M3 had finished a "cursory global survey" of the moon before contact was lost. In September 2009 NASA scientists announced that

M3 had found evidence of water on the moon's surface. Scientists have long speculated that the lunar surface contains water molecules, but this is the first evidence of their existence.

In 2013 ISRO launched its first mission to Mars, which is described in Chapter 7. As of February 2016, the agency (http://www.isro.gov.in/chandrayaan-2) was developing its Chandrayaan 2 mission, which will include an orbiter, lander, and rover that will investigate Earth's moon. Although an official launch date had not been announced, the spacecraft was expected to launch in 2017 or 2018.

SPACE SCIENCE ASSETS THAT ARE NOT IN OUTER SPACE

Many assets that are devoted to space science are not actually in outer space. These include ground-based telescopes and assets that fly aboard balloons and airplanes.

Ground-Based Telescopes

Glass lenses have been used to make eyeglasses for several centuries. When telescopes were first developed in Europe around the beginning of the 1600s, they included a series of lenses within a long slender tube that magnified distant objects. In 1610 the Italian scientist Galileo (1564–1642) popularized the telescope when he published *Sidereus Nuncius* (*Starry Messenger*). The book described Galileo's astronomical observations, including the discovery of four moons around Jupiter.

OPTICAL TELESCOPES. The earliest telescopes were optical telescopes. Like the human eye, they can discern only visible light. In 2007 Spanish astronomers unveiled the largest terrestrial optical telescope to date: Gran Telescopio Canarias. The observatory is located on the Canary Islands off the coast of Morocco in northern Africa. That same year the Large Binocular Telescope (LBT) also became fully operational. The LBT features twin mirrors and is located near Safford, Arizona, as part of the Mount Graham International Observatory. The LBT is a collaboration between U.S., Italian, and German astronomy organizations and educational institutions. Both telescopes are primarily used to search for exoplanets, brown dwarfs, nearby stars, and other celestial phenomena.

RADIO TELESCOPES. During the 1930s the American engineer Karl Jansky (1905–1950) built an instrument that was capable of detecting radio waves. His radio telescope picked up waves that were generated by thunderstorms and by an unknown source from outer space. Over time, scientists constructed larger and more powerful radio telescopes to pick up the radio waves that arrive on Earth from outer space.

During the 1980s dozens of radio telescopes in New Mexico were linked together to enhance their capability. This was called the Very Large Array. In 1993 the National Science Foundation created a more powerful system called the Very Long Baseline Array. It includes 10, 82-foot (25-m) radio antennas that are located across the United States, from Hawaii in the west to the Virgin Islands in the east. The telescopes are connected by a computer network and provide very high-quality radio wave images.

The largest single-dish radio telescope on Earth is the Arecibo Observatory (https://www.naic.edu/) in Puerto Rico. The observatory was built during the 1960s and as of February 2016 was operated by a multinational team of organizations led by SRI International for the National Science Foundation. On November 16, 1974, scientists used the observatory to send a radio message out into the galaxy. The message was coded in binary, meaning that a series of zeroes and ones were transmitted by shifting frequencies. The total broadcast took less than three minutes. It was a pictorial message. The message showed the numbers 1 through 10, the atomic numbers of five chemical elements, the chemical formula of deoxyribonucleic acid, information about the human form and Earth's population, a stick-figure person, the location of Earth in relation to the sun and the other planets, a representation of the Arecibo telescope, and information about its size.

INFRARED OBSERVATORIES. During the 1800s scientists first developed detectors for infrared radiation that was coming from outer space. Because the water vapor in Earth's atmosphere blocks out most infrared radiation, these detectors were placed atop mountains, where the air is thinner. Later, they were sent up in high-altitude balloons and even airplanes.

Mauna Kea is an extinct volcano in Hawaii that rises 14,000 feet (4,270 m) above the Pacific Ocean. A number of observatories have been erected atop Mauna Kea, because the thin atmosphere allows detection of near-infrared radiation. The Keck Observatory is operated by NASA in conjunction with the California Institute of Technology and the University of California. The Subaru Telescope is a project of the National Institutes of Natural Science of Japan. The Gemini Observatory (which includes optical and infrared capabilities) is an international collaboration between the United States, the United Kingdom, Canada, Chile, Australia, Argentina, and Brazil. Other universities and institutions from around the world also operate telescopes atop Mauna Kea.

In 2014 construction began on yet another observatory at the site, the Thirty-Meter Telescope (TMT; http://www.tmt.org/science-case). The powerful telescope will operate at wavelengths from the ultraviolet to the mid-infrared spectrum. However, the project has attracted fierce opposition from native Hawaiians who consider Mauna Kea to be sacred ground. Nola Taylor Redd notes in "Third Observatory to Close on Sacred Hawaiian Mountain" (Space.com, October 29, 2015) that the state's governor asked researchers to "shut down 25 percent of the telescopes on the mountain" before the TMT begins operations. By October 2015, four of the 13 observatories on Mauna Kea had agreed to close down. However, in December 2015 the Hawaii Supreme Court invalidated the construction permit for the TMT. As a result, its status was uncertain as of February 2016.

The Atacama Desert in Chile is another popular outpost for terrestrial telescopes. This arid region lies on a plateau west of the Andes Mountains. According to Priit J. Vesilind, in "The Driest Place on Earth" (NationalGeographic.com, August 2003), the desert is the driest place on Earth. The dry air and high altitude provide ideal conditions for space viewing. In "Why Chile Is an Astronomer's Paradise" (BBC.co.uk, July 24, 2011), Gideon Long indicates that some of the world's most powerful terrestrial telescopes are located in the Atacama Desert, including the Paranal Observatory, which is operated by the European Southern Observatory.

NEW TECHNOLOGIES. As noted earlier in this chapter, the shielding effects of Earth's atmosphere limit the ability of ground-based telescopes to accurately discern the electromagnetic spectrum. Scientists rely on technologies called "adaptive optics" to compensate for these effects and make the images clearer. Elizabeth Howell notes in "'The New Cool': How These Sharp Space Pictures Were Snapped from a Ground Telescope" (UniverseToday.com, July 3, 2013) that new advanced optics technologies are enabling ground-based telescopes to capture clearer images than ever before. Such innovations are driving great interest in the development and construction of new ground-based telescopes. For example, the Giant Magellan Telescope (http://www.gmto.org/overview/) is to be constructed in Chile by 2021 and will provide near-infrared and visible light capabilities.

CHAPTER 7
MARS

Mars moves through our skies in its stately dance, distant and enigmatic, a world awaiting exploration.

—Carl Sagan, "Mars: A New World to Explore" (*National Geographic*, December 1967)

Mars has been a mystery to humans for thousands of years. Although much is known about it, there is still much more to learn. Mars is the fourth planet from the sun and the planet most like Earth in the solar system. It is named after the mythical god of war whom the Greeks called Ares and the Romans called Mars. Mars is also known as the Red Planet because it looks reddish from Earth. Mars is a dusty, cold world. Rays of ultraviolet radiation beat down on its surface continuously, and its atmosphere is nearly all carbon dioxide.

People on Earth have long been fascinated with the idea of life on Mars. Ancient people could see Mars as a pale reddish light in the nighttime sky. They believed it was stained with the blood of fallen warriors. Once telescopes were invented, people had a better view of the planet, but many still thought it was inhabited. Patterns of straight lines could be seen on its surface. To some, these were evidence of water canals dug into the ground by Martians. This notion lingered for decades in the public imagination.

At the dawn of the space age, robotic probes were sent to Mars to settle the question once and for all. They found a frozen wasteland of fine powdery dust. Neither canals nor Martians could be located. There was some water vapor in the atmosphere and some frozen water at the planet's poles. The frozen water intrigued scientists, because where there is water, there is the potential for Earth-like life.

As of February 2016, robotic explorers continued to uncover the mysteries of Mars, and there were long-range plans for human missions to the Red Planet.

EARLY TELESCOPIC VIEWS OF MARS

The Italian scientist Galileo (1564–1642) was probably the first to see Mars through a telescope. He noticed that sometimes it appeared larger than at other times. He believed that its distance from Earth was changing over time. During the 17th century the German astronomer Johannes Kepler (1571–1630) studied Mars's movement for years. His observations helped him to develop the laws of planetary motion for which he would become famous.

As telescopes improved, astronomers reported seeing dark and light patches on Mars that also varied in size over time. Some people thought these were patches of vegetation changing in response to the changing seasons. Others believed they represented contrasting areas of land and sea. Over the centuries, a belief grew that Mars was populated. This notion gained strength during the late 1800s thanks to the Italian astronomer Giovanni Schiaparelli (1835–1910). He created some of the first maps of the planet and assigned names to prominent features. His naming system relied on place names taken from the Bible and ancient mythology. Schiaparelli said he saw straight lines on the Martian surface and called them *canali*, which translates as "channels" or "canals." Many people interpreted the word to mean that there were artificial canals on Mars.

In August 1877 the American astronomer Asaph Hall (1829–1907) discovered that Mars has two moons. He named them Phobos (meaning "fear") and Deimos (meaning "flight" or "panic"). These are two characters mentioned in ancient Greek mythology as being servants to the god Mars. A few decades later Percival Lowell (1855–1916), an American mathematician and amateur astronomer, greatly popularized the idea that Mars was inhabited by intelligent beings. In 1894 he founded the Lowell Observatory in Flagstaff, Arizona, and studied the Red Planet for years. He believed he saw artificial canals on Mars and wrote several books and magazine articles about his theories.

Lowell's ideas were not shared by most astronomers of the time. For example, in 1907 the British naturalist Alfred Russel Wallace (1823–1913) wrote the book *Is Mars Habitable?*, which examined Lowell's claims one by one and attacked them with scientific data and reasoning. The book is considered to be a pioneering work in the field of exobiology (the investigation of possible life beyond Earth). Wallace argued that Mars was a frozen desert and ended the book with the definitive statement: "Mars, therefore, is not only uninhabited by intelligent beings such as Mr. Lowell postulates, but is absolutely UNINHABITABLE."

SCIENTIFIC FACTS ABOUT MARS

Table 7.1 lists facts about Mars and compares the planet to Earth. Mars is a small planet. Its diameter (twice the radius) is about half that of Earth. The force of gravity is much weaker on Mars than it is on Earth. An astronaut standing on Mars would feel only 38% as much gravity as on Earth. Mars's surface temperature can range between −225 and 70 degrees Fahrenheit (−153 and 20 degrees C). Like Earth, Mars has different seasons throughout its orbit because it is tilted.

Marking the passage of time on a planet other than Earth is complicated. Timekeeping measurements are based on Earth's movements in relation to other celestial bodies. A "day" on Earth is the amount of time it takes for Earth to rotate once about its axis; that is, when the sun reappears in the same position in the sky. Astronomers call this a "solar day." In addition, time divisions are based on long-standing, but arbitrary, decisions. For example, a day is divided into 24 hours. An hour is divided into 60 minutes, and so forth.

It can be argued that using the same units and divisions to mark time passage on other planets is illogical because each planet moves differently in relation to other celestial bodies than does Earth. Timekeeping on Mars has posed a challenge for scientists since the first spacecraft traveled there. In "Technical Notes on Mars Solar Time as Adopted by the Mars24 Sunclock" (June 30, 2015, http://www.giss.nasa.gov/tools/mars24/help/notes.html), Michael Allison and Robert Schmunk note that "each Mars lander mission project has adopted a different reference for its solar timekeeping and mission clock."

To keep things simple, Martin time can be defined using Earthly time terms, such as hours and minutes. That is the convention used in this chapter. Allison and Schmunk indicate that a Martian day (a solar day or "sol") lasts 24 hours, 39 minutes, and 35 seconds. Thus, a sol is slighter longer than a solar day on Earth.

Martian Geology and Atmosphere

Mars is called a terrestrial planet because it is composed of rocky material, like Mercury, Venus, and Earth.

TABLE 7.1

Facts about Mars

Discovery
Date of discovery: Unknown
Discovered by: Known by the ancients

Orbit size around sun
Metric: 227,943,824 km
English: 141,637,725 miles
Scientific notation: 2.2794382 x 10^8 km
Astronomical units: 1.523662 A.U.
By comparison: 1.524 x Earth

Mean orbit velocity
Metric: 86,677 km/h
English: 53,858 mph
Scientific notation: 2.4077 x 10^4 m/s
By comparison: 0.808 x Earth

Orbit eccentricity
0.0933941
By comparison: 5.589 x Earth

Equatorial inclination
25.2

Equatorial radius
Metric: 3,389.5 km
English: 2,106.1 miles
Scientific notation: 3.3895 x 10^3 km
By comparison: 0.5320 x Earth

Equatorial circumference
Metric: 21,296.9 km
English: 13,233.3 miles
Scientific notation: 2.12969 x 10^4 km

Volume
Metric: 163,115,609,799 km^3
English: 39,133,515,914 $miles^3$
Scientific notation: 1.63116 X 10^{11} km^3
By comparison: 0.151 x Earth

Mass
Metric: 641,693,000,000,000,000,000,000 kg
Scientific notation: 6.4169 x 10^{23} kg
By comparison: 0.107 x Earth

Density
Metric: 3.934 g/cm^3
By comparison: 0.714 x Earth

Surface area
Metric: 144,371,391 km^2
English: 55,742,106 square miles
Scientific notation: 1.4437 x 10^8 km^2
By comparison: 0.283 x Earth

Surface gravity
Metric: 3.71 m/s^2
English: 12.2 ft/s^2
By comparison: If you weigh 100 pounds on Earth, you would weigh 38 pounds on Mars.

Escape velocity
Metric: 18,108 km/h
English: 11,252 mph
Scientific notation: 5.030 x 10^3 m/s
By comparison: Escape velocity of Earth is 25,030 mph.

Sidereal rotation period
1.026 Earth days
24.623 hours
By comparison: Earth's rotation period is 23.934 hours.

Surface temperature
Metric: −153 to +20 °C
English: −225 to +70 °F
Scientific notation: 120 to 293 K

Atmospheric constituents
Carbon dioxide, nitrogen, argon
Scientific notation: CO_2, N_2, Ar
By comparison: CO_2 is responsible for the Greenhouse Effect and is used for carbonation in beverages. N_2 is 80% of Earth's air and is a crucial element in DNA. Ar is used to make blue neon light bulbs.

Notes: A.U. = astronomical units. cm = centimeters. ft = feet. g = grams. h = hour. kg = kilograms. km = kilometers. m = meters. mi = miles. mph = miles per hour. s = second. °C = degrees Celsius. °F = degrees Fahrenheit. K = degrees Kelvin.

SOURCE: "Mars: By the Numbers," in *Solar System Exploration: Planets*, National Aeronautics and Space Administration, 2015, http://solarsystem.nasa.gov/planets/mars/facts (accessed December 8, 2015)

Mars has some of the same geological features as Earth, including volcanoes, valleys, ridges, plains, and canyons.

Most Martian features have two-word names. One of the words is a geological term and is usually from Latin or Greek, for example, *mons* for "mountain," *planitia* for "low plain," *planum* for "high plain," and *vallis* for "valley." The other word comes from the classical naming system begun by Schiaparelli during the 1800s or from later astronomers, including the Greek astronomer Eugène Michael Antoniadi (1870–1944). Beginning in 1919 the International Astronomical Union (IAU) became the official designator of names for celestial objects and the features on them. Only the IAU has this authority.

There are two particularly prominent features on Mars. The first is the volcano Olympus Mons that is about 16 miles (26 km) high. This is three times higher than Mount Everest on Earth. In English Olympus Mons means "Mount Olympus." This was the home of the gods in ancient Greek mythology. The other notable geological feature on Mars is the canyon Valles Marineris (Mariner Valleys). This enormous canyon is about 1,900 miles (3,000 km) long and up to 370 miles (600 km) wide and 5 miles (8 km) deep in places. It is named after *Mariner 9*, which photographed it during the early 1970s.

The surface of Mars is covered with a fine powdery dust that has a pale reddish tint. This is due to the presence of oxidized iron minerals (like rust) on the planet's surface. The Martian atmosphere is thin and contains more than 95% carbon dioxide. There is a tiny amount of oxygen, but not enough for humans to breathe. It is windy on Mars. Strong winds sometimes engulf the planet in dust storms that turn the atmosphere a hazy yellowish-brown color. The wind also blows clouds around the sky.

The Martian poles are covered by solid carbon dioxide (dry ice) that is layered with dust and water ice. These polar caps change in size as the seasons change. Sometimes during the summer the uppermost dry ice evaporates away, only to reform when the weather turns cold again.

Martian Moons

The two Martian moons Phobos and Deimos are not round spheres like Earth's moon. They are shaped like lopsided potatoes. Phobos is 17 miles (27 km) long and 12 miles (19 km) wide and is approximately 5,800 miles (9,300 km) from Mars; Deimos is 10 miles (16 km) long and 6 miles (10 km) wide and is nearly 15,000 miles (24,100 km) from Mars.

The Martian moons are small compared with other moons in the solar system. Many scientists believe Phobos and Deimos are actually asteroids that wandered too close to Mars and were captured by its gravity. There is a large asteroid belt located between the orbits of Mars and Jupiter. This could be where Phobos and Deimos originated.

Mars in Orbit and Opposition

Earth has a nearly circular orbital path around the sun, whereas Mars's path is slightly more elliptical (oval shaped). (See Figure 7.1.) Earth takes 365 days to complete an orbit around the Sun. By contrast, a Martian orbit lasts nearly 687 days. Thus, Mars and Earth are constantly changing position in relation to each other. At their most distant point the two planets are 233 million miles (375 million km) apart. At their closest point they are less than 35 million miles (56 million km) apart. This explains why in some years Mars looks closer to Earth than in other years.

FIGURE 7.1

Orbits of Earth and Mars around the Sun

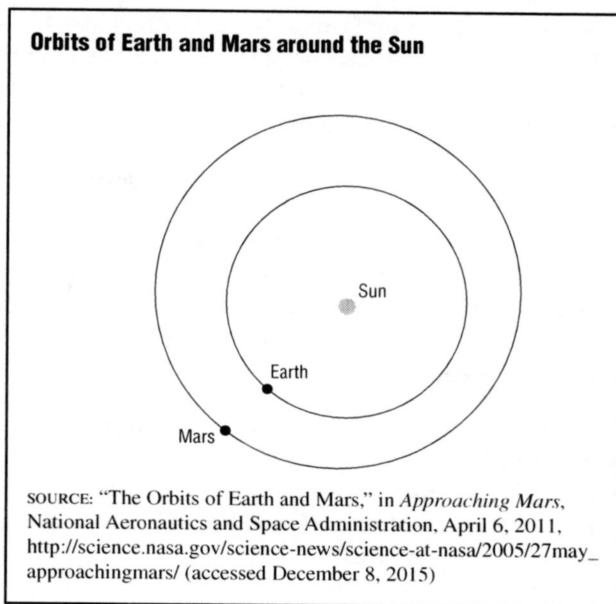

SOURCE: "The Orbits of Earth and Mars," in *Approaching Mars*, National Aeronautics and Space Administration, April 6, 2011, http://science.nasa.gov/science-news/science-at-nasa/2005/27may_approachingmars/ (accessed December 8, 2015)

About every 26 months the sun, Earth, and Mars line up in a row with Earth lying directly in the middle. This configuration is called Mars in opposition. It means that Mars is closer to Earth than usual and is easier to observe. The most recent opposition of Mars occurred in April 2014. The next two oppositions will be in May 2016 and July 2018. Oppositions are good times to send spacecraft to Mars, because the planet is closer to Earth than usual.

Perihelion is the innermost point of an orbit around the sun. Mars is closest to Earth when Mars is at perihelion and in Earth opposition at the same time. This occurs approximately every 15 years. Most of the historic discoveries about Mars occurred when the planet was in opposition, particularly perihelic opposition. The most recent perihelic opposition occurred in August 2003, when Mars was only 34.7 million miles (55.8 million km) from Earth. The next perihelic opposition will occur in July 2018.

MISSIONS TO MARS

After the moon the planet Mars was the destination of choice during the early days of space travel. The Soviet Union was particularly eager to reach the Red Planet before the United States.

Many Failures

Historically, spacecraft have had a difficult time making it to Mars in working order and staying that way. Many missions intended for Mars have failed for one reason or another. Some were plagued by launch problems, whereas others suffered malfunctions during flight, descent, or landing.

Mars missions undertaken during the 1960s by the Soviet Union were particularly trouble-prone. All seven of them failed. Although the next decade showed some improvement, little usable data were obtained from the spacecraft that reached their destination. The one attempt to reach Mars by the Russian Space Agency (of the former Soviet Union, or the Russian Federation), in 1996, failed when the spacecraft was unable to leave Earth orbit.

The Mars missions of the National Aeronautics and Space Administration (NASA) fared better. Most of them conducted during the 1960s and 1970s achieved their objectives. Then there was a long lull in NASA's Mars Exploration Program.

During the 1990s NASA launched five separate spacecraft to Mars: *Mars Observer* (1992), *Mars Global Surveyor* (1996), *Mars Pathfinder* (1996), *Mars Climate Orbiter* (1998), and *Mars Polar Lander* (1999). Three of them were lost on arrival.

NASA lost contact with the *Mars Observer* just before it was to go into orbit around Mars. It is believed that some kind of fuel explosion destroyed the spacecraft as it began its maneuvering sequence. The *Mars Observer* carried a highly sophisticated gamma-ray spectrometer that was designed to map the Martian surface composition from orbit. The failure of the mission resulted in a loss estimated at $1 billion. This was by far the most expensive of NASA's failed Mars missions.

In September 1999 the *Mars Climate Orbiter* was more than 60 miles (97 km) off course when it ran into the Martian atmosphere and was destroyed. The loss of the spacecraft was particularly embarrassing for NASA, because it was due to human error. An investigation revealed that flight controllers had mistakenly converted between metric and English units. This caused erroneous steering commands to be sent to the spacecraft. Outside investigators complained that the problem was larger than a few mathematical errors. They blamed overconfidence and poor mission oversight by NASA management.

The agency's embarrassment deepened a few months later when the *Mars Polar Lander* was lost. A software problem caused the spacecraft to think it had touched down on the surface although it had not. The computer apparently shut down the engines during descent and let the spacecraft plummet at high speed into the ground, where it was destroyed.

The Russian and Japanese space agencies suffered their own failures during the 1990s with spacecraft bound for Mars. In 1996 Russia's *Mars 96* spacecraft was lost when the launch vehicle failed. Likewise, in 1998 a Japanese spacecraft named *Nozomi* suffered radiation damage during its flight and never made it into orbit around Mars. Instead, it was put into a heliocentric (sun-centered) orbit.

Since the start of the 21st century NASA's spacecraft to Mars have been very successful, with no failures as of February 2016. India's space agency also achieved success by putting an orbiter around Mars. Other agencies have not been as fortunate. In 2003 the European Space Agency (ESA) launched the *Mars Express*, an orbiter that carried the lander *Beagle 2*. Although the *Mars Express* successfully made it into Martian orbit, the lander was lost during its descent. In November 2011 a Chinese-Russian collaboration called *Yinghuo-1/Phobos-Grunt* was launched. The combined orbiter/lander failed to leave Earth orbit and was torn apart when it reentered Earth's atmosphere in January 2012.

THE MARINER PROGRAM

The Mariner Program included a series of NASA spacecraft launched between 1962 and 1973 to explore the inner solar system (Mercury, Venus, and Mars). These were relatively low-cost missions conducted with small spacecraft. Six of the Mariner spacecraft were designed for Mars missions. Two of these spacecraft failed. In 1964 *Mariner 3* malfunctioned after takeoff and never made it to Mars. In 1971 *Mariner 8* failed during launch. This left four successful Mariner Mars spacecraft: *Mariner 4*, *Mariner 6*, *Mariner 7*, and *Mariner 9*.

In July 1965 *Mariner 4* achieved the first successful flyby of Mars. A planetary flyby mission is one in which a spacecraft is put on a trajectory that takes it near enough to a planet for detailed observation, but not close enough to be pulled in by the planet's gravity. During its flyby, *Mariner 4* took 21 photos, the first close-ups ever obtained of Mars. They showed a world pockmarked with craters, probably from meteor strikes.

In 1969 *Mariner 6* and *Mariner 7* conducted a dual mission to Mars. Both spacecraft flew by the planet and together sent back 201 photos. These photos revealed that the features once thought to be canals were not canals after all. Instead, it appears that a number of small features or shadows on Mars only looked like they were aligned when viewed through Earth-based telescopes. The illusion was perpetuated by a human tendency to see order in a random collection of shapes. The mystery of the canals had finally been solved.

Mariner 9 turned out to be the most fruitful of the Mariner missions. In November 1971 the spacecraft went into orbit around Mars after a five-and-a-half-month flight from Earth. It was the first artificial satellite ever to be placed in orbit around the planet. It remained in orbit for nearly a year and returned more than 7,300 photos of the planet's surface. For the first time, scientists got a good look at Mars's surface features, such as volcanoes and valleys. *Mariner 9* showed geological features that looked like dry flood channels. It also captured the first close-up photos of Phobos and Deimos.

Scientists learned from Mariner data that Mars has virtually no magnetic field and is bombarded with ultraviolet radiation. Earth's extensive magnetic field (or magnetosphere) helps protect the planet from dangerous electromagnetic radiation traveling through space. Scientists know that lack of such protection on Mars makes it exceedingly difficult for life to exist on the planet.

THE VIKING MISSION

In August and September 1975 NASA launched a pair of spacecraft to Mars under its Viking mission. Each contained an identical orbiter and lander. It took them nearly a year to reach their destination. On July 20, 1976, the *Viking 1* lander set down on the western slope of Chryse Planitia (Plains of Gold). On September 3, 1976, the *Viking 2* lander set down in Utopia Planitia (Plains of Utopia). The general locations of the landing sites are shown in Figure 7.2.

In "Viking Mission to Mars" (September 29, 2015, http://nssdc.gsfc.nasa.gov/planetary/viking.html), Ed Grayzeck of NASA describes the findings from the mission. The orbiters imaged the entire planetary surface. The landers also took images and collected and analyzed surface samples. They measured seismic phenomena and provided NASA with constant weather reports. Scientists learned that Mars has volcanoes, enormous canyons, craters, lava plains, and surface features that appear to have been formed from wind-blown sand or from the action of surface water. The images revealed seasonal dust storms and movement of atmospheric gases between the polar caps. The landing sites were found to contain iron-rich clays; however, no signs of life were detected.

The Viking images also revealed that Mars has a light yellowish-brown atmosphere due to the presence of airborne dust. In other words, the Red Planet is actually more the color of butterscotch.

The *Viking 2* orbiter continued functioning until July 1978, and its lander ended communications in April 1980. The *Viking 1* orbiter was powered down in August 1980, and its lander continued to make transmissions to Earth until November 1982.

THE MARS GLOBAL SURVEYOR MISSION

More than 20 years passed between the launch of the highly productive Viking mission and another successful

FIGURE 7.2

Locations of landing sites for past Mars rovers and landers

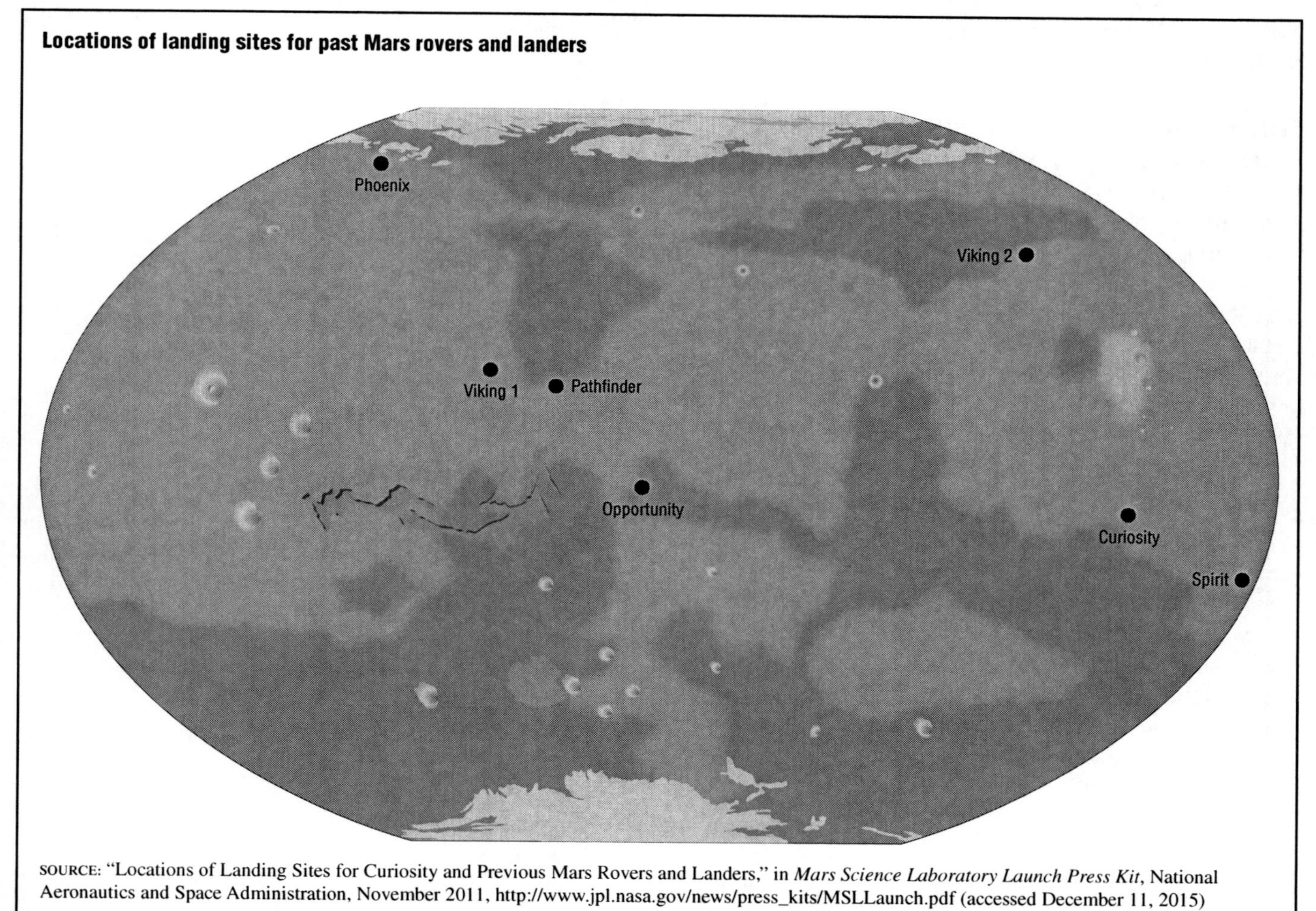

SOURCE: "Locations of Landing Sites for Curiosity and Previous Mars Rovers and Landers," in *Mars Science Laboratory Launch Press Kit*, National Aeronautics and Space Administration, November 2011, http://www.jpl.nasa.gov/news/press_kits/MSLLaunch.pdf (accessed December 11, 2015)

mission to Mars, which was the *Mars Global Surveyor* (*MGS*) launched in 1996. The spacecraft arrived near the planet 10 months later and was slowly put into a Martian orbit.

In 1999 the spacecraft began its mapping mission. NASA summarizes the mission's major findings in "Mars Global Surveyor: Important Discoveries" (January 26, 2010, http://mars.jpl.nasa.gov/mgs/mission/MGSdiscoveries.html). *MGS* images showed previously unknown canyons within the planet's polar ice caps and new craters. Scientists also learned that Mars might once have had a global magnetic field around it, like the one that exists around Earth. Images revealed gullies that appeared to have had water flowing through them—some quite recently. A "fan-shaped area of interweaving, curved ridges" was discovered that might be an ancient river delta. In addition, evidence of the mineral hematite was found. Hematite is a fine-grained mineral known to form in wet conditions.

In November 2006 the *MGS* suddenly went silent. NASA notes in "Mars Global Surveyor (MGS) Spacecraft Loss of Contact" (April 13, 2007, http://mars.jpl.nasa.gov/mgs/mission/mgs_white_paper_20070413.pdf) that a computer error caused the spacecraft to orient itself toward the sun incorrectly. It overheated and lost all battery power. The *MGS* had operated for 9 years and 52 days, the longest Mars mission to that time.

THE MARS PATHFINDER MISSION

On December 4, 1996, NASA launched the spacecraft *Mars Pathfinder*, which traveled for seven months before it entered the gravitational influence of Mars. It parachuted to the surface and deployed air bags to cushion its landing. It landed on July 4, 1997, in a rocky flood plain called Ares Vallis (Valley of Ares). (See Figure 7.2.) Once on the surface of Mars the spacecraft was referred to as the lander. It was renamed the Carl Sagan Memorial Station, in memory of the American astronomer Carl Sagan (1934–1996), who died while the spacecraft was en route to Mars.

The lander unfolded three hinged solar panels onto the ground. (See Figure 7.3.) It then released a small six-wheeled rover named *Sojourner*. (See Figure 7.4.) The name resulted from a NASA contest in which schoolchildren proposed names of historical heroines for the mission. The winning entry suggested Sojourner Truth (c. 1797–1883), an African American woman who crusaded for human rights during the 1800s. According to NASA, in "Rover Sojourner: Mission Overview" (2016, http://marsprogram.jpl.nasa.gov/MPF/rover/mission.html), the tiny rover was 24.5 inches (62.3 cm) long, 18.7 inches (47.5 cm) wide, and 10.9 inches (27.7 cm) high.

For two and a half months *Sojourner* collected data about Martian soil, radiation levels, and rocks. Meanwhile, the lander collected images and relayed data back to Earth. It also measured the amount of dust and water vapor in the atmosphere and collected dust particles for analysis.

On September 27, 1997, the lander relayed its last data transmission back to Earth. In "Mars Pathfinder Winds down after Phenomenal Mission" (November 4, 1997, http://marsprogram.jpl.nasa.gov/MPF/mpf-pressrel.html), NASA reports that the lander and *Sojourner* greatly exceeded their expected lifetimes of 30 days and 7 days, respectively, before NASA lost contact with them. They returned more than 16,000 images during the mission. In addition, scientists learned a great deal about the Martian atmosphere and weather. They found that Martian dust contains magnetic particles and that the planet experiences frequent dust devils due to sudden wind gusts. One of the most exciting discoveries was the presence of rounded pebbles and depressions in rocks that likely formed because of running water on the surface of the planet sometime in the past.

THE 2001 MARS ODYSSEY MISSION

On April 7, 2001, the *Mars Odyssey Orbiter* was launched by NASA. It reached Mars six months later and was slowly put into orbit. In February 2002 it began mapping the planet's surface. A schematic of the spacecraft is shown in Figure 7.5. It includes three scientific instruments: a thermal imaging system, a gamma-ray spectrometer, and the Mars Radiation Environment Experiment.

The thermal imaging system collects surface images in the infrared portion of the electromagnetic spectrum. (See Figure 6.8 in Chapter 6.) Scientists use *Odyssey*'s images to identify and map minerals in the surface soils and rocks. The spacecraft's gamma-ray spectrometer can detect the presence of various chemical elements on the planet's surface. This is particularly useful for finding water ice that is buried beneath the surface and for detecting salty minerals. The Mars Radiation Environment Experiment also collects radiation data that will be useful for planning any future Mars expeditions by humans.

In December 2012 the orbiter achieved a milestone when it became the longest-serving Mars mission in history. As of February 2016, NASA (http://marsprogram.jpl.nasa.gov/odyssey/) reported that the spacecraft was still operational and functioning well.

THE PERIHELIC OPPOSITION OF 2003

Scientists knew that 2003 was going to be a good year to go to Mars, because Mars would be in perihelic opposition to Earth. This meant that less fuel and flight time than usual would be required to send a spacecraft from Earth to Mars. During the late 1990s the Japan Aerospace Exploration Agency (JAXA), the ESA, and NASA began planning Mars missions to coincide with

FIGURE 7.3

Mars Pathfinder

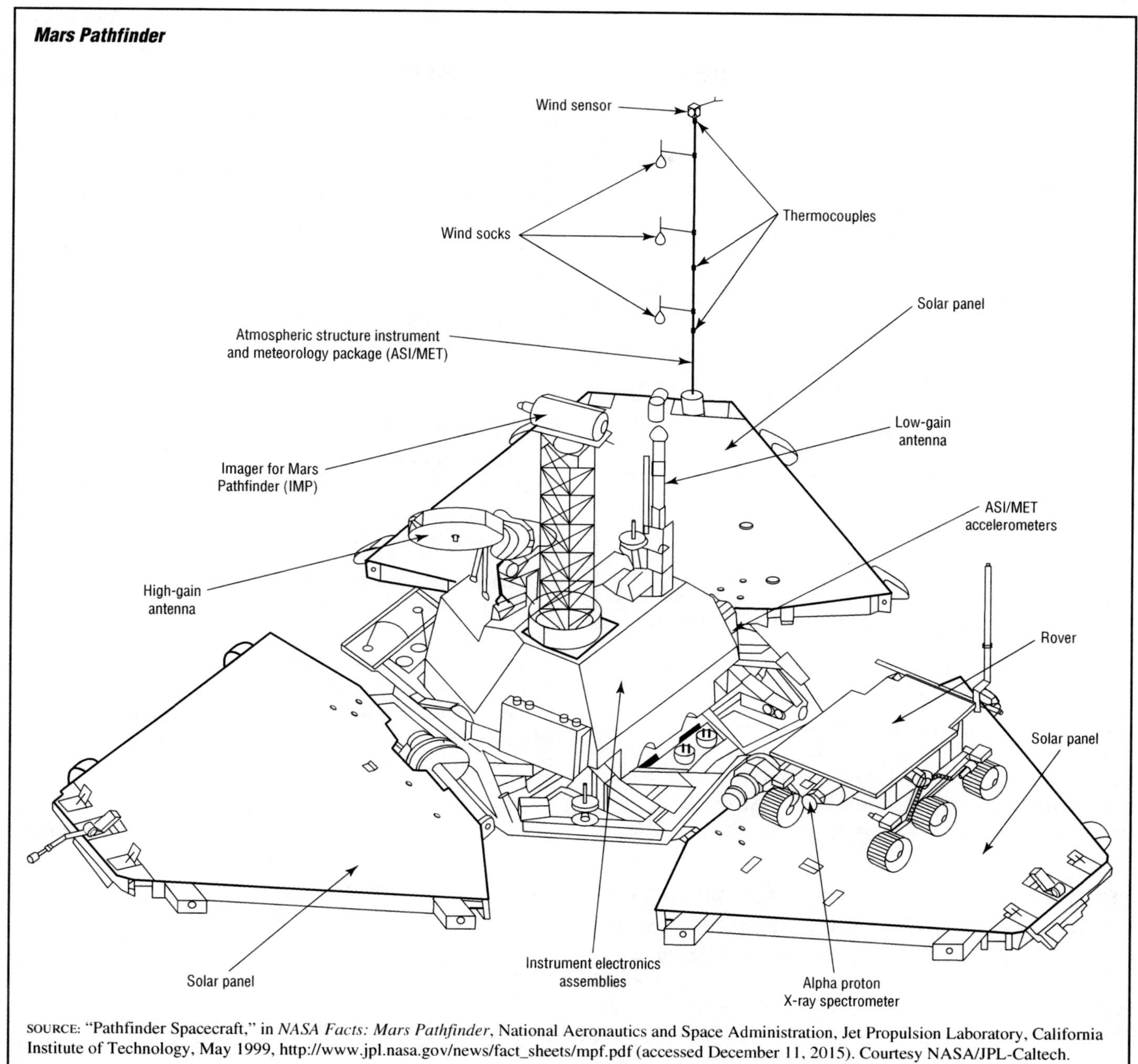

SOURCE: "Pathfinder Spacecraft," in *NASA Facts: Mars Pathfinder*, National Aeronautics and Space Administration, Jet Propulsion Laboratory, California Institute of Technology, May 1999, http://www.jpl.nasa.gov/news/fact_sheets/mpf.pdf (accessed December 11, 2015). Courtesy NASA/JPL-Caltech.

the perihelic opposition of 2003. As noted earlier, a JAXA orbiter named *Nozomi* failed to reach Mars.

THE MARS EXPRESS MISSION

The Mars Express mission is the first mission to Mars by the ESA. In June 2003 the *Mars Express Orbiter* was launched from the Baikonur Cosmodrome in Kazakhstan. The spacecraft included a lander named *Beagle 2*. The name was chosen in honor of the ship on which the British naturalist Charles Darwin (1809–1882) traveled during the 1830s while exploring South America and the Pacific region.

In late November 2003 the *Mars Express Orbiter* reached the planet's vicinity and prepared to go into orbit. On December 19, 2003, *Beagle 2* was released from the orbiter. Six days later the lander entered the Martian atmosphere on its way to a landing site at Isidis Planitia (Plains of Isis). The ESA lost contact with *Beagle 2* as it descended toward the planet. Repeated attempts to reestablish contact were made over the next few months, but they were unsuccessful.

Once the *Mars Express Orbiter* reached its orbit, it began collecting planetary data. As of February 2016, the ESA (http://www.esa.int/esaMI/Mars_Express/index.html)

FIGURE 7.4

***Sojourner* rover**

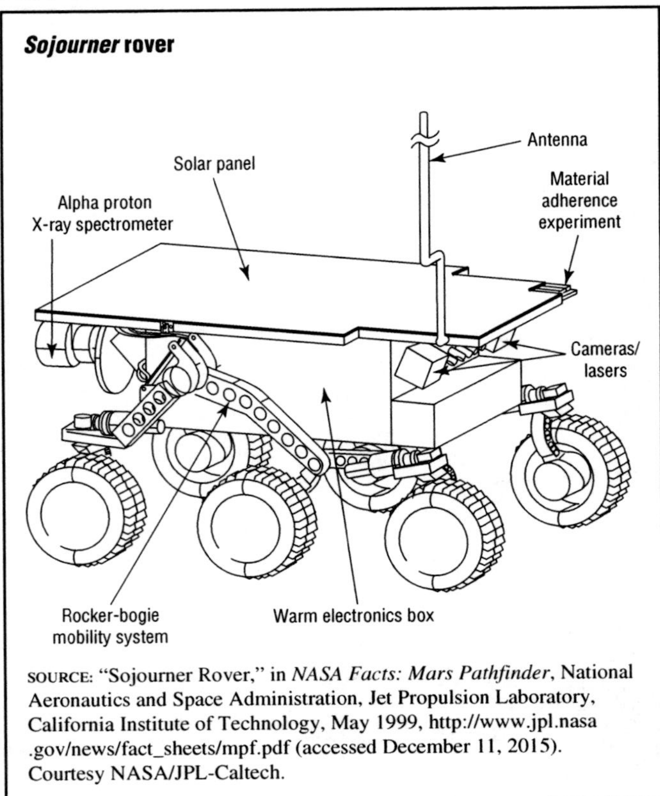

SOURCE: "Sojourner Rover," in *NASA Facts: Mars Pathfinder*, National Aeronautics and Space Administration, Jet Propulsion Laboratory, California Institute of Technology, May 1999, http://www.jpl.nasa.gov/news/fact_sheets/mpf.pdf (accessed December 11, 2015). Courtesy NASA/JPL-Caltech.

reported that the spacecraft was still operating. The orbiter carries seven instruments that investigate the Martian atmosphere and geological structure and search for subsurface water. One of the instruments, the Analyzer of Space Plasmas and Energetic Atoms 3, was supplied by NASA.

THE MARS EXPLORATION ROVERS MISSION

NASA's Mars Exploration Rover (MER) mission began in 2003 and included twin spacecraft, each carrying a lander. Inside each lander was a golf cart–sized rover that was designed to explore the Martian surface. The rovers were named *Spirit* and *Opportunity*. The names were the winning entries in a naming contest NASA held in 2002. The winning entry came from a third-grade student living in Scottsdale, Arizona. She was born in Russia and adopted by an American family. She chose the names to honor her feelings about the United States.

Spirit launched first on June 10, 2003. *Opportunity* launched four weeks later on July 7, 2003. The launch dates were chosen to put the spacecraft in flight near the time of Mars's perihelic opposition.

On January 4, 2004, the *Spirit* landed on Mars. The landing site was in Gusev Crater, which was named in honor of the Russian astronomer Matvei Gusev (1826–1866). The crater is about 100 miles (160 km) in diameter and lies at the end of a long valley known as Ma'adim Vallis. This translates as Mars Valley, because *Ma'adim* is Hebrew for "Mars." Major valleys on the Red Planet are named after Mars in different languages.

On January 25, 2004, the *Opportunity* set down near Mars's equator in an area called Meridiani Planum (Meridian Plain), which is considered to be the site of 0 degrees longitude on Mars. This is the longitude arbitrarily selected by astrogeologists to be the prime meridian for the rest of the planet. *Opportunity*'s landing site was nearly halfway around Mars from that of *Spirit*'s landing site. (See Figure 7.2.)

The landing sites were chosen for their flat terrain and because they were considered prime locations to look for evidence of ancient water.

Roving *Spirit* and *Opportunity*

The components of each rover are labeled in Figure 7.6. Each rover is just over 5 feet (1.5 m) long and weighed about 380 pounds (172 kg) on Earth. The panoramic cameras sit about 5 feet above the ground atop a mast.

Each rover carries a package of science instruments called an Athena science payload. Each payload includes two survey instruments, three instruments for close-up investigation of rocks, and a tool for scraping off the outer layer of rocks. The rovers were designed to move at a top speed of 2 inches (5 cm) per second. An average speed of 0.4 of an inch (1 cm) per second was expected for a rover traveling over rougher terrain.

The rovers were designed to operate independently of their landers. Each rover carries its own telecommunications equipment, camera, and computer. The electronic equipment receives power from batteries that are repeatedly recharged by solar arrays. It was late summer on Mars when the rovers began their mission. Scientists expected that power generation would taper off after about 90 sols (or 92 Earth days) and eventually stop as the arrays became too dust-coated to harness solar power. However, scientists were surprised when dust devils kept sweeping by the rovers and blowing the dust off the arrays. These periodic cleanings allowed the rovers to keep operating for much longer than expected.

The Name Game

Only the IAU has the authority to assign official names to planetary features. Major features, such as mountains, valleys, and large craters, have already been named. The IAU naming process can take many months and even years to accomplish. NASA scientists handling images from the MERs have to quickly assign temporary working names to the many new smaller features that are being revealed. The evolution of this process is described in the article "Naming Mars: You're in Charge" (Astro-Bio.net, June 20, 2004).

FIGURE 7.5

Scientific instruments on *2001 Mars Odyssey* orbiter

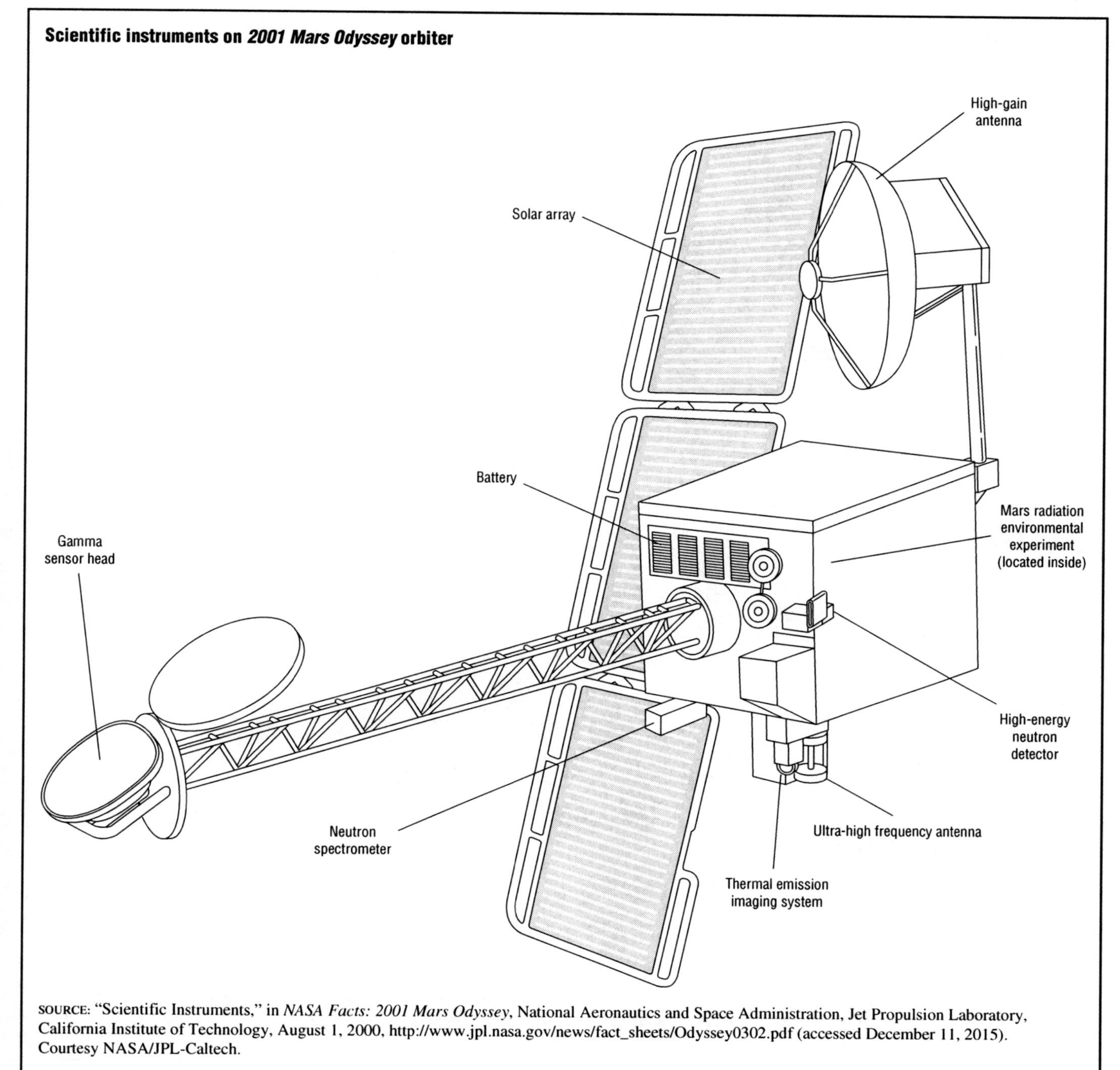

SOURCE: "Scientific Instruments," in *NASA Facts: 2001 Mars Odyssey*, National Aeronautics and Space Administration, Jet Propulsion Laboratory, California Institute of Technology, August 1, 2000, http://www.jpl.nasa.gov/news/fact_sheets/Odyssey0302.pdf (accessed December 11, 2015). Courtesy NASA/JPL-Caltech.

Most of the names are picked arbitrarily by whatever scientist first views an incoming image. Features are named after people, places, sailing ships, or other things the scientist fancies. For example, *Opportunity* landed within a tiny crater dubbed Eagle Crater in honor of the *Apollo 11* spacecraft that carried the first men to Earth's moon.

Water and Blueberries

On March 2, 2004, NASA scientists announced that *Opportunity* had uncovered strong evidence that the Meridiani Planum had been "soaking wet" in the past.

The claim was based on examination of the chemical composition and structure of rocks that were found in an outcrop in the area. The rocks contained minerals such as sulfate salts that are known to form in watery areas on Earth. The rocks also had niches in which crystals appear to have grown in the past. These empty niches are called vugs and are a strong indicator that the rocks sat in water for some time. Finally, there are spherules (small spheres) about the size of ball bearings embedded in the rocks. Scientists nicknamed the spherules blueberries. The iron-rich composition of the blueberries and the

FIGURE 7.6

Mars Exploration Rover

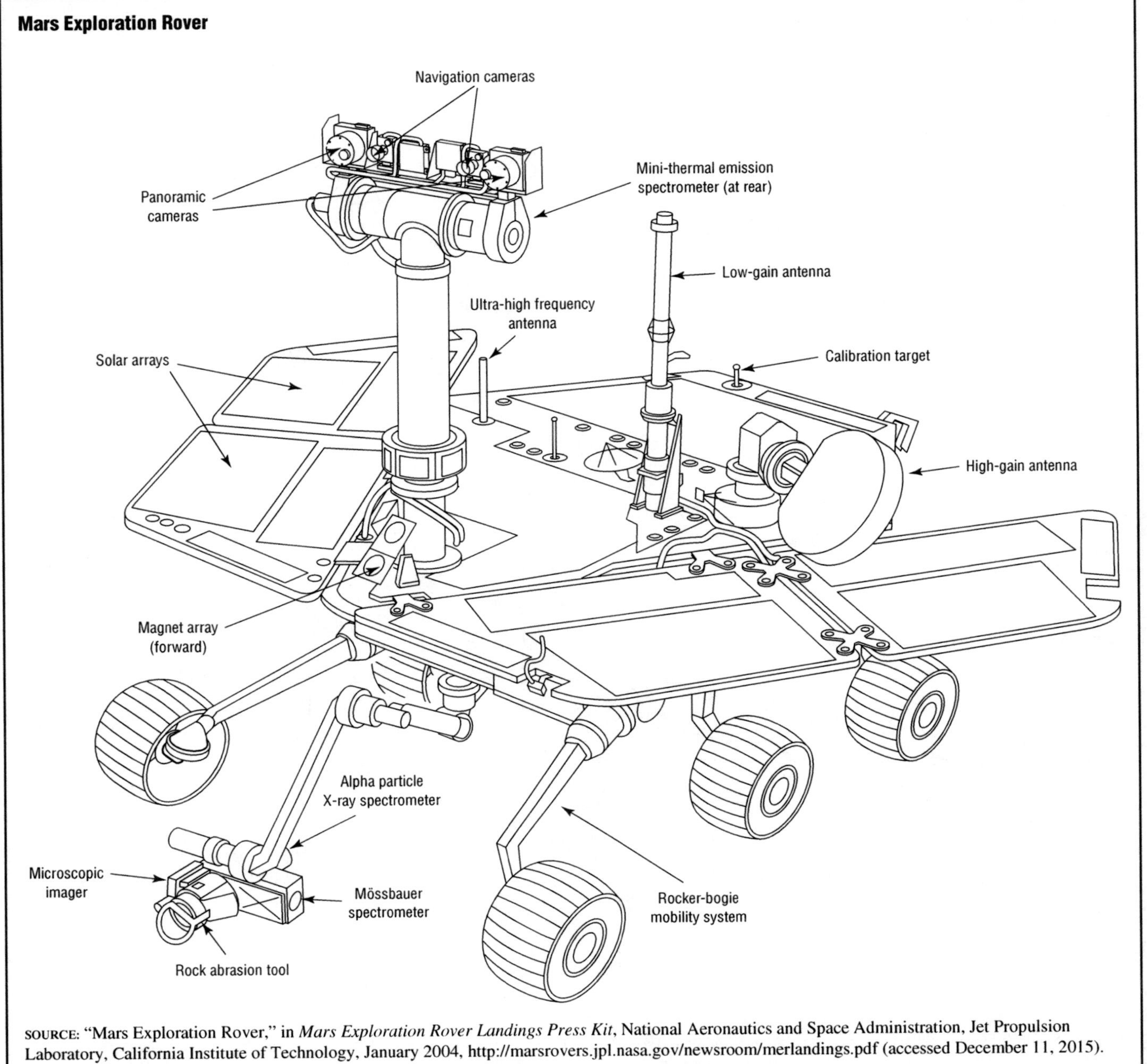

SOURCE: "Mars Exploration Rover," in *Mars Exploration Rover Landings Press Kit*, National Aeronautics and Space Administration, Jet Propulsion Laboratory, California Institute of Technology, January 2004, http://marsrovers.jpl.nasa.gov/newsroom/merlandings.pdf (accessed December 11, 2015). Courtesy NASA/JPL-Caltech.

way they are embedded in the rocks hint that water acted against the rocks in the past.

Spirit Is Lost

In April 2009 *Spirit* became stuck in sand and could not be freed. NASA reports in the press release "NASA Trapped Mars Rover Finds Evidence of Subsurface Water" (October 28, 2010, http://marsrovers.jpl.nasa.gov/newsroom/pressreleases/20101028a.html) that *Spirit* became stuck in soft sand after breaking through the surface crust in an area dubbed "Troy." Repeated attempts to free the rover were unsuccessful. With the bitterly cold Martian winter approaching, NASA failed to orient the rover's solar panels toward the sun. As a result, *Spirit* was put in a silent hibernation mode in which its radio and heaters were turned off to allow all possible energy collected from the solar panels to be devoted to charging the rover's batteries.

On March 22, 2010, *Spirit* sent out its last communication. Although scientists hoped the rover might "reawaken" during warmer weather, it did not. As of February 2016, no further transmission had been received from *Spirit*. In "Where Are the Rovers Now?" (2016, http://marsrover.nasa.gov/mission/traverse_maps.html), NASA indicates that *Spirit* traveled a total of 4.8 miles (7.7 km) during its mission on the Martian surface.

Status Update for *Opportunity*

In January 2016 NASA (http://mars.nasa.gov/mer/mission/status_opportunityAll.html) celebrated the 12th anniversary of the landing of the rovers on Mars. At that time, *Opportunity* was exploring a depression dubbed Marathon Valley. The nickname was selected because as the rover was traversing the valley it topped 26.2 miles (42.2 km). This is the official distance of a running marathon on Earth. *Opportunity* was using its abrasion tool to grind into a rock and chemically analyze its composition. Scientists believe that Marathon Valley is rich in clay minerals and hope to learn more about them using data collected by the rover.

THE MARS RECONNAISSANCE ORBITER MISSION

On August 12, 2005, NASA launched the *Mars Reconnaissance Orbiter* (*MRO*) toward the Red Planet. A powerful Atlas V two-stage rocket was used to hoist the heavy orbiter into space.

The *MRO*'s mission is to collect data that can be analyzed to determine if water has ever existed on Mars long enough for life to evolve. The orbiter is equipped with sophisticated atmospheric, mineralogy, and radar probes that investigate the atmosphere, terrain, and subsurface of the planet. (See Figure 7.7.) It also carries a high-resolution camera to take detailed images of the Martian surface. NASA calls the spacecraft its "eyes in the sky." The *MRO* entered Mars orbit in March 2006, and after several months of maneuvering began its science mission. The mission was scheduled to last for one Martian year (687 Earth days), but was eventually extended. As of February 2016, the *MRO* (http://mars.jpl.nasa.gov/mro/) was still in orbit around Mars. The data it collects are used to set sampling targets for the *Opportunity* rover.

FIGURE 7.7

Mars Reconnaissance Orbiter

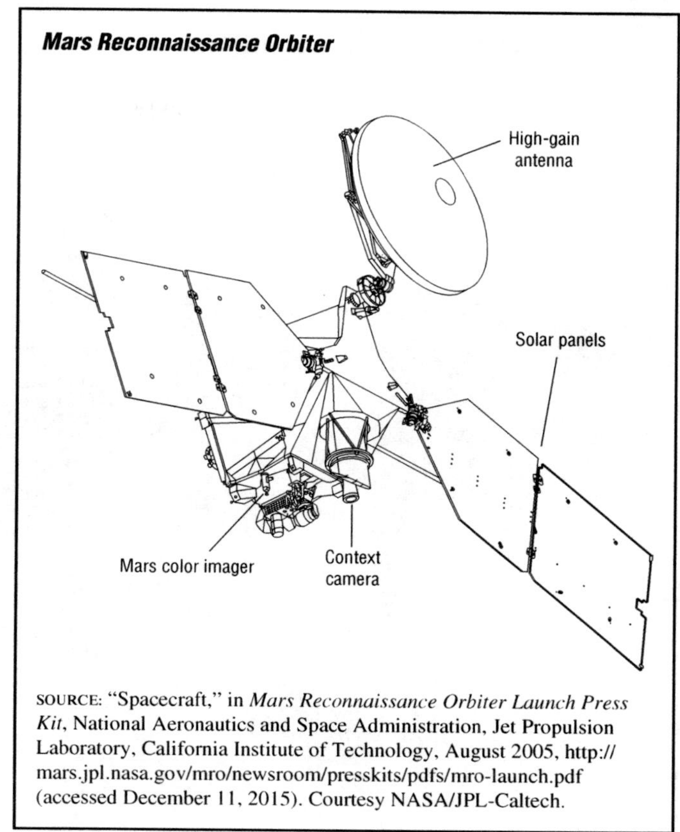

SOURCE: "Spacecraft," in *Mars Reconnaissance Orbiter Launch Press Kit*, National Aeronautics and Space Administration, Jet Propulsion Laboratory, California Institute of Technology, August 2005, http://mars.jpl.nasa.gov/mro/newsroom/presskits/pdfs/mro-launch.pdf (accessed December 11, 2015). Courtesy NASA/JPL-Caltech.

THE PHOENIX MARS LANDER MISSION

On August 4, 2007, NASA launched the *Phoenix Mars Lander*. It set down on Mars's northern polar region in May 2008. (See Figure 7.2.) *Phoenix*'s landing site was in Vastitas Borealis (Northern Plains), a relatively flat landscape that is believed to contain water ice close to the surface.

Phoenix carried seven science instruments, including a robotic arm for digging and collecting soil and ice samples. (See Figure 7.8.) Samples were analyzed by onboard instruments for water and carbon-containing compounds. *Phoenix* also included a stereoscopic imager that recorded full-color panoramic views of the environment, gas and soil analyzers, a meteorological station that tracked daily and seasonal weather changes, and a descent imager that photographed Mars during the spacecraft's descent.

The primary mission duration was projected to be 90 to 150 sols (approximately 92 to 154 Earth days). Scientists knew that *Phoenix* would cease operations when winter set in at Mars's north polar region, because there would be no sunlight to capture on the solar arrays and recharge the spacecraft's batteries. In November 2008 NASA received the last signal from *Phoenix*. As of February 2016, repeated attempts by *Odyssey* to contact *Phoenix* had been unsuccessful.

Raw data collected by NASA during the mission have been made available through the agency's Planetary Data System (http://pds.jpl.nasa.gov/), which is the data release portal for all of NASA's planetary missions. As of February 2016, scientists continued to analyze the data.

THE MARS SCIENCE LABORATORY/ CURIOSITY MISSION

On November 26, 2011, NASA launched the *Mars Science Laboratory* (*MSL*). It carried a rover named *Curiosity*. According to NASA, in "NASA Selects Student's Entry as New Mars Rover Name" (May 27, 2009,

FIGURE 7.8

Configuration of *Phoenix Mars Lander* after landing

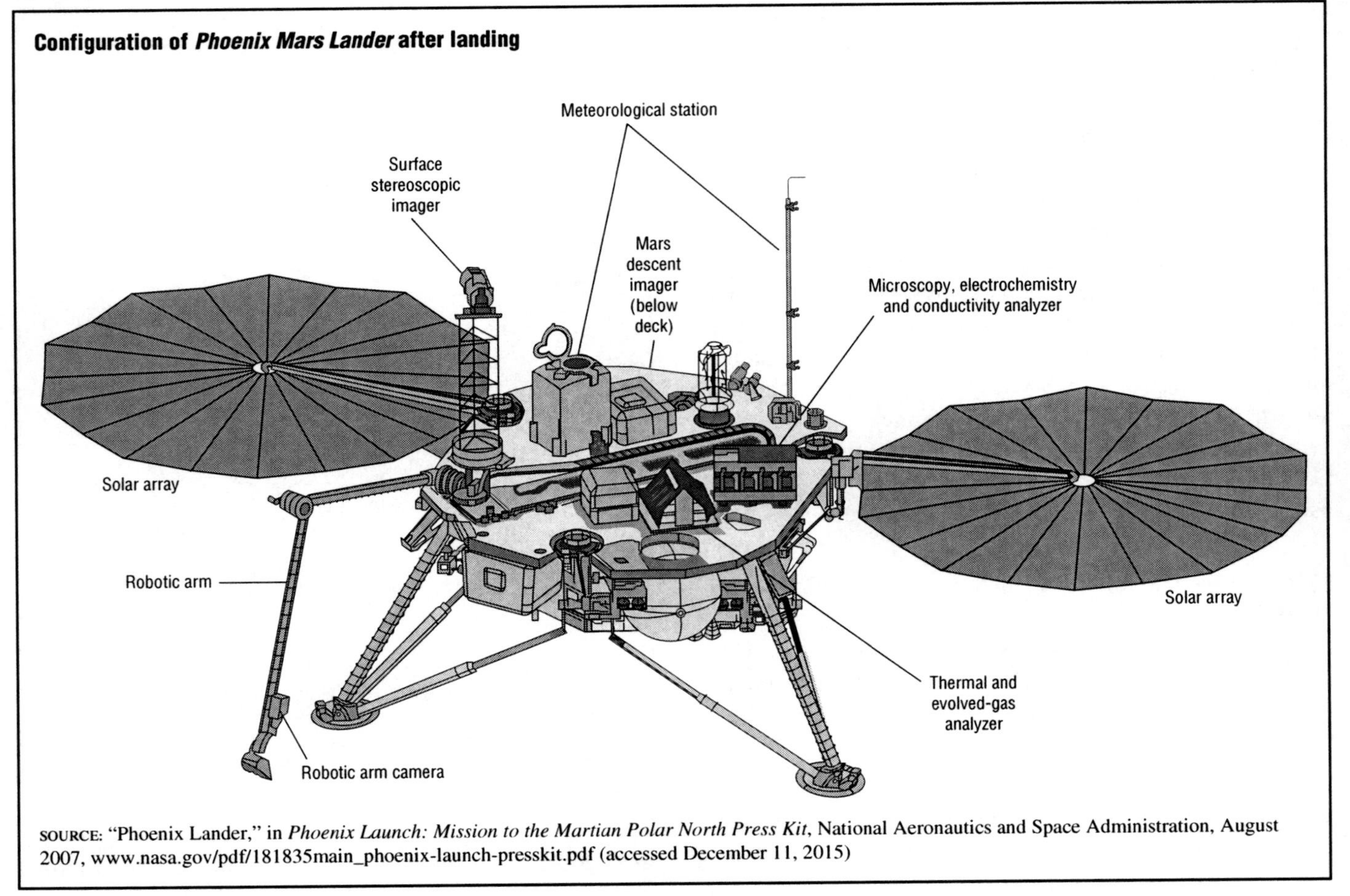

SOURCE: "Phoenix Lander," in *Phoenix Launch: Mission to the Martian Polar North Press Kit*, National Aeronautics and Space Administration, August 2007, www.nasa.gov/pdf/181835main_phoenix-launch-presskit.pdf (accessed December 11, 2015)

http://www.nasa.gov/mission_pages/msl/msl-20090527.html), the rover's name was chosen by a sixth-grade student from Lenexa, Kansas.

As shown in Figure 7.9, the rover is equipped with various science instruments and a mast camera. It was designed to collect soil and rock samples and subject them to detailed chemical analysis. Scientists are looking for evidence suggesting that microbial life once existed on Mars. The vehicle is about 9 feet (2.7 m) long and weighs approximately four times as much as one of the MERs. In "Mars Science Laboratory/Curiosity" (November 2014, http://www.jpl.nasa.gov/news/fact_sheets/mars-science-laboratory.pdf), NASA notes that the *MSL* includes a sophisticated suite of instruments that were developed by scientists in the United States, Canada, Spain, and Russia. The *MSL* also includes a Radiation Assessment Detector. Radiation data are considered essential for the planning of any future human missions to Mars.

The *MSL*, with the *Curiosity* rover on board, landed on Mars in August 2012. NASA (http://www.nasa.gov/mission_pages/msl/news/msl20120822.html) nicknamed the touchdown site Bradbury Landing in honor of the American writer Ray Bradbury (1920–2012). During the 1950s he published a series of stories called *The Martian Chronicles*, in which well-intentioned humans travel to Mars and accidentally spread deadly Earth germs among the Martian population.

Curiosity set off on its journey across Gale Crater, which is located not far from the final resting place of the *Spirit* rover. (See Figure 7.2.) In "Curiosity's Landing Site: Gale Crater" (2016, http://mars.nasa.gov/msl/mission/timeline/prelaunch/landingsiteselection/aboutgalecrater/), NASA indicates that Gale Crater is approximately 96 miles (154 km) in diameter. The agency explains, "Scientists chose Gale Crater as the landing site for Curiosity because it has many signs that water was present over its history." These signs include clays and sulfate minerals that formed in water.

Gale Crater holds within it a mountain that rises approximately 3 miles (5 km) above the crater's floor. The mountain bears the official name of Aeolis Mons (Mount Aeolis), as designated by the IAU. However, NASA commonly calls it Mount Sharp. According to the agency (March 28, 2012, http://www.jpl.nasa.gov/news/news.php?release=2012-090), the name honors the American geologist Robert Phillip Sharp (1911–2004), who

FIGURE 7.9

Mars *Curiosity* rover

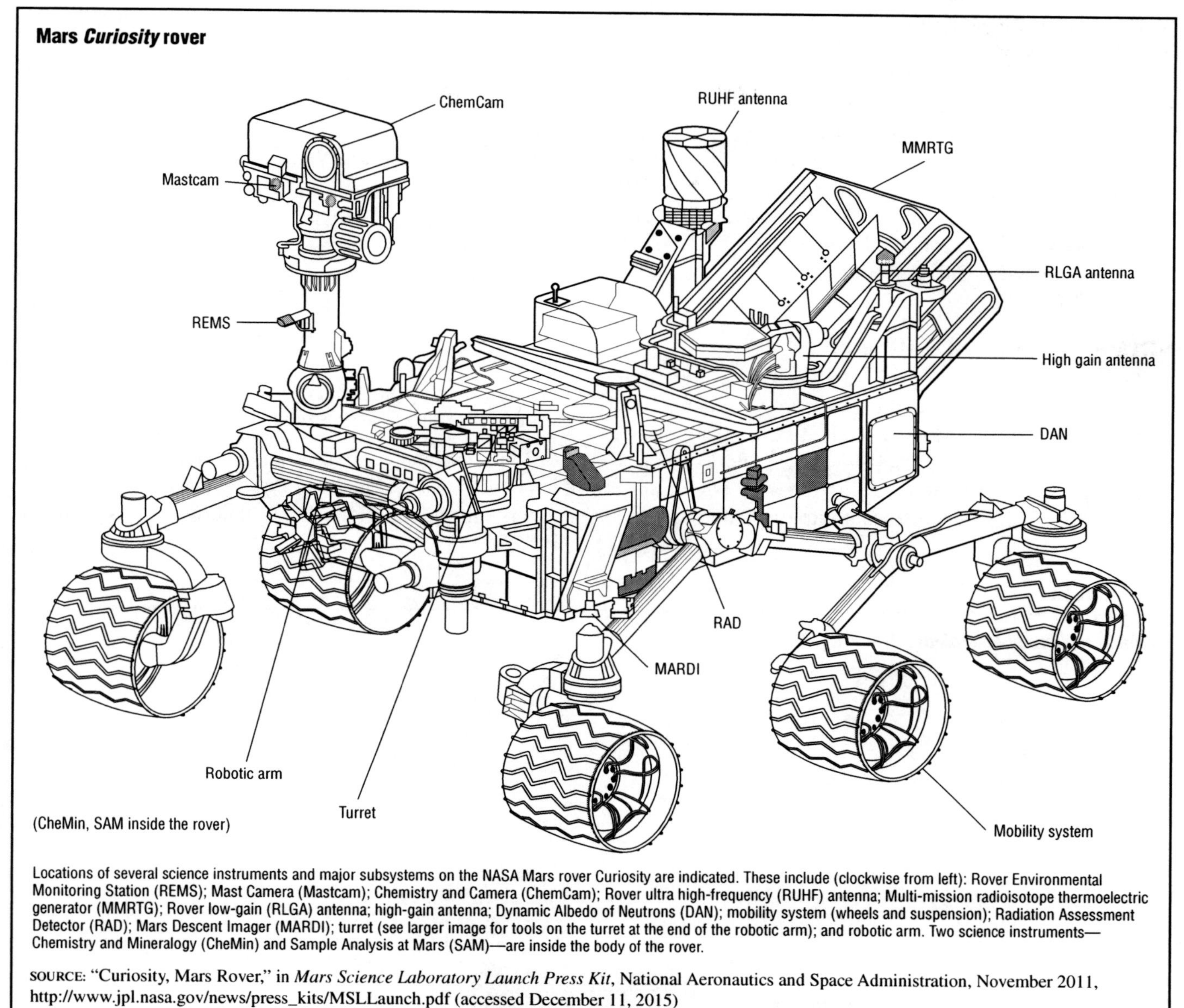

Locations of several science instruments and major subsystems on the NASA Mars rover Curiosity are indicated. These include (clockwise from left): Rover Environmental Monitoring Station (REMS); Mast Camera (Mastcam); Chemistry and Camera (ChemCam); Rover ultra high-frequency (RUHF) antenna; Multi-mission radioisotope thermoelectric generator (MMRTG); Rover low-gain (RLGA) antenna; high-gain antenna; Dynamic Albedo of Neutrons (DAN); mobility system (wheels and suspension); Radiation Assessment Detector (RAD); Mars Descent Imager (MARDI); turret (see larger image for tools on the turret at the end of the robotic arm); and robotic arm. Two science instruments—Chemistry and Mineralogy (CheMin) and Sample Analysis at Mars (SAM)—are inside the body of the rover.

SOURCE: "Curiosity, Mars Rover," in *Mars Science Laboratory Launch Press Kit*, National Aeronautics and Space Administration, November 2011, http://www.jpl.nasa.gov/news/press_kits/MSLLaunch.pdf (accessed December 11, 2015)

provided input for early NASA missions to Mars. The rover's long-term goal is to investigate the lower elevations of Mount Sharp.

In the press release "Rover Rounds Martian Dune to Get to the Other Side" (January 4, 2016, http://mars.jpl.nasa.gov/msl/news/whatsnew/index.cfm?FuseAction=ShowNews&NewsID=1882), NASA indicates that *Curiosity* reached the base of Mount Sharp in 2014. Since then, it has slowly worked its way up to higher layers of terrain. As of January 2016, the rover was exploring Bagnold Dunes, a band of dark sand dunes along the northwestern edge of Mount Sharp. According to NASA, "The mission's dune-investigation campaign is designed to increase understanding about how wind moves and sorts grains of sand, in an environment with less gravity and much less atmosphere than well-studied dune fields on Earth."

The Curiosity mission is considered a "flagship" mission by NASA. As explained in Chapter 6, flagship missions are characterized by significant "scope, strategic objectives, and technology challenges." They are also inherently expensive. NASA notes in "Mars Science Laboratory/Curiosity: By the Numbers" (2016, http://solarsystem.nasa.gov/missions/marsscilab/facts) that the MSL/Curiosity mission has a total estimated lifetime cost of $2.5 billion. In "Mars Science Laboratory/Curiosity," the agency explains that the primary mission duration was set at 24 months, a milestone that was reached during the summer of 2014.

In July 2014 an expert panel convened by NASA (http://www.lpi.usra.edu/pss/sep2014/Senior-Review-2014-Report.pdf) conducted a senior review of seven of the agency's planetary missions, including the MSL/Curiosity mission.

One of the panel's goals was to assess the "science per dollar" value of extending the missions. In this respect, the panel found the MSL/Curiosity mission lacking because relatively few samples were taken during its prime mission or planned for its extended mission. The panel complains, "This [is] a poor science return for such a large investment in a flagship mission." Despite this criticism, the mission was extended. NASA's budget request for fiscal year 2016 (http://www.nasa.gov/sites/default/files/atoms/files/fy2016_budget_book_508_tagged_0.pdf) includes $58 million in funding for the MSL/Curiosity mission.

INDIA'S MARS ORBITER MISSION

On November 5, 2013, the Indian Space Research Organisation, an agency of the Indian government, successfully launched its Mars orbiter mission. The spacecraft is known informally as *Mangalyaan*, which means "Mars craft" in Hindi. It is the nation's first interplanetary mission. The spacecraft went into orbit around Mars in September 2014. As of February 2016, it continued to operate. According to the Indian Space Research Organisation (2016, http://www.isro.gov.in/pslv-c25-mars-orbiter-mission/payloads), the orbiter carries a payload that conducts imaging, maps the planet's surface and mineralogy, detects methane (an organic or carbon-containing gas), and analyzes the Martian atmosphere.

THE MARS ATMOSPHERE AND VOLATILE EVOLUTION MISSION

On November 18, 2013, less than two weeks after the *Mangalyaan* orbiter left Earth, NASA launched the *Mars Atmosphere and Volatile Evolution Mission* (*MAVEN*). Figure 7.10 shows the spacecraft and its instruments. It went into orbit around Mars in September 2014 and was still operational as of February 2016.

The Laboratory for Atmospheric and Space Physics at the University of Colorado, Boulder, is one of NASA's partners in the mission. In "Exploring Mars' Climate

FIGURE 7.10

***Mars Atmosphere and Volatile Evolution Mission* (MAVEN)**

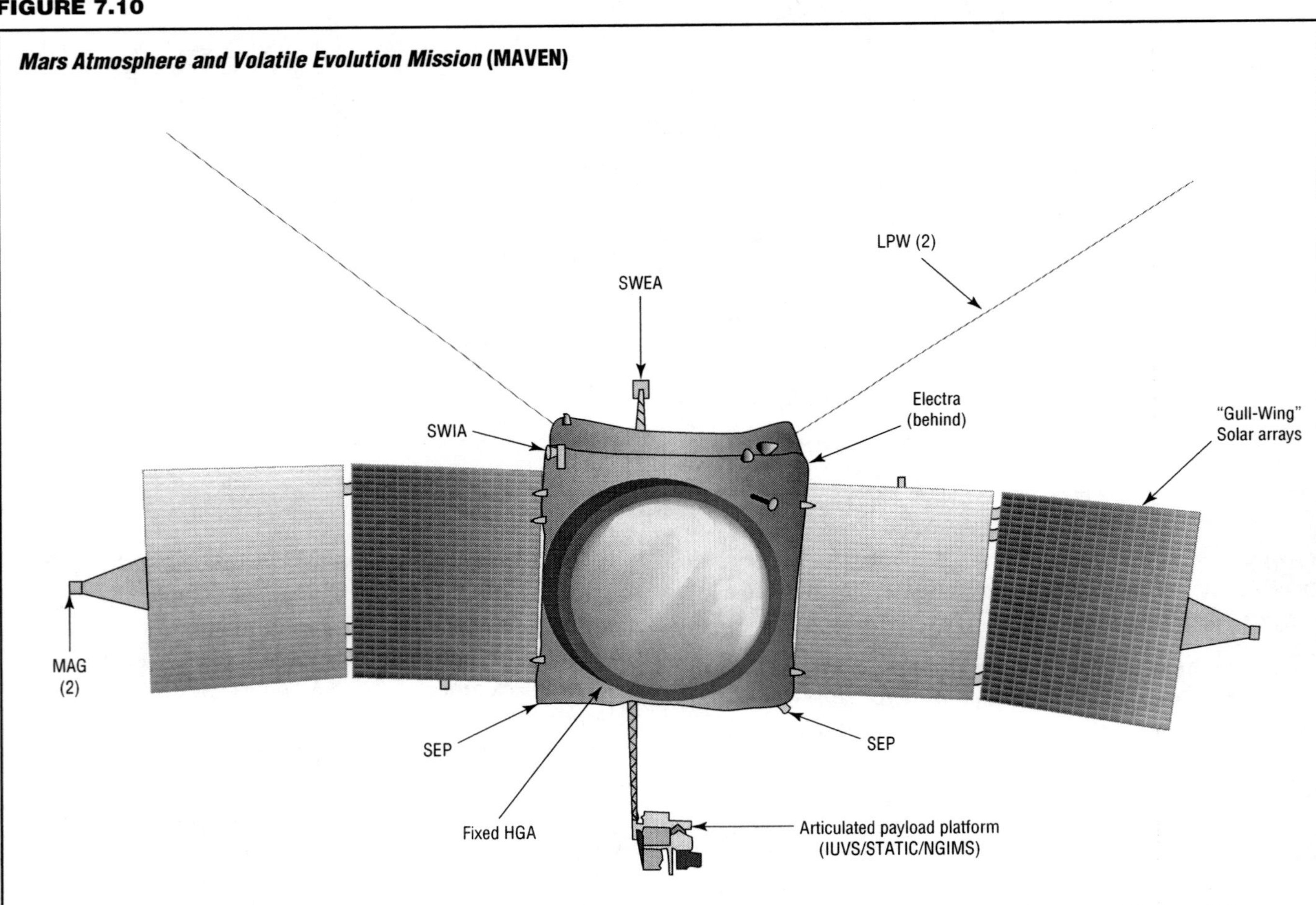

Notes: MAVEN = *Mars Atmosphere and Volatile Evolution Mission.* HGA = High-Gain Antenna. IUVS = Imaging UltraViolet Spectrograph. LPW = Langmuir Probe and Waves. MAG = Magnetometer. NGIMS = Neutral Gas and Ion Mass Spectrometer. SEP = Solar Energetic Particle. STATIC = SupraThermal and Thermal Ion Composition. SWEA = Solar Wind Electron Analyzer. SWIA = Solar Wind Ion Analyzer.

SOURCE: "MAVEN's Science Payload Consists of Eight Instruments in Three Packages," in *Press Kit: MAVEN*, National Aeronautics and Space Administration, November 2013, http://mars.nasa.gov/files/resources/MAVEN_PressKit_Final.pdf (accessed December 8, 2015)

History" (2016, http://lasp.colorado.edu/home/maven/), the laboratory explains that "the mission's goal is to explore the planet's upper atmosphere, ionosphere, and interactions with the Sun and solar wind." The laboratory notes in "MAVEN Reveals Speed of Solar Wind Stripping Martian Atmosphere" (November 5, 2015, http://lasp.colorado.edu/home/maven/2015/11/05/maven-reveals-speed-of-solar-wind-stripping-martian-atmosphere/) that data reveal that the Martian atmosphere "is losing gas to space via stripping by the solar wind." As explained in Chapter 6, the solar wind is a flow of energetic particles that constantly blows from the sun. Over millennia the stripping process is believed to have removed from Mars a "warm and wet environment that might have supported surface life."

THE FUTURE OF MARS EXPLORATION

Mars has become one of the most popular destinations for robotic spacecraft. As of February 2016, there were five spacecraft in orbit and two operational rovers on the Martian surface.

Funding constraints have seriously thwarted the United States' capability to develop and lead future Mars missions. During the first decade of the 21st century NASA intended to collaborate with the ESA in its ExoMars Programme with mission launches beginning during the late 2010s. However, in February 2012 NASA announced that it was pulling out of the project, much to the disappointment of U.S. space scientists.

Instead, NASA developed its own Mars mission, Interior Exploration Using Seismic Investigations, Geodesy, and Heat Transport (InSight). It was designed to investigate Mars's interior structure, such as crust, mantle, and core, using a lander. The *InSight* spacecraft was slated to launch in March 2016, when Mars would be in opposition. However, in December 2015 NASA called off the planned launch. In the press release "NASA Suspends 2016 Launch of InSight Mission to Mars, Media Teleconference Today" (December 22, 2015, http://www.nasa.gov/press-release/nasa-suspends-2016-launch-of-insight-mission-to-mars), the agency indicates that an instrument called the Seismic Experiment for Interior Structure was found to be faulty. The instrument was built by the French space agency the Centre National d'Etudes Spatiales (National Space Study Center). NASA notes that the favorable 2016 launch window only lasts for a few weeks; thus, the launch could not be postponed to later in the year. As of February 2016, future plans for the InSight mission were unknown. However, NASA continued to plan a 2020 mission to Mars that will include a rover.

Internationally, the ESA has an ExoMars Programme launch scheduled for March 2016. The mission is devoted to exobiology. According to the agency (February 17, 2016, http://exploration.esa.int/mars/46124-mission-overview/), the spacecraft will include an orbiter and a lander named *Schiaparelli* (after the Italian astronomer). The primary mission goals are "to search for evidence of methane and other trace atmospheric gases that could be signatures of active biological or geological processes and to test key technologies in preparation for ESA's contribution to subsequent missions to Mars." Russia will provide the launch vehicle for the spacecraft, which is expected to go into orbit around the Red Planet in October 2016.

In "ExoMars Mission (2018)" (October 28, 2015, http://exploration.esa.int/mars/48088-mission-overview/), the ESA indicates that a follow-up ExoMars mission scheduled for 2018 will include a Russian-built lander and a European-built rover. As noted earlier, 2018 is an auspicious year for Mars exploration because the planet will be in perihelic opposition with Earth and thus closer than usual.

Crewed Missions to Mars?

Several space agencies have conceptual plans to send crewed missions to Mars in the distant future. As noted in Chapter 2, in 2004 President George W. Bush (1946–) proposed that astronauts travel to the moon and then to Mars. However, in 2010 President Barack Obama (1961–) laid out a different agenda that called for a crewed mission to an asteroid and then to Mars, possibly sometime during the 2030s.

Srinivas Laxman notes in "China's Space Mission: The Long March to the Moon and Mars" (AsianScientist.com, June 27, 2011) that China's long-range plans for space exploration include a crewed mission to the Red Planet between 2040 and 2060.

Getting humans safely to Mars poses numerous technological and biological challenges that remain to be solved. One problem is launch vehicle capability. Crewed spacecraft that are designed to spend more than just a few days in space are heavier than robotic spacecraft because humans require so many resources. The Saturn V rockets described in Chapter 2 were the most powerful rockets ever built. They lofted the Apollo spacecraft toward the moon during the late 1960s and early 1970s. Since then, no rockets have been launched with so great a lifting capacity. However, the Space Launch System (SLS) vehicles being developed by NASA are expected to be even more powerful than the Saturn V rockets. The first SLS test flight is expected in 2017.

As of 2016, it took a robotic spacecraft around six months, at best, to travel from Earth to Mars. This means that astronauts on a round trip to Mars would be in weightless conditions aboard a spacecraft for at least a year. Chapter 5 describes the medical problems, such as bone loss and high intracranial pressure within the skull, that are associated with prolonged exposure to microgravity. Another very worrisome challenge is radiation exposure. In "Calculated Risks: How Radiation Rules

Manned Mars Exploration" (Space.com, February 18, 2014), Sheyna E. Gifford states that "in open space, human beings continuously contend with intense solar and cosmic background radiation. Solar energetic particles (SEPs) and galactic cosmic rays (GCRs) turn a trip to Mars into a six-month radiation shower." According to Gifford, the first radiation readings obtained by the *MSL*'s Radiation Assessment Detector were published by NASA in January 2014. They indicate that a human on a six-month voyage to Mars would be exposed to more radiation than a human spending the equivalent time on the *International Space Station.* Crewed missions to Mars will likely require spacecraft that are much quicker and much better shielded against radiation than those currently available.

Some private organizations are proposing crewed missions to Mars and have captured significant public attention. For example, Mars One is a nonprofit organization headquartered in the Netherlands. In 2013 it made headlines after it invited people around the world to apply to take a one-way trip to the Red Planet to establish a colony there during the early 2020s. Approximately 100,000 people applied and paid the required application fees. The organization indicated in "Will Mars One Meet the Exact Time Schedule?" (http://www.mars-one.com/faq/finance-and-feasibility/will-mars-one-meet-the-exact-time-schedule) that as of February 2016 it was trying to raise $6 billion to send its first four-person team to Mars.

CHAPTER 8
THE FAR PLANETS AND BEYOND

The farther we penetrate the unknown, the vaster and more marvelous it becomes.

—Charles A. Lindbergh Jr., *Autobiography of Values* (1978)

Beyond Mars lie the far planets: Jupiter, Saturn, Uranus, and Neptune. Although they are a great distance from the sun, they are not even close to the edge of the solar system. Beyond Neptune is a large icy area called the Kuiper Belt that extends outward 7 billion miles (11.3 billion km). Within it there are untold numbers of celestial bodies orbiting the sun. One of these Kuiper Belt Objects is Pluto, formerly considered a full-fledged planet, but now considered to be a dwarf planet. Figure 8.1 shows the relative locations of the far planets and Pluto. They are far from the sun, in a cold and dark part of the solar system.

In ancient times people noticed that some lights in the sky followed odd paths around the heavens. The Greeks called them *asteres planetos* (wandering stars). Later, they would be called planets. The ancients could see only two of the far planets in the nighttime sky: Jupiter and Saturn.

Jupiter was named for the mythical Roman god of light and sky. He was the supreme god also known as Jove or *dies pater* (shining father). His counterpart in Greek mythology was named Zeus. Saturn was named after the god of agriculture, who was also Jupiter's father. His Greek counterpart was called Kronos.

Following the invention of the telescope, Uranus, Neptune, and Pluto were discovered. Uranus was named for the father of the god Saturn. Neptune was the god of the sea and Jupiter's brother in Roman mythology. Pluto was named after the Greek god of the underworld.

When the space age began, humans sent robotic spacecraft to investigate the far planets. They have returned images of strange and marvelous worlds that are composed of gas and slush instead of rock. Many new moons have been revealed. Some of these moons are covered with ice and have atmospheres. There could be liquid water teeming with life beneath that ice.

THREE CENTURIES OF DISCOVERY

It took three centuries for humans to uncover the far planets in the solar system. During the 1600s the telescope opened up new opportunities for observation. People learned that Jupiter and Saturn had moons and that Saturn had rings. The telescope also showed that the wandering stars were not stars at all, because they did not generate their own light, but reflected light from the sun.

No new planets were discovered during the 1600s. The far planets were still too distant and fuzzy to be recognized for what they were. Uranus was discovered during the late 1700s. Another century passed before the discovery of Neptune. Pluto was discovered in 1930.

Astronomers categorize planets based on geology and composition. Mercury, Venus, Earth, and Mars are called the terrestrial planets because they are made of rock and metal. Jupiter, Saturn, Uranus, and Neptune are called the gas giants. Some scientists think they may have solid cores, but the exterior of these planets consists of huge clouds of gas. These planets are also known as the Jovian planets (after Jove or Jupiter). All of them have ring systems.

Pluto is a different story. It is a small ice world. For decades astronomers argued whether it was even a planet. In 2006 the debate was ended by a decision from the International Astronomical Union (IAU), the body that is responsible for naming celestial objects. In August 2006 the IAU proclaimed that Pluto was officially a dwarf planet. By official definition, a dwarf planet is planetlike in that it orbits the sun and has sufficient mass and self-gravity to be nearly round in shape. However, the IAU explains in the press release

FIGURE 8.1

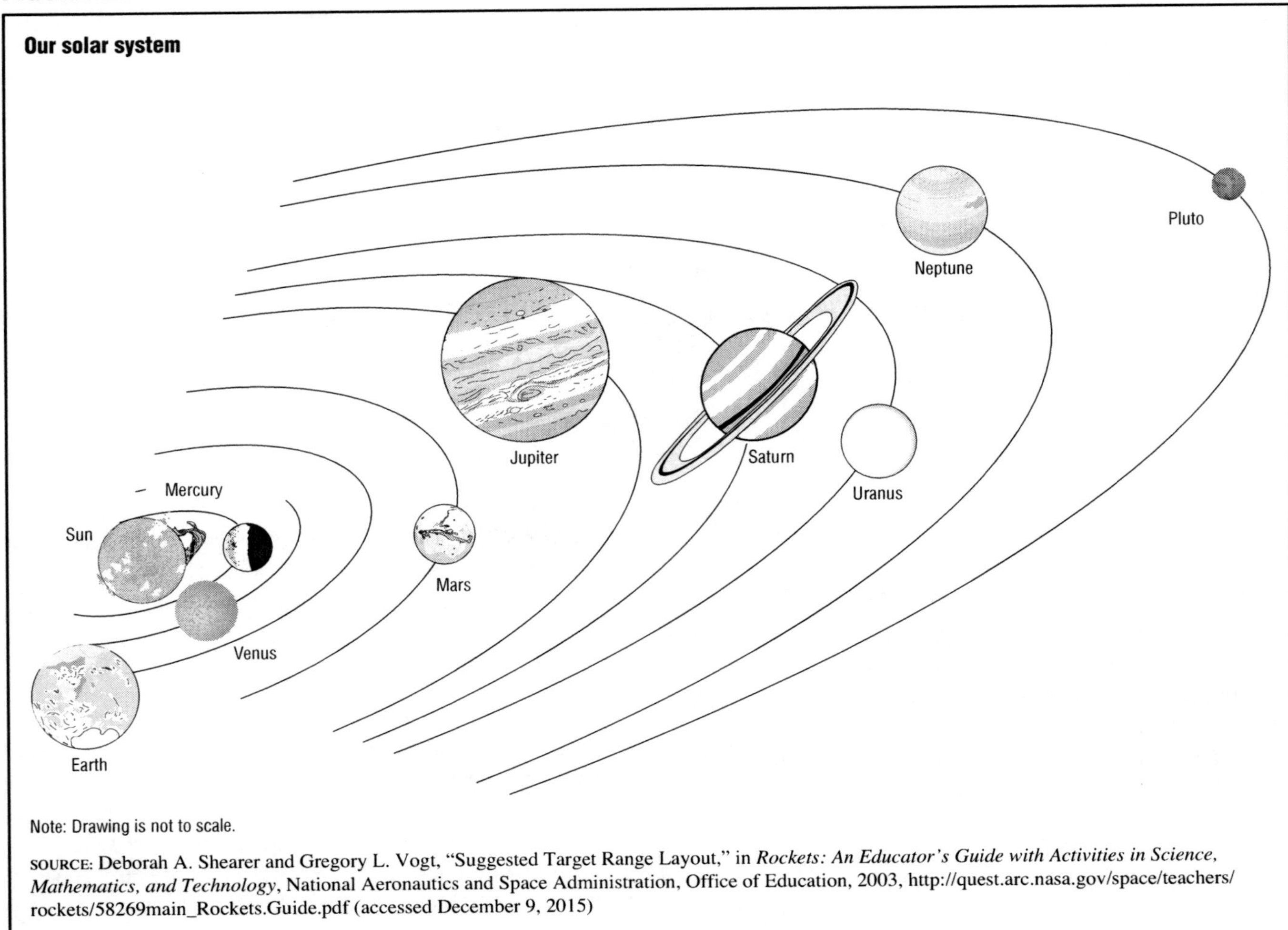

Note: Drawing is not to scale.

SOURCE: Deborah A. Shearer and Gregory L. Vogt, "Suggested Target Range Layout," in *Rockets: An Educator's Guide with Activities in Science, Mathematics, and Technology*, National Aeronautics and Space Administration, Office of Education, 2003, http://quest.arc.nasa.gov/space/teachers/rockets/58269main_Rockets.Guide.pdf (accessed December 9, 2015)

"IAU 2006 General Assembly: Result of the IAU Resolution Votes" (August 24, 2006, http://www.iau.org/iau0603.414.0.html) that unlike a planet, a dwarf planet "has not cleared the neighbourhood around its orbit." In other words, a dwarf planet, unlike a planet, does not clear smaller objects out of its orbit.

JUPITER

Jupiter is the fifth planet from the sun and the largest planet in the solar system. It takes the planet nearly 12 Earth years to make one orbit around the sun. Jupiter is just over 11 times larger than Earth. The planet is bright enough to be seen with the naked eye and appears yellowish from Earth. It is similar in composition to a small star and has an incredibly powerful magnetic field that stretches out millions of miles. The poles experience dazzling auroras that are many times more powerful and brighter than the auroras (northern and southern lights) on Earth.

Jupiter's atmosphere is approximately 90% hydrogen, and the remaining 10% is mostly helium, with traces of methane, water, and ammonia. Its sky is streaked with clouds and often with lightning. A gigantic hurricane-like storm has raged on the planet for untold centuries. It is a cold high-pressure area that is two to three times wider than Earth. The storm is nicknamed the Great Red Spot. The red color is probably due to the presence of certain chemical elements within the storm. Scientists believe that Jupiter's surface is slushy rather than solid.

Jupiter's Moons

On January 7, 1610, the Italian astronomer Galileo (1564–1642) was looking through his homemade telescope and discovered four celestial objects near Jupiter. At first he thought they were stars. After watching them for a week, he realized they were satellites in orbit around Jupiter. Two months later Galileo published his findings in *Sidereus Nuncius* (*Starry Messenger*).

That same year the German astronomer Simon Marius (1573–1624) published *Mundus Iovialis* (*The Jovian World*), in which he claimed that he discovered the satellites before Galileo. Marius did not provide any observational data in his book, and Galileo was better respected. As a result, the credit was given to Galileo.

Eventually, the four moons were named Io, Europa, Ganymede, and Callisto.

In the years following Galileo's discovery, other observers discovered many smaller moons around Jupiter. The pace of these discoveries accelerated greatly during the late 20th and early 21st centuries as better equipment was developed. In 2003 astronomers at an observatory atop Mauna Kea in Hawaii spotted 24 previously unknown moons around Jupiter. These moons have been named after the lovers, favorites, or descendants of Zeus in accordance with IAU guidelines. The National Aeronautics and Space Administration (NASA) notes in "Planetary Satellite Discovery Circumstances" (http://ssd.jpl.nasa.gov/?sat_discovery) that as of March 2015 Jupiter had 67 known moons that had officially been named by the IAU.

FIGURE 8.2

Size comparison of some of the moons of Saturn

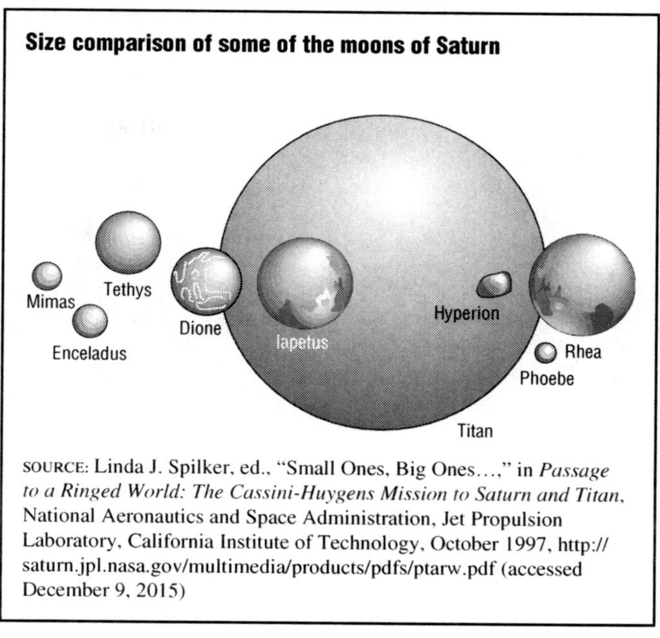

SOURCE: Linda J. Spilker, ed., "Small Ones, Big Ones...," in *Passage to a Ringed World: The Cassini-Huygens Mission to Saturn and Titan*, National Aeronautics and Space Administration, Jet Propulsion Laboratory, California Institute of Technology, October 1997, http://saturn.jpl.nasa.gov/multimedia/products/pdfs/ptarw.pdf (accessed December 9, 2015)

SATURN

Saturn is the sixth planet from the sun and the second-largest planet in the solar system. It takes 29.5 Earth years to orbit around the sun. Saturn's atmosphere is mostly hydrogen, with traces of helium and methane. It is a hazy yellow color. The planet is very windy, with wind speeds reaching 1,000 miles per hour (1,600 km/h).

Saturn is flat at the poles. The planet is surrounded by several thin rings of orbiting material that circle near its equator. In 1610, when Galileo first saw Saturn through his telescope, its rings appeared to him to be two dim stars on either side of the planet. He described these stars as "handles." In the following years other astronomers saw the strange shapes around Saturn. They were variously described as ears or arms that extended from the planet's surface. It would take an improvement in telescopic power before their true nature was revealed.

Moons and Rings

Christiaan Huygens (1629–1695) was a Dutch astronomer who became famous for his observations of Saturn. He and his brother Constantyn built new and more powerful telescopes that were greatly admired by astronomers of the time.

In 1655 Huygens discovered a satellite around Saturn. This turned out to be the planet's largest moon. In 1656 he wrote about his discovery in *De Saturni Luna Observatio Nova* (*New Observation of a Moon of Saturn*). Huygens referred to his discovery as simply Saturn's moon. Later, it would be called Titan.

Over the centuries, many more moons were discovered around Saturn. Figure 8.2 compares the size of Titan to some other Saturnian moons. According to NASA, in "Planetary Satellite Discovery Circumstances," as of March 2015 Saturn was known to have 62 moons. Other moons orbiting Saturn are possible but have yet to be confirmed by astronomers.

Huygens was the first person to figure out that the mysterious shapes near Saturn were not stars, arms, or ears, but a ring of material around the planet. Huygens mistakenly thought the ring was one solid object. During the 1600s Giovanni Domenico Cassini (1625–1712) discovered a major gap in the ring around Saturn. This proved that the structure was not one solid object as Huygens had thought. The gap would later be called the Cassini Division. Cassini believed that Saturn's rings were composed of millions of small particles. This view was shared by the French astronomer Jean Chapelain (1595–1674). However, it was not generally accepted until the 18th century.

Modern astronomers believe the rings are composed of chunks of ice. These chunks range in size from tiny particles to as large as automobiles. (See Figure 8.3.) In the 21st century scientists know that Saturn's ring system is actually many ringlets of different sizes that are nestled within each other with gaps or divisions between ring systems.

URANUS

Uranus is the seventh planet from the sun and the third-largest planet in the solar system. It looks featureless through even the most powerful telescopes. Scientists believe the planet is shrouded in clouds that hide it from view. The presence of methane in the upper atmosphere is believed to account for the planet's light blue-green color. It takes 84 Earth years for Uranus to orbit around the sun. Uranus is unique in the solar system

FIGURE 8.3

Saturn ring composition

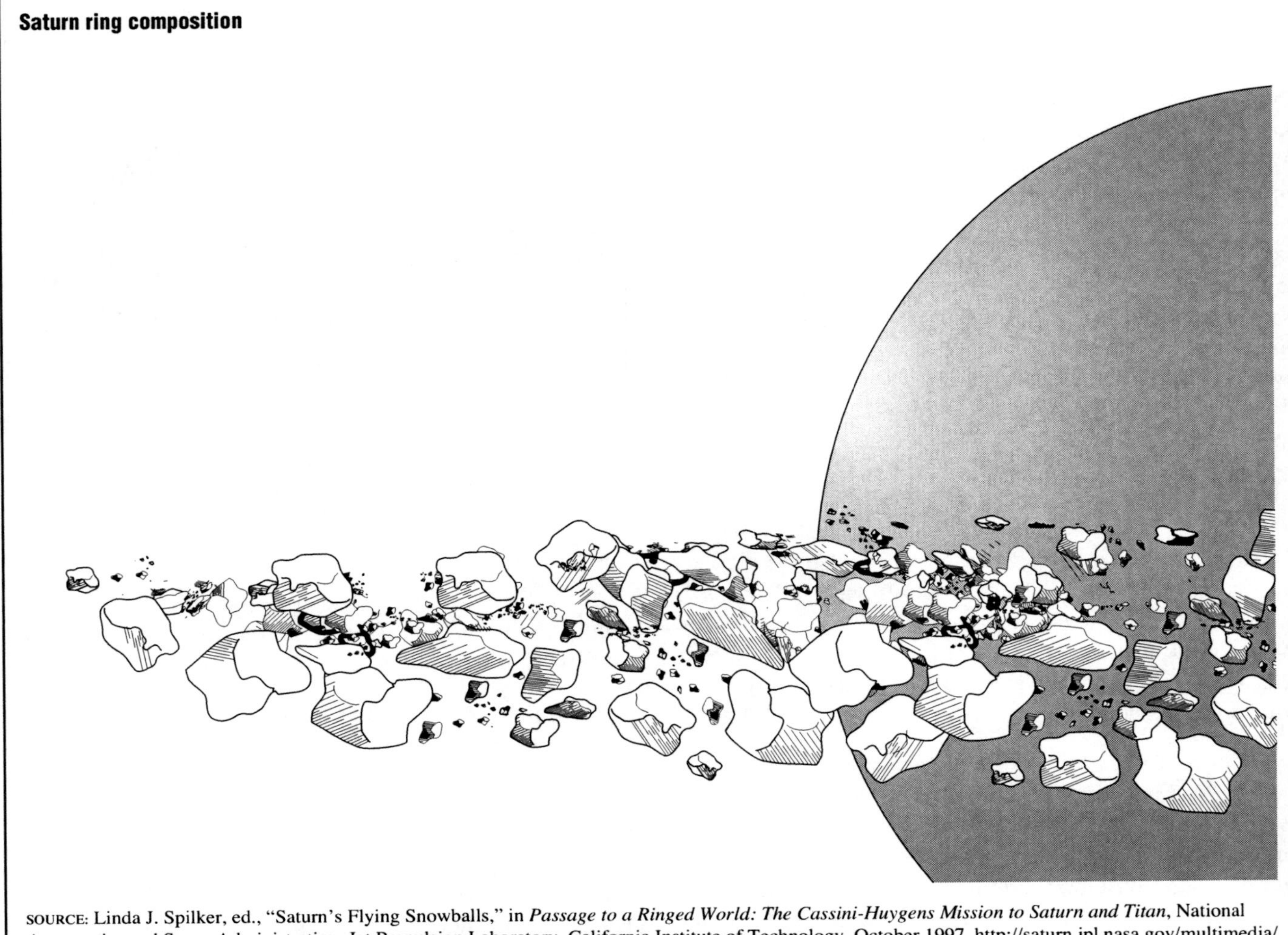

SOURCE: Linda J. Spilker, ed., "Saturn's Flying Snowballs," in *Passage to a Ringed World: The Cassini-Huygens Mission to Saturn and Titan*, National Aeronautics and Space Administration, Jet Propulsion Laboratory, California Institute of Technology, October 1997, http://saturn.jpl.nasa.gov/multimedia/products/pdfs/ptarw.pdf (accessed December 9, 2015). Courtesy NASA/JPL-Caltech.

because its axis is tilted so far from its orbital plane. The planet lies on its side as it orbits with a pole pointed toward the sun.

Herschel Discovers Uranus and Two of Its Moons

The astronomer William Frederick Herschel (1738–1822) was born in Germany, but lived and worked in Britain. On March 13, 1781, he was searching the sky with his telescope when he discovered Uranus. Herschel wanted to name the planet Georgium Sidus in honor of King George III (1738–1820) of Britain. However, the name Uranus was selected from ancient mythology.

Six years later, in 1787, Herschel was the first to spot satellites around the planet. He discovered the two largest moons: Titania and Oberon. Since then, more than two dozen additional moons have been discovered. In "Planetary Satellite Discovery Circumstances," NASA states that Uranus had 27 known satellites as of March 2015. The moons are named after characters from the plays of William Shakespeare (1564–1616) and from the poem "The Rape of the Lock" by Alexander Pope (1688–1744).

NEPTUNE

Neptune is the eighth planet from the sun. The planet is far from Earth and extremely difficult to observe. According to NASA, in "Neptune: In Depth" (2016, http://solarsystem.nasa.gov/planets/neptune/indepth), it takes the planet 165 Earth years to make one orbit around the sun. Neptune is an icy giant and extremely windy. It is a vivid shade of bright blue due to the presence of methane and other chemicals in its atmosphere. The planet has six rings that are not uniform in thickness all the way around, but thin to nonexistent in places.

Neptune's Controversial Discovery

Neptune's discovery is a twisted tale of mathematics, bureaucrats, and international competition. Following the discovery of Uranus in 1781, astronomers watched the planet for several decades. They were puzzled because its orbit did not follow the expected path. Some astronomers began to suspect that there might be another planet beyond Uranus. The effect of its gravity would explain the irregularities that astronomers saw in Uranus's orbit.

During the 1840s John Couch Adams (1819–1892) of Britain and Urbain-Jean-Joseph Leverrier (1811–1877) of France used mathematics to plot the location of this mystery planet. Adams presented his theory to George Biddell Airy (1801–1892), the Astronomer Royal of Britain. For some reason, Airy failed to pursue the matter right away. Meanwhile, Leverrier submitted his theory to Johann Gottfried Galle (1812–1910), the director of the Berlin Observatory. On the night of September 23, 1846, Galle used Leverrier's notes to locate the planet in the sky.

However, it soon came out that Adams had described the location of the planet months before Leverrier. A heated argument erupted between France and Britain. Astronomers decided to split the credit for the planet's discovery between Adams and Leverrier. Galle is considered the first to observe the planet. However, a review of Galileo's notes from the 1600s revealed that Galileo actually spotted the planet centuries before, but he thought it was a fixed star.

Neptune's Moons

Only weeks after Neptune's discovery its first moon was discovered. In early October 1846 the British astronomer William Lassell (1799–1880) spotted the moon. It was named Triton, after the son of Poseidon, the sea god. The name was suggested by the French astronomer Nicolas Camille Flammarion (1842–1925). In 1949 the Dutch American astronomer Gerard Peter Kuiper (1905–1973) found another moon around Neptune. According to NASA, in "Planetary Satellite Discovery Circumstances," in 1989 images from the spacecraft *Voyager 2* revealed six previously unknown moons. Five more moons were discovered between 2002 and 2003. They are named after characters associated with Neptune or Poseidon (his Greek counterpart) or other sea-related individuals in ancient mythology. In 2013 a 14th moon was discovered; as of February 2016, it had not been officially named.

DWARF PLANETS

Dwarf planets are a new category of celestial bodies. The designation was created in 2006 by IAU resolution. Although Pluto is the best-known dwarf planet, it was not the first one discovered. This distinction goes to Ceres, a small world named after a Roman goddess. Ceres was discovered in 1801 by the Italian astronomer Giuseppe Piazzi (1746–1826). He found it in the massive asteroid belt lying between Mars and Jupiter. Another dwarf planet is Eris, which was named after a Greek goddess. It was discovered in July 2005 by astronomers at the California Institute of Technology in Pasadena, California. Like Pluto, Eris is a trans-Neptunian object, meaning it lies beyond Neptune. According to NASA, in "Plutoids" (June 13, 2008, http://solarsystem.nasa.gov/galleries/plutoids), trans-Neptunian dwarf planets are called plutoids. Some dwarf planets have moons. For example, in September 2005 scientists at the Keck Observatory in Hawaii discovered that Eris has a moon. It was named Dysnomia after the daughter of the goddess Eris.

As of February 2016, NASA (http://solarsystem.nasa.gov/planets/dwarf) noted that two other dwarf planets—Makemake in the asteroid belt and the plutoid Haumea—have been confirmed by the IAU. Many other celestial objects may also qualify as dwarf planets once detailed data are obtained about them. For example, as of November 2015, the astronomer Mark Brown of the California Institute of Technology maintained a website (http://www.gps.caltech.edu/~mbrown/dps.html) listing more than 500 objects that he believes could be designated as dwarf planets. One of these objects, dubbed 2012VP$_{113}$ by its discoverers, was first described by Chadwick A. Trujillo and Scott S. Sheppard in "A Sedna-Like Body with a Perihelion of 80 Astronomical Units" (*Nature*, vol. 507, no. 7493, March 2014). According to the researchers, the trans-Neptunian object orbits far from the sun in an area known as the Oort cloud. In "Oort Cloud: In Depth" (2016, http://solarsystem.nasa.gov/planets/oort/indepth), NASA explains that the Oort cloud is named after the Dutch astronomer Jan Hendrik Oort (1900–1992) and is "believed to be a thick bubble of icy debris that surrounds our solar system."

Tombaugh Discovers Pluto

The American astronomer Clyde William Tombaugh (1906–1997) is credited with discovering the dwarf planet Pluto. Tombaugh made the discovery on February 18, 1930, while working at the Lowell Observatory in Flagstaff, Arizona. The name Pluto was finally selected for the planet from many suggestions made by the public. The name was originally suggested by an 11-year-old British girl named Venetia Burney. NASA reports in "Students Send an Instrument to Space" (2016, https://discoverynewfrontiers.nasa.gov/missions/missions_nh.cfml) that Burney (who had since married and taken the last name Phair) died at the age of 90 in 2009.

Pluto was the Greek god of the underworld and was able to make himself invisible. The name seemed appropriate for the darkest planet in the solar system that had been so difficult to find. It takes Pluto 248 Earth years to make one revolution of the sun. It has a highly elliptical orbit, in that sometimes it is closer to the sun than Neptune. This last occurred between 1979 and 1999. Pluto is a dark and icy world with a surface of frozen nitrogen, methane, and carbon dioxide. For decades it was observed and photographed only from great distances. However, NASA's *New Horizons* spacecraft, which is described later in this chapter, flew by Pluto in 2015 and collected data about the dwarf planet and its moons.

Pluto's Moons

Pluto's primary moon is Charon. It is named after a character in Greek mythology who ferried the souls of the dead across the river Styx to the underworld. On June 22, 1978, Charon was discovered by James W. Christy (1938–) at the U.S. Naval Observatory in Washington, D.C. Christy suggested the name that is now assigned to the moon. NASA notes in "Planetary Satellite Discovery Circumstances" that as of March 2015 Pluto had five known moons (Charon, Nix, Hydra, Kerberos, and Styx).

THE PIONEER MISSIONS

During the early 1970s the United States began a series of interplanetary missions that were designed to explore the far planets. The first of these missions was aptly named Pioneer. As described in Chapter 2, the first four Pioneer missions were launched during the 1950s and were devoted to lunar research. During the 1960s *Pioneer 5* through *Pioneer 9* collected solar data. During the 1970s NASA launched four Pioneer missions: *Pioneer 10* and *Pioneer 11* explored the outer planets, and *Pioneer 12* and *Pioneer 13* (the last of the series) explored Venus.

NASA (June 30, 1996, http://apod.nasa.gov/apod/ap960630.html) notes that a 6-inch (15.2-cm) by 9-inch (22.9-cm) metal plaque was mounted on each spacecraft. The plaque included illustrations of a human man and woman, the spacecraft's silhouette, and some mathematical, chemical, and astronomical data represented in binary code symbols. An image of the solar system at the bottom of the plaque shows a Pioneer spacecraft leaving Earth and passing between Jupiter and Saturn on its way out of the solar system.

Pioneer 10

On March 3, 1972, *Pioneer 10* was launched atop an Atlas-Centaur rocket from the Cape Canaveral Air Force Station in Florida. It was the first mission ever sent to the outer solar system. Ultimately, it became the first human-made object to travel past the planets of the solar system and head toward interstellar space.

Pioneer 10 was the first spacecraft to travel through the asteroid belt between Mars and Jupiter. Scientists had feared that this would be a dangerous area of space. They learned that the asteroids in the belt are spread far apart and do not pose a significant hazard to spacecraft flying through it.

In December 1973 *Pioneer 10* was the first spacecraft to investigate Jupiter. Its closest approach came within 124,000 miles (200,000 km) of the planet. *Pioneer 10* carried various instruments to study the solar wind, magnetic fields, cosmic radiation and dust, and hydrogen concentrations in space. Its Jupiter studies focused on the planet's magnetic effects, radio waves, and atmosphere. The atmospheres of Jupiter's satellites (particularly Io) were also investigated.

Over the years the instruments aboard the spacecraft began to fail or were turned off by NASA to conserve power. In 1997 NASA ceased routine tracking of the spacecraft because of budget reasons. The spacecraft was the most distant human-made object in space until February 1998, when it was passed (in total distance from Earth) by an even faster spacecraft called *Voyager 1* (which is traveling in an opposite direction). NASA notes in the press release "Pioneer 10 Spacecraft Sends Last Signal" (February 25, 2003, http://www.nasa.gov/centers/ames/news/releases/2003/03_25HQ.html) that the last detected signal from *Pioneer 10* was in January 2003. The signal was "very weak," and NASA engineers believe that the spacecraft lacks enough power to send another signal back to Earth. At the time, *Pioneer 10* was approximately 7.6 billion miles (12.2 billion km) away from Earth heading toward the star Aldebaran (the eye in the constellation Taurus), which is 82 light-years away. It will take the spacecraft over 2 million years to reach the star.

Pioneer 11

On April 6, 1973, the *Pioneer 11* spacecraft was launched into space by an Atlas-Centaur rocket. A year and a half later it flew by Jupiter on its way to Saturn. The spacecraft approached within 21,000 miles (33,800 km) of Jupiter. It was the first spacecraft to observe the planet's polar regions. It also returned detailed images of the Great Red Spot. Like its sister spacecraft, *Pioneer 11* investigated solar and cosmic phenomena and interplanetary and planetary magnetic fields during its journey.

In September 1979 *Pioneer 11* flew within 13,000 miles (20,900 km) of Saturn and returned the first close-up pictures of the planet and its rings. It continued past the planet toward the edge of the solar system. Routine mission operations were ended in September 1995, and NASA received the last transmission from the spacecraft the following month. According to NASA, in "The Pioneer Missions" (March 26, 2007, http://www.nasa.gov/centers/ames/missions/archive/pioneer.html), the spacecraft may still be sending out transmissions, but it cannot be maneuvered to point those signals back to Earth. Thus, no further transmission reception is expected. NASA notes that by the end of 1995 *Pioneer 11* was approximately 4 billion miles (6.4 billion km) from Earth. It is heading in the direction of the constellation of Aquila. It will reach one of the stars in the constellation in about 4 million years.

Pioneer and Plutonium

The Pioneer spacecraft were equipped with radioisotope power systems to take advantage of the heat that is released during the natural radioactive decay of a plutonium pellet. These power systems provided heat and electricity for the spacecraft. Although sending plutonium into space is controversial, NASA continues to use this power source on all of its missions to the far planets. The planets are too far from the sun to make solar power a feasible and reliable choice for these spacecraft.

THE VOYAGER PROGRAM

In 1977 NASA began another bold mission to investigate Jupiter and Saturn. The program was called Voyager, and it included twin robotic spacecraft named *Voyager 1* and *Voyager 2*. An illustration of a Voyager spacecraft is shown in Figure 8.4.

The various instruments on board were designed to detect and measure the solar wind and other charged particles, cosmic radiation, magnetic field intensities, and plasma waves. The original five-year Voyager mission was so successful that it was extended to include flybys of Uranus and Neptune.

The Voyager Interplanetary Mission

Both spacecraft were launched into space atop Titan rockets. *Voyager 2* was the first to launch, on August 20, 1977. It was followed on September 5, 1977, by *Voyager 1*. Both spacecraft traveled for two years to fly by Jupiter. They made several scientific observations as they passed Jupiter and continued on to Saturn. *Voyager 1* was on a faster trajectory than *Voyager 2*, so it reached the planet first. *Voyager 2* was directed to fly by Uranus and Neptune. It was the first spacecraft to do so.

The Voyager spacecraft were two of the most successful in NASA's history. In "Voyager Project Information" (March 2, 2015, http://nssdc.gsfc.nasa.gov/planetary/voyager.html) and "Fact Sheet: The Voyager Interplanetary Mission" (2016, http://voyager.jpl.nasa.gov/news/factsheet.html), NASA describes a number of discoveries that were revealed by the Voyager spacecraft about the gas giants in the outer solar system:

- Jupiter, Uranus, and Neptune have faint ring systems.
- Jupiter has a complicated atmosphere in which auroras are common.
- Jupiter's moon Io has active volcanoes.
- Jupiter's moon Europa has a smooth surface composed of water ice.
- Jupiter and its moons lie within an intense radiation belt.
- Saturn's ringlets are not as uniform and separate as expected—some are kinked or braided together, and additional gaps between rings were discovered.

FIGURE 8.4

The *Voyager* spacecraft

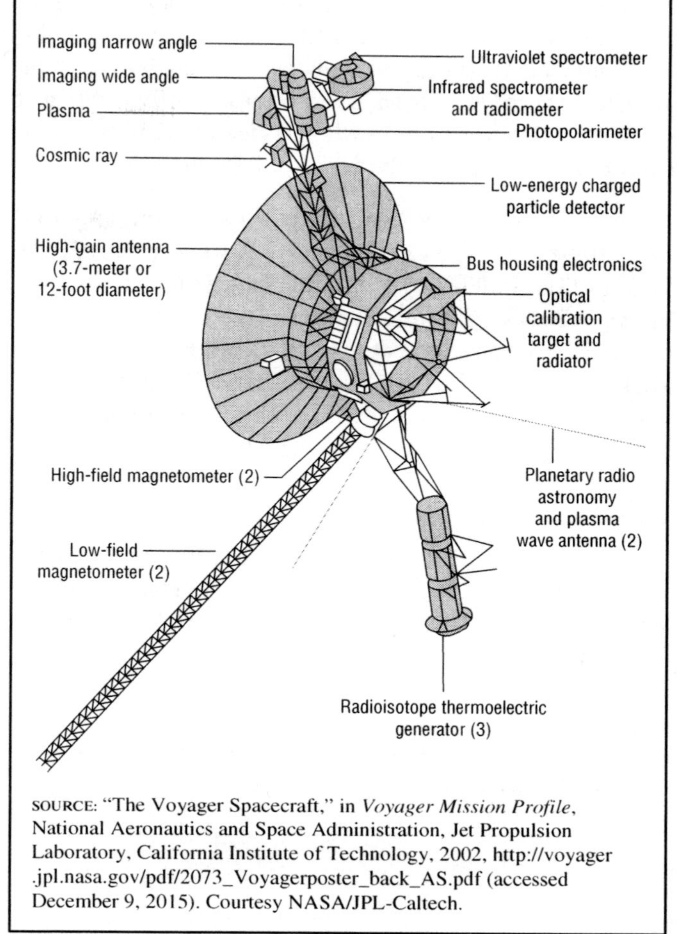

SOURCE: "The Voyager Spacecraft," in *Voyager Mission Profile*, National Aeronautics and Space Administration, Jet Propulsion Laboratory, California Institute of Technology, 2002, http://voyager.jpl.nasa.gov/pdf/2073_Voyagerposter_back_AS.pdf (accessed December 9, 2015). Courtesy NASA/JPL-Caltech.

- Saturn's weather is relatively tame compared with Jupiter's weather.
- Saturn's largest moon, Titan, has a dense smoggy atmosphere that contains nitrogen and carbon-containing compounds.
- Saturn's moon Mimas has a massive impact crater.
- Jupiter, Saturn, and Neptune have auroral zones (areas in the upper atmosphere in which magnetic phenomena cause light displays to appear, such as Earth's northern and southern lights, which are known as aurora borealis and aurora australis, respectively).
- Uranus and Neptune have magnetospheres.
- Neptune's atmosphere experiences large-scale storms, including one massive storm dubbed "the Great Dark Spot."
- Neptune's moon Triton has "geyser-like structures" on its surface and an atmosphere.

The discovery of water ice on the surface of Europa was particularly exciting, because it raises the possibility that there is liquid water underneath the ice.

The Voyager Interstellar Mission

The Voyager spacecraft proved to be so hardy after completing their planetary missions that they were sent on a new mission in 1989 called the Voyager Interstellar Mission (VIM). *Voyager 1* became the most distant human-made object in space in February 1998, when it reached a distance of 6.5 billion miles (10.5 billion km) from Earth, surpassing the record of *Pioneer 10*.

According to NASA (http://voyager.jpl.nasa.gov/where/index.html), as of February 2016 *Voyager 1* was 12.5 billion miles (20.1 billion km) from Earth, and *Voyager 2* was 10.3 billion miles (16.6 billion km) from Earth. The agency indicates in "Operations Plan to End Mission" (May 18, 2015, http://voyager.jpl.nasa.gov/science/thirty.html) that the spacecraft are expected to continue operating until at least 2020.

Messages from Earth

In "Golden Record" (2016, http://voyager.jpl.nasa.gov/spacecraft/goldenrec.html), NASA explains that the Voyager spacecraft carry written and recorded messages from Earth, in case they come across any intelligent life. Attached to each spacecraft is a 12-inch (30.5-cm) gold-plated copper disk inside a protective aluminum case. The cover of the protective case has symbolic instructions for playing the disc and a diagram of Earth's location in the solar system carved into it. The disks contain recorded greetings in 55 different languages and various other sounds, including bits of music and natural and human-made sounds. There are 115 images that are encoded in analog form on the disks of various Earth scenes. These include pictures of people, objects, and places from around the world. The disks carry printed messages from President Jimmy Carter (1924–) and Kurt Waldheim (1918–2007), the secretary-general of the United Nations.

THE GALILEO MISSION

NASA's Galileo mission was the first to put a spacecraft in orbit around one of the far planets. The mission to Jupiter included a scientific probe that left the orbiter and plunged into the planet's atmosphere. A diagram of the spacecraft including the descent probe is provided in Figure 8.5. The probe was 4 feet (1.2 m) in diameter and 3 feet (0.9 m) long. The mission was operated by NASA's Jet Propulsion Laboratory in Pasadena, California.

FIGURE 8.5

The *Galileo* spacecraft

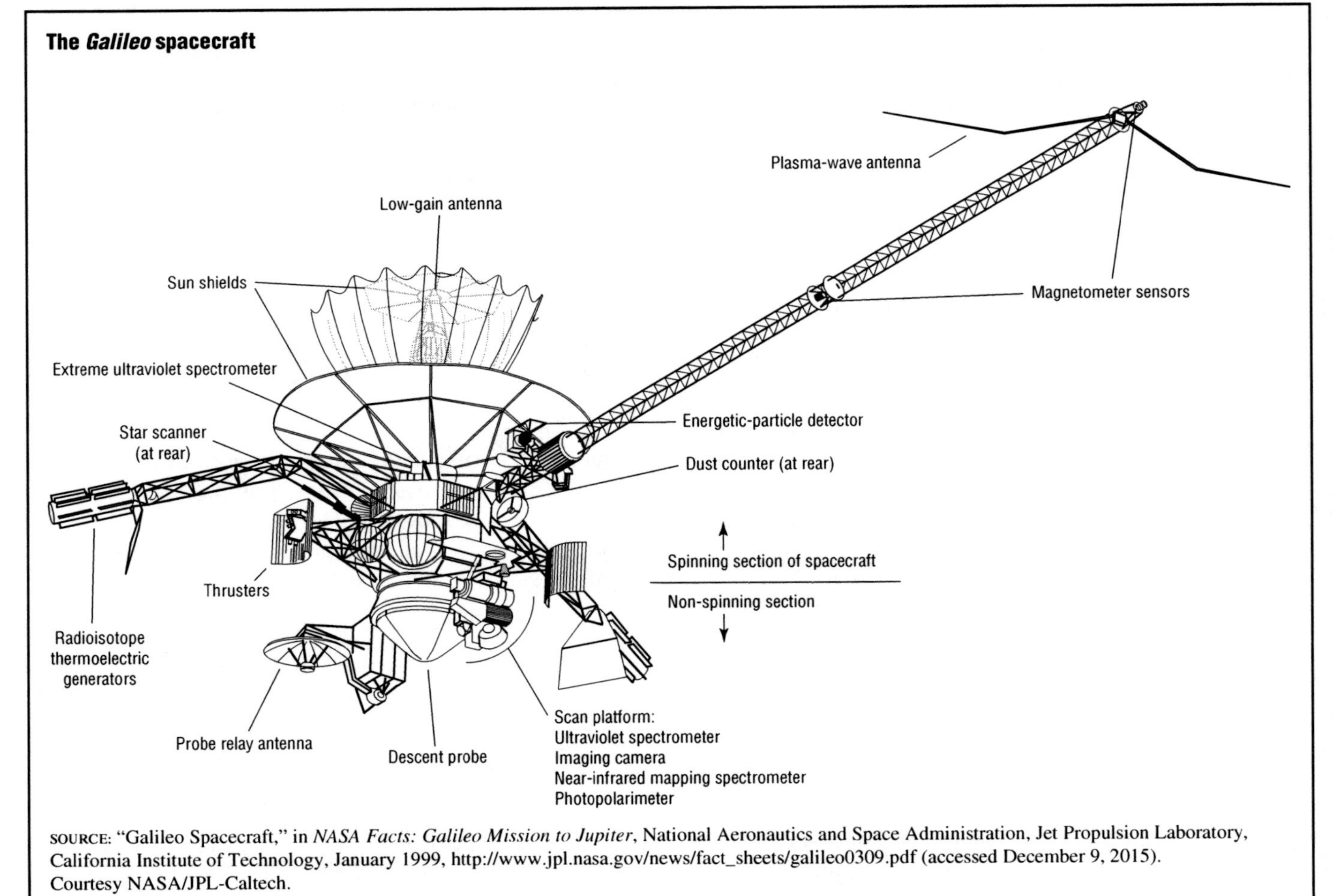

SOURCE: "Galileo Spacecraft," in *NASA Facts: Galileo Mission to Jupiter*, National Aeronautics and Space Administration, Jet Propulsion Laboratory, California Institute of Technology, January 1999, http://www.jpl.nasa.gov/news/fact_sheets/galileo0309.pdf (accessed December 9, 2015). Courtesy NASA/JPL-Caltech.

On October 18, 1989, the space shuttle *Atlantis* lifted off from the Kennedy Space Center in Florida with the *Galileo* spacecraft on board. The shuttle astronauts released *Galileo* in Earth orbit, then the craft used its two-stage inertial upper stage rocket to boost itself toward Venus.

The spacecraft swung by Venus once and Earth twice as part of gravity assist maneuvers. These are maneuvers in which a spacecraft flies in close enough to a planet to get a boost from the orbital momentum of a planet traveling around the sun. In "Gravity Assists/Flybys: A Quick Gravity Assist Primer" (2016, http://saturn.jpl.nasa.gov/mission/missiongravityassistprimer/), NASA compares a gravity assist to throwing a table tennis ball to skim along the top of one of the moving blades of an electric fan. The blades circle the fan's motor at a high rate of speed. The ball gets close enough to one of the blades to pick up momentum and shoot off in a different direction. Using gravity assists during space flight saves on fuel. This is particularly important for long journeys to the outer solar system.

By July 1995 *Galileo* was nearing Jupiter. It released the probe, which began a five-month plunge toward the planet. On December 7, 1995, the orbiter was in position when the probe began its final descent at more than 106,000 miles per hour (171,000 km/h). According to NASA, in "Solar System Exploration: Galileo" (July 9, 2010, https://solarsystem.nasa.gov/galileo/mission/journey-probe.cfm), for 58 minutes the heavily protected probe transmitted data about Jupiter's atmosphere, temperature, and weather. It was eventually destroyed by the intense heat and pressure surrounding the planet. It had penetrated 124 miles (200 km) into the violent atmosphere.

The orbiter spent the next eight years in orbit around Jupiter. It conducted many flybys of the moons Europa, Ganymede, and Callisto and used its 11 scientific instruments to collect data about radiation, magnetic fields, charged particles, and cosmic dust. By September 2003 the orbiter was running low on propellant, so NASA scientists decided to destroy the spacecraft by purposely plunging it into Jupiter's atmosphere. The scientists feared that it could run out of fuel and crash into one of Jupiter's moons. This could contaminate environments that might contain water and life forms.

The Galileo mission was hugely successful. NASA indicates in *Galileo End of Mission Press Kit* (September 2003, http://www.jpl.nasa.gov/news/press_kits/galileo-end.pdf) that the spacecraft traveled more than 2.8 billion miles (4.6 billion km) during its long journey, capturing thousands of detailed images of the planet and its largest moons and collecting a wealth of data about these celestial objects. According to NASA, the top-10 science results of the mission were:

- Information about Jupiter's evolution since the planet formed
- First observation of ammonia clouds in the atmosphere of another planet
- Details about extensive volcanic activity on Jupiter's moon Io
- Information about Io's atmospheric plasma interactions
- Evidence of liquid oceans on Jupiter's moon Europa
- Evidence of a magnetic field on Jupiter's moon Ganymede
- Evidence of a saltwater layer on Jupiter's moons Europa, Callisto, and Ganymede
- Evidence of a thin atmospheric layer on Jupiter's moons Europa, Callisto, and Ganymede
- Details about Jupiter's ring system formation
- Information about Jupiter's magnetosphere

In addition, the mission provided the first flyby and images of the asteroids Gaspra and Ida and showed scientists the first-known moon around an asteroid (Dactyl, which orbits Ida). It also provided the only direct observation of the 1994 impact of the comet Shoemaker-Levy 9 with Jupiter.

THE CASSINI-HUYGENS MISSION

In 1997 NASA collaborated with the European Space Agency (ESA) and the Agenzia Spaziale Italiana (Italian Space Agency) to launch the Cassini-Huygens mission to Saturn. The *Cassini* orbiter was designed to orbit the planet for four years and to release a probe that would land on Titan, Saturn's largest moon. (See Figure 8.6.) Specific mission objectives were to investigate Saturn's magnetosphere and atmosphere, determine the structure and behavior of its rings, and characterize the composition, weather, and geological history of its moons.

On October 15, 1997, the spacecraft was launched atop a Titan IV Centaur rocket from the Cape Canaveral Air Force Station. Over the next three years it received two gravity assists from Venus and one each from Earth and Jupiter. *Cassini* arrived at Saturn in July 2004, becoming the first spacecraft ever to orbit the planet.

The *Cassini* orbiter is equipped with 12 scientific instruments. It also carried the *Huygens* probe with six instruments of its own. The probe was released on December 25, 2004, and began its three-week journey to the surface of Titan. It penetrated the thick cloud cover that hides the moon and touched down on January 14, 2005. The probe sampled Titan's atmosphere and provided the first photographs ever of its surface. The probe was active for nearly two and a half hours during its

FIGURE 8.6

The *Cassini* spacecraft

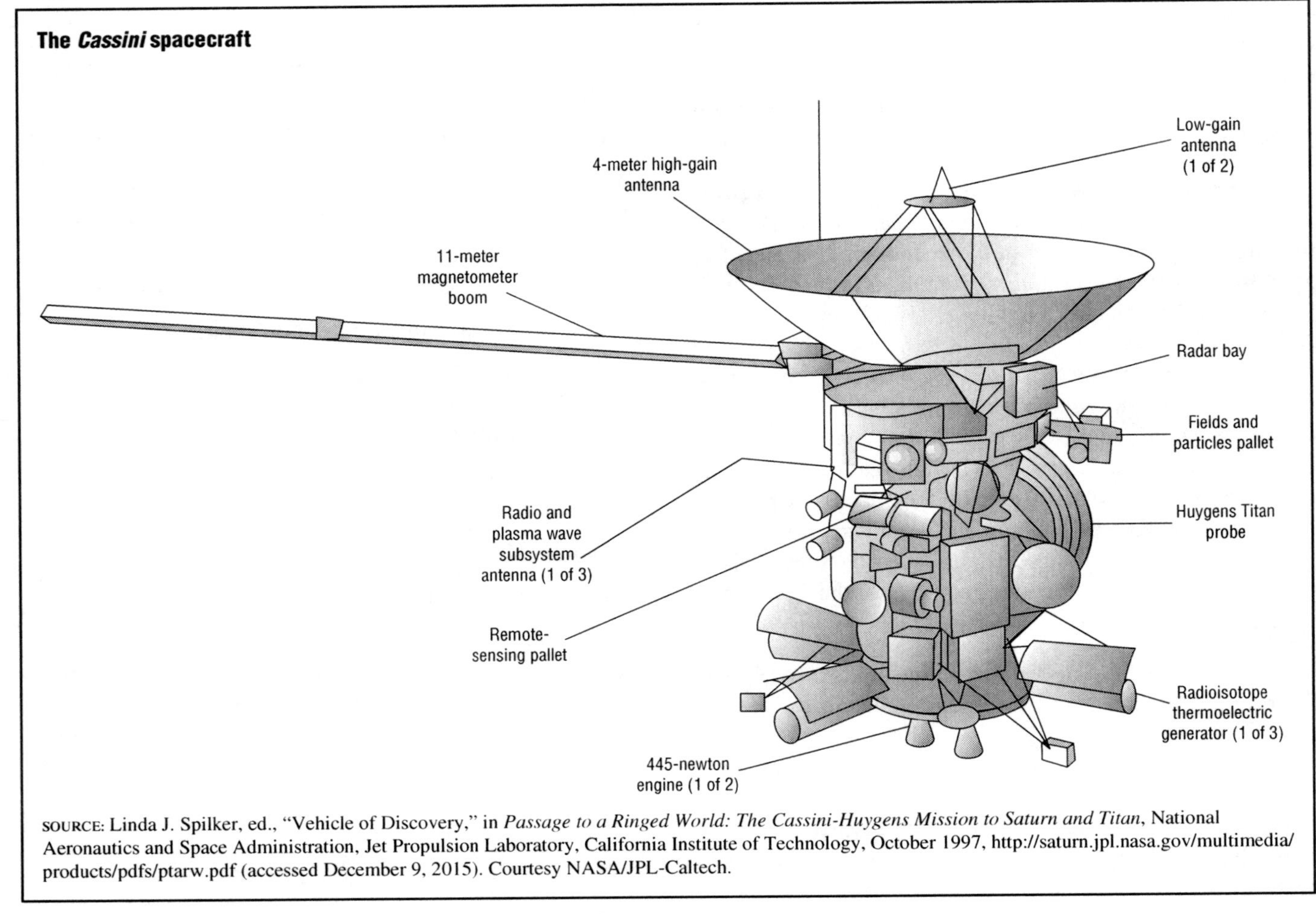

SOURCE: Linda J. Spilker, ed., "Vehicle of Discovery," in *Passage to a Ringed World: The Cassini-Huygens Mission to Saturn and Titan*, National Aeronautics and Space Administration, Jet Propulsion Laboratory, California Institute of Technology, October 1997, http://saturn.jpl.nasa.gov/multimedia/products/pdfs/ptarw.pdf (accessed December 9, 2015). Courtesy NASA/JPL-Caltech.

descent and another hour and 12 minutes after landing before its battery power ceased.

In "Cassini-Huygens Top 10 Science Highlights" (July 20, 2005, http://www.nasa.gov/mission_pages/cassini/whycassini/science-top-ten_p1of3_20050720.html), NASA notes that the *Cassini* orbiter revealed important findings during its first few months of operation. Chief among them were surprises about Titan's surface and organic atmosphere. The surface does not include global oceans as scientists expected, but is Earthlike in some ways. There is evidence of volcanoes, erosion, craters, dunes, and dry and wet lake beds. Titan's atmosphere contains organic chemicals, such as benzene and methane. Scientists believe the moon experiences methane showers from clouds sweeping overhead. The first detailed images of Phoebe reveal it is scarred with many large craters. There is evidence of water ice and silicate and organic materials on the surface.

Mission Extensions

In 2008 the Cassini-Huygens mission was extended through 2010 and called the Cassini Equinox mission (February 25, 2011, http://sci.esa.int/cassini-huygens/43181-cassini-tour-equinox-mission/). Scientists used the orbiter to study Saturn's equinox, which occurred in August 2009. At that time the sun began to shine on the Northern Hemisphere of the planet.

In 2010 NASA began another extension called the Cassini Solstice mission (2016, http://saturn.jpl.nasa.gov/mission/introduction/), which will extend through September 2017 and allow scientists to capture data during the Saturnian summer solstice in May 2017. Other research targets include Saturn's magnetosphere and rings and further studies of the moons Titan, Enceladus, Dione, Rhea, and Mimas.

Mission Highlights

NASA provides in "Cassini 10 Years at Saturn Top 10 Discoveries" (June 25, 2014, http://saturn.jpl.nasa.gov/news/cassinifeatures/10thannivdiscoveries/) a list of mission highlights from the date of launch through 2014. Specifically, the agency notes the discovery of icy plumes on Saturn's moon Enceladus and the presence of rain, rivers, lakes, and seas on Saturn's moon Titan.

In "2016 Saturn Tour Highlights" (2016, http://saturn.jpl.nasa.gov/mission/saturntourdates/), NASA provides a detailed timeline of the mission events that are scheduled for 2016, such as flybys of Saturn's moon Titan. The agency indicates the types of observations that will be made and the data that will be collected. NASA (2016, http://saturn.jpl.nasa.gov/photos/halloffame/) also maintains a photo gallery containing dozens of "Hall of Fame" images chosen from the hundreds of thousands of images taken during the mission.

The Cassini mission is called a "flagship" mission by NASA. As noted in Chapter 6, flagship missions are particularly expensive missions, because they feature significant "scope, strategic objectives, and technology challenges." In "Mission Overview" (2016, http://saturn.jpl.nasa.gov/mission/quickfacts/), the agency indicates that the mission's total lifetime cost to the United States is estimated at $2.6 billion. European partners spent another $660 million in development and operation costs.

THE NEW HORIZONS MISSION

As mentioned earlier, the Kuiper Belt consists of many icy worlds in a vast region that lies beyond Neptune in the solar system. (See Figure 8.7.) Astronomers refer to celestial bodies in this region as Kuiper Belt Objects (KBOs). Since the first discovery of a KBO in 1992, scientists have determined that there are many thousands of these objects. Pluto is now considered a KBO, as are the recently discovered objects Eris, Quaoar, Orcus, and Varuna. These objects lie far from Earth, and little is known about them.

FIGURE 8.7

Locations of the Kuiper Belt and the orbit of Eris in the outer solar system

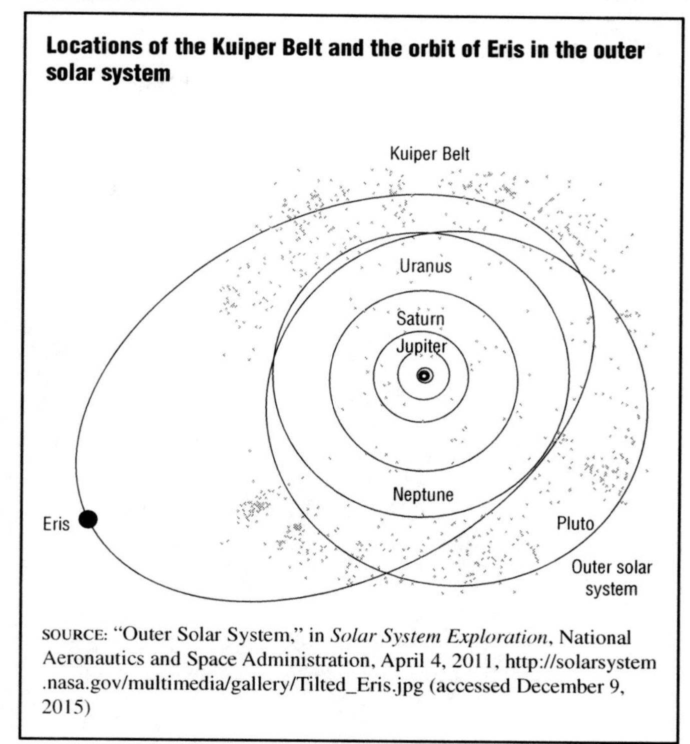

SOURCE: "Outer Solar System," in *Solar System Exploration*, National Aeronautics and Space Administration, April 4, 2011, http://solarsystem.nasa.gov/multimedia/gallery/Tilted_Eris.jpg (accessed December 9, 2015)

The *New Horizons* spacecraft was launched toward Pluto and the Kuiper Belt on January 19, 2006, aboard an Atlas V rocket. (See Figure 8.8.) More than a year later the spacecraft passed by Jupiter. It is the first mission to be conducted under NASA's New Frontiers Program. The spacecraft is operated for NASA by the Johns Hopkins University Applied Physics Laboratory in Laurel, Maryland.

In July 2015 *New Horizons* made its closest approach to Pluto and its moons. According to NASA, in "NASA's New Horizons Spacecraft Begins Intensive Data Downlink Phase" (September 4, 2015, http://www.nasa.gov/feature/nasa-s-new-horizons-spacecraft-begins-intensive-data-downlink-phase), the spacecraft was approximately 3 billion miles (4.8 billion km) from Earth at that time. Its scientific instruments assessed the geology and atmosphere of Pluto and its primary moon Charon and mapped their surface compositions. In "New Horizons Finds Blue Skies and Water Ice on Pluto" (October 8, 2015, https://www.nasa.gov/nh/nh-finds-blue-skies-and-water-ice-on-pluto), NASA states that Pluto's atmosphere contains tiny soot-like particles called tholins that are reddish in color. They scatter incoming sunlight, which gives the planet's sky a bluish haze. In addition, *New Horizons* data indicate there are "numerous small, exposed regions of water ice on Pluto." These icy regions are also reddish in color.

According to NASA (http://pluto.jhuapl.edu/Mission/Where-is-New-Horizons/index.php), as of February 2016 *New Horizons* had moved beyond Pluto and was scheduled to perform flybys of objects in the Kuiper Belt.

THE JUNO MISSION

On August 5, 2011, NASA launched the spacecraft *Juno* atop an Atlas rocket on its way toward Jupiter. (See Figure 8.9.) The mission is part of NASA's New Frontiers Program. In 2016 the orbiter will assume a polar orbit around Jupiter and study the planet's geology, atmosphere, and climate. NASA notes in "Juno Overview" (September 21, 2015, http://www.nasa.gov/mission_pages/juno/overview/index.html) that the mission's primary goal is "to understand the origin and evolution of Jupiter." The spacecraft is designed to do tasks such as "investigate the existence of a solid planetary core, map Jupiter's intense magnetic field, measure the amount of water and ammonia in the deep atmosphere, and observe the planet's auroras."

According to NASA, in "Juno Spacecraft and Instruments" (July 30, 2015, http://www.nasa.gov/mission_pages/juno/spacecraft/#.Uw4cKp5dWrY), the orbiter is carrying dozens of sensors and scientific instruments. One of the instruments—the JunoCam—will capture images that will be disseminated to the public. As of February 2016, NASA (http://www.nasa.gov/mission_pages/juno/

FIGURE 8.8

***New Horizons* spacecraft**

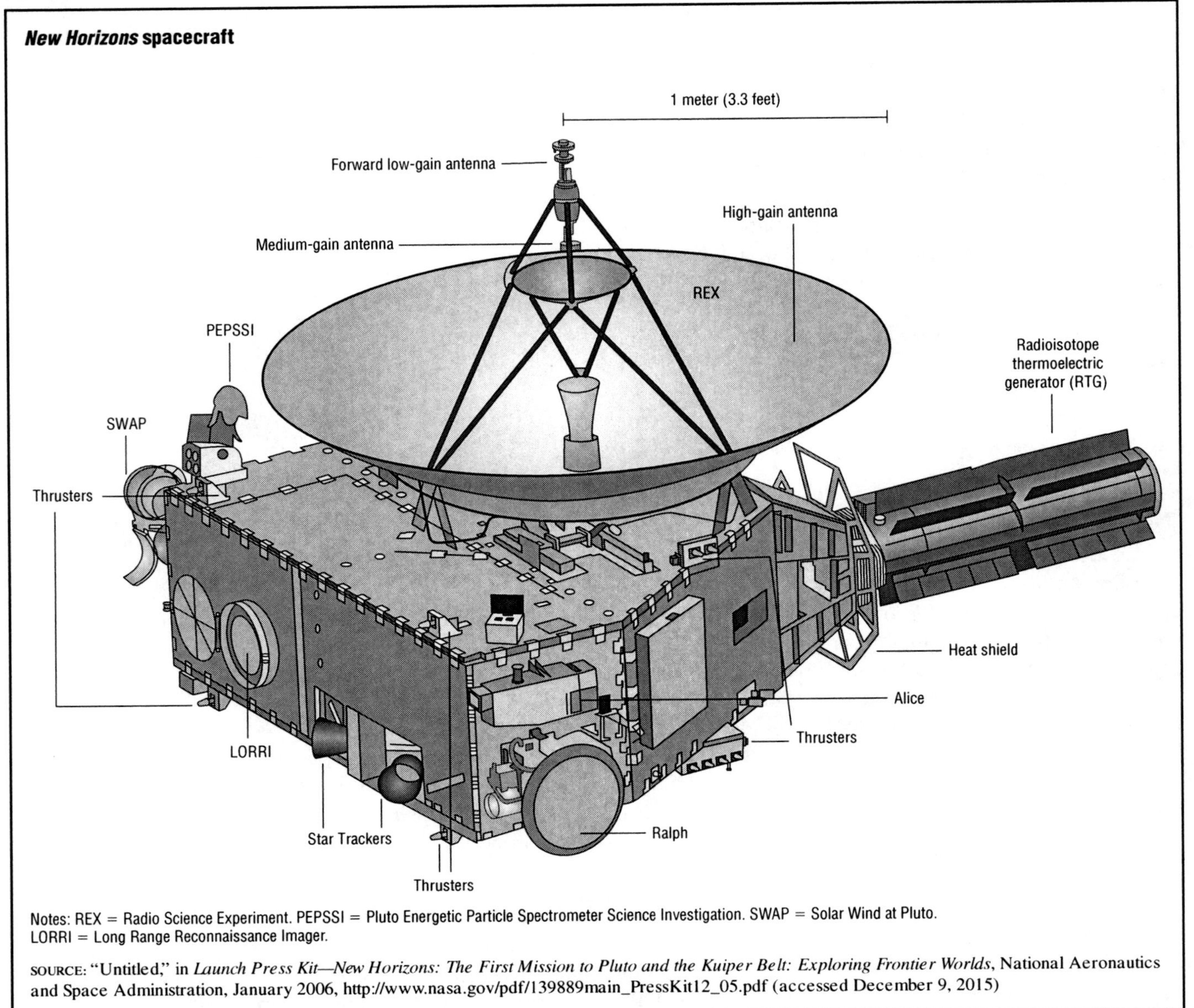

Notes: REX = Radio Science Experiment. PEPSSI = Pluto Energetic Particle Spectrometer Science Investigation. SWAP = Solar Wind at Pluto. LORRI = Long Range Reconnaissance Imager.

SOURCE: "Untitled," in *Launch Press Kit—New Horizons: The First Mission to Pluto and the Kuiper Belt: Exploring Frontier Worlds*, National Aeronautics and Space Administration, January 2006, http://www.nasa.gov/pdf/139889main_PressKit12_05.pdf (accessed December 9, 2015)

where) reported that *Juno* was approximately 48 million miles (77 million km) from Jupiter and was expected to reach the planet in July 2016.

OCEAN WORLDS

Multiple space missions have revealed (or hinted at) the presence of water or water ice on various far planets and their moons and some dwarf planets. Figure 8.10 compares Earth to other worlds believed to have watery oceans. In "Ocean Worlds" (2015, http://www.jpl.nasa.gov/infographics/uploads/infographics/print/11262.pdf), NASA calls these worlds "the best known candidates in our search for life in the solar system." The agency explains in "The Solar System and Beyond Is Awash in Water" (April 7, 2015, http://www.jpl.nasa.gov/news/news.php?feature=4541) that that there are "three ingredients needed for life as we know it: liquid water, essential chemical elements for biological processes, and sources of energy that could be used by living things." In this respect, NASA believes that the most promising worlds are Jupiter's moon Europa and Saturn's moon Enceladus.

FUTURE MISSIONS TO THE FAR PLANETS AND BEYOND

Chapter 6 describes NASA's Science Mission Directorate and its planetary science program, which includes various mission lines. These include relatively low-cost missions operated under the Discovery Program or other lines and the much more expensive flagship missions. During the early 2010s budget constraints forced NASA to limit its participation in flagship planetary missions. A joint NASA-ESA mission called Europa Jupiter System

FIGURE 8.9

***Juno* spacecraft**

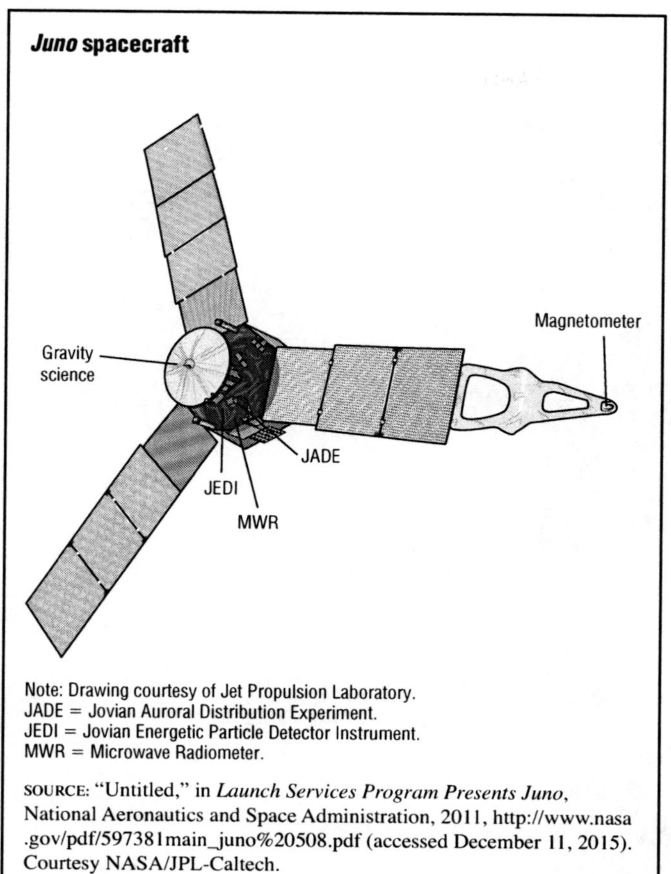

Note: Drawing courtesy of Jet Propulsion Laboratory.
JADE = Jovian Auroral Distribution Experiment.
JEDI = Jovian Energetic Particle Detector Instrument.
MWR = Microwave Radiometer.

SOURCE: "Untitled," in *Launch Services Program Presents Juno*, National Aeronautics and Space Administration, 2011, http://www.nasa.gov/pdf/597381main_juno%20508.pdf (accessed December 11, 2015). Courtesy NASA/JPL-Caltech.

Mission–Laplace (EJSM-Laplace) was expected to be NASA's next flagship effort and launched sometime during the 2020s. However, this plan was abandoned in 2011 because of funding concerns. As of February 2016, the agency was planning a smaller-scale mission that will launch during the early 2020s. Table 8.1 lists the instruments that have been selected to fly on the spacecraft. The orbiter will circle Jupiter and conduct multiple flybys of its moon Europa.

In January 2012 the ESA (February 25, 2016, http://sci.esa.int/juice/) announced that the EJSM-Laplace mission had been reformulated into a European-led mission called Jupiter Icy Moons Explorer. As of February 2016, the mission was planned for launch in 2022. The orbiter will arrive near Jupiter in 2030 and collect data about the planet and three of its icy moons: Callisto, Europa, and Ganymede.

FIGURE 8.10

Some planetary bodies with known or suspected watery oceans

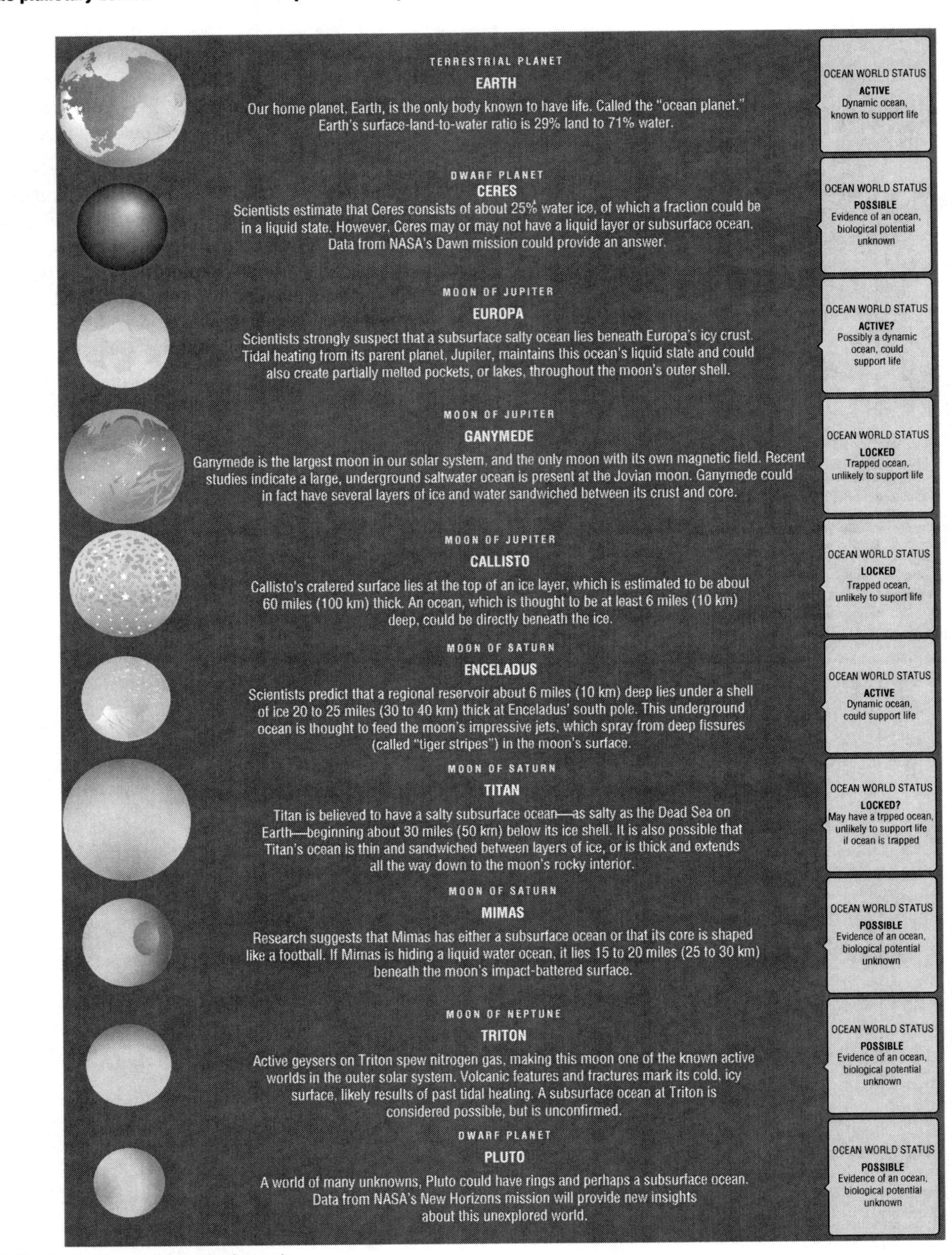

Note: Planetary bodies are not shown exactly to scale.

SOURCE: Adapted from "Ocean Worlds," in *JPL Infographics*, National Aeronautics and Space Administration, Jet Propulsion Laboratory, California Institute of Technology, undated, http://www.jpl.nasa.gov/infographics/infographic.view.php?id=11262 (accessed December 9, 2015)

TABLE 8.1

Instruments selected for NASA spacecraft under development to travel to Saturn's moon Europa in the 2020s

Instrument	Description
Plasma Instrument for Magnetic Sounding (PIMS)	This instrument works in conjunction with a magnetometer and is key to determining Europa's ice shell thickness, ocean depth, and salinity by correcting the magnetic induction signal for plasma currents around Europa.
Interior Characterization of Europa using Magnetometry (ICEMAG)	This magnetometer will measure the magnetic field near Europa and—in conjunction with the PIMS instrument—infer the location, thickness and salinity of Europa's subsurface ocean using multi-frequency electromagnetic sounding.
Mapping Imaging Spectrometer for Europa (MISE)	This instrument will probe the composition of Europa, identifying and mapping the distributions of organics, salts, acid hydrates, water ice phases, and other materials to determine the habitability of Europa's ocean.
Europa Imaging System (EIS)	The wide and narrow angle cameras on this instrument will map most of Europa at 50 meter (164 foot) resolution, and will provide images of areas of Europa's surface at up to 100 times higher resolution.
Radar for Europa Assessment and Sounding: Ocean to Near-surface (REASON)	This dual-frequency ice penetrating radar instrument is designed to characterize and sound Europa's icy crust from the near-surface to the ocean, revealing the hidden structure of Europa's ice shell and potential water within.
Europa Thermal Emission Imaging System (E-THEMIS)	This "heat detector" will provide high spatial resolution, multi-spectral thermal imaging of Europa to help detect active sites, such as potential vents erupting plumes of water into space.
MAss SPectrometer for Planetary EXploration/Europa (MASPEX)	This instrument will determine the composition of the surface and subsurface ocean by measuring Europa's extremely tenuous atmosphere and any surface material ejected into space.
Ultraviolet Spectrograph/Europa (UVS)	This instrument will adopt the same technique used by the Hubble Space Telescope to detect the likely presence of water plumes erupting from Europa's surface. UVS will be able to detect small plumes and will provide valuable data about the composition and dynamics of the moon's rarefied atmosphere.
Surface Dust Mass Analyzer (SUDA)	This instrument will measure the composition of small, solid particles ejected from Europa, providing the opportunity to directly sample the surface and potential plumes on low-altitude flybys.

SOURCE: "The NASA selectees are," in *NASA's Europa Mission Begins with Selection of Science Instruments*, National Aeronautics and Space Administration, May 26, 2015, http://www.nasa.gov/press-release/nasa-s-europa-mission-begins-with-selection-of-science-instruments (accessed December 9, 2015)

CHAPTER 9
PUBLIC OPINION ABOUT SPACE EXPLORATION

Space... the expensive frontier.

—Adam Anderson (*Bizmology*, July 21, 2009)

Humans seem to have an inherent desire to surmount great obstacles and push into new frontiers. There have always been brave people willing to risk their lives on bold and dangerous journeys into uncharted territory. They have climbed Mount Everest, traversed wild jungles, crossed barren deserts, and sailed stormy seas. Successful explorers become popular heroes. Their achievements thrill and delight people who do not have the ability, resources, or courage to go themselves.

The U.S. space program taps into this spirit of adventure. Astronauts became the heroic explorers of the 20th century. They opened new frontiers and set foot on the moon. However, these successes were achieved at a high price. They cost the country human lives and billions of dollars that some critics say could have been spent feeding the poor, healing the sick, and housing the homeless. Was it worth it?

Space exploration is appealing on a psychological level. It is awesome, daring, and closely associated with American can-do optimism and patriotic pride. A robust space program also showcases and strengthens U.S. capabilities in science, engineering, and technology. These are powerful motivations to keep venturing out into space.

However, the United States faces a number of expensive problems: incurable diseases, crime, poverty, pollution, climate change, unemployment, terrorism, and war. People who are concerned with poor social conditions resent the billions of dollars that are spent on exploring outer space. Within the scientific community, many respected researchers would rather see scarce funds devoted to Earth-related research than to space science. There are promising scientific and medical frontiers on this planet that still need exploring.

In a democratic society the public gets to weigh the relative costs and benefits of national goals and decide which ones to pursue. Public opinion polls show that most Americans have an uneasy devotion to the nation's space travel agenda. They love the idea but hate paying the bill. Sometimes they wonder if money that is spent on space exploration might be better spent on Earth-based issues. It is an ongoing debate that has raged since the earliest days of space exploration.

IS SPACE EXPLORATION IMPORTANT TO SOCIETY?

In November 1999, as the 20th century came to a close, Frank Newport, David W. Moore, and Lydia Saad of the Gallup Organization asked people to rank 18 specific events of the 20th century in order of importance and reported the results in *The Most Important Events of the Century from the Viewpoint of the People* (December 6, 1999, http://www.gallup.com/poll/3427/Most-Important-Events-Century-From-Viewpoint-People.aspx). Landing a man on the moon ranked seventh in the list of "most important events of the century." (See Table 9.1.) This put it behind major events associated with World War I (1914–1918), World War II (1939–1945), and important social milestones that granted rights to women and minorities.

The last space shuttle flight took place during the summer of 2011. This event prompted researchers to poll Americans on their views regarding the nation's space program. In June 2011 the Pew Research Center for the People and the Press (http://people-press.org/files/legacy-questionnaires/June11%20space%20topline%20for%20release.pdf) conducted a political survey that included questions about the space program. Poll participants were asked whether they believed it is "essential or not essential that the United States continue to be a world leader in space exploration." More than half (58%) said it is essential, whereas 38% said it is not essential. The remaining 4% did not know or refused to answer.

TABLE 9.1

Public ranking of the most important events of the 20th century, 1999

	(MI) %	(I) %	(SI) %	(NI) %	(DK) %
1. World War II	71	21	5	2	1
2. Women gaining the right to vote in 1920	66	20	11	3	0
3. Dropping the atomic bomb on Hiroshima in 1945	66	20	9	4	1
4. The Nazi Holocaust during World War II	65	20	9	5	1
5. Passage of the 1964 Civil Rights Act	58	26	13	2	1
6. World War I	53	28	11	5	3
7. Landing a man on the moon in 1969	50	30	15	5	*
8. The assassination of President Kennedy in 1963	50	29	15	5	1
9. The fall of the Berlin Wall in 1989	48	30	19	3	0
10. The U.S. Depression in the 1930s	48	29	18	3	2
11. The breakup of the Soviet Union in the early 1990s	46	31	19	3	1
12. The Vietnam War in the 1960s and early 1970s	37	31	20	11	1
13. Charles Lindbergh's transatlantic flight in 1927	27	33	28	11	1
14. The launching of the Russian Sputnik satellites in the 1950s	25	34	28	12	1
15. The Korean War in the early 1950s	21	36	32	10	1
16. The Persian Gulf War in 1991	18	38	32	11	1
17. The impeachment of President Bill Clinton in 1998	15	25	24	35	1
18. The Watergate scandal involving Richard Nixon in the 1970s	14	32	34	19	1

(MI) = Most important.
(I) = Important.
(SI) = Somewhat important.
(NI) = Not important.
(DK) = Don't know.
* = Less than 0.5%.

SOURCE: Frank Newport, David W. Moore, and Lydia Saad, "The Results," in *The Most Important Events of the Century from the Viewpoint of the People*, The Gallup Organization, December 6, 1999, http://www.gallup.com/poll/3427/Most-Important-Events-Century-From-Viewpoint-People.aspx (accessed December 9, 2015).

A similar question was posed by ORC International (http://i2.cdn.turner.com/cnn/2011/images/07/21/poll.july21.pdf) in a poll that was conducted in July 2011 for CNN. When asked "How important do you think it is for the United States to be ahead of Russia and other countries in space exploration?," the respondents were mixed in their opinions. More than a third (38%) said "very important," and a quarter (26%) said "fairly important." However, 36% said it is "not too important" to them. The remaining 1% had no opinion on the matter.

The Pew Research Center for the People and the Press asked poll participants how much they thought the U.S. space program had contributed to the following: "Scientific advances that all Americans can use," "This country's national pride and patriotism," and "Encouraging people's interest in science and technology." The results indicate that just over one-third of respondents each thought the U.S. space program had contributed "a lot" or "some" to these purposes. Another 16% to 20% said the space program had contributed "not much." Between 5% and 8% indicated the space program had contributed "nothing at all." The remaining 4% to 5% either did not know or refused to answer.

In "The Space Shuttle Program" (CBSNews.com, July 8, 2011), CBS News provides the results of a poll it conducted on the same issue in June 2011. Poll participants were asked the following question: "Since the moon landing more than 40 years ago, do you think the U.S. space program has accomplished more than you expected, less than you expected, or about what you expected it to accomplish by now?" Only 23% said the space program had accomplished "more" than they expected, 35% said it had accomplished "about" what they expected, and 36% said it had accomplished "less" than they expected. The remaining 6% did not know or did not answer the question.

SHOULD SPACE TRAVEL BE A SPENDING PRIORITY?

History shows that space travel was a high priority during the 1960s. President John F. Kennedy (1917–1963) and Vice President Lyndon B. Johnson (1908–1973) were convinced that putting a man on the moon was vital to U.S. political interests during the cold war. They persuaded Congress to devote billions of dollars to the effort. At the time, the public was not enthusiastic about the idea. According to the Gallup Organization, most polls it conducted during the 1960s showed that less than 50% of Americans considered the endeavor worth the cost.

In *Where Do We Go from Here: Chaos or Community?* (1967), the civil rights leader Martin Luther King Jr. (1929–1968) said, "Without denying the value of scientific endeavor, there is a striking absurdity in committing billions to reach the moon where no people live, while only a fraction of that amount is appropriated to service the densely populated slums." King's sentiment sums up a moral question that has plagued the space program since its inception. Is it right for a nation to spend its money on space travel while there are people suffering on Earth?

The National Aeronautics and Space Administration (NASA) would argue that its budget accounts for only a tiny fraction of the nation's total spending. As shown in Table 2.7 in Chapter 2, NASA's federal allocations totaled $17.6 billion in fiscal year (FY) 2014. The Office of Management and Budget indicates in *Fiscal Year 2014 Budget of the U.S. Government* (February 2014, http://www.whitehouse.gov/sites/default/files/omb/budget/fy2014/assets/budget.pdf) that total federal spending in FY 2014 was $3.6 trillion; thus, NASA's spending accounted for less than 0.5% of total federal spending.

The University of Chicago's National Opinion Research Center conducts the General Social Survey (GSS). The GSS covers a variety of subjects, including the nation's spending priorities.

The GSS asks respondents to indicate their level of support for 23 national spending priorities. The list includes specific programs, such as welfare and Social Security, and more generic priorities, such as the environment, crime, health, and space exploration. The results of the most recent polling were reported by Tom W. Smith of the National Opinion Research Center in "Trends in National Spending Priorities, 1973–2014" (March 2015, http://www.norc.org/PDFs/GSS%20Reports/GSS_Trends%20in%20Spending_1973-2014.pdf). Smith notes that space exploration "has always finished close to the bottom." In 2014 space exploration ranked 20th out of the 23 categories. Only assistance to big cities, welfare programs, and foreign aid received lower spending priorities. Nearly half (44.4%) of those asked in 2014 said national spending on space exploration is "about right." Nearly a third (30%) believed the United States spends "too much," and 25.5% believed "too little" is spent. This question has been part of the GSS poll in various years dating back to 1973. Discontent with spending on space exploration peaked in 1974, when 62.9% of those asked said "too much" was being spent. Favor for additional spending on space exploration was at its highest point in 2014.

SHOULD SPACE TRAVEL BE A SCIENCE PRIORITY?

In the 1960s television series *Star Trek*, space was called "the final frontier." While this may be true from a philosophical viewpoint, it does not apply as well to the realm of science. Geneticists, oceanographers, geologists, and biologists have long maintained that there are still many scientific and medical frontiers to be explored on Earth.

In the United States scientific research is funded by private and government sources. NASA, the National Institutes of Health, the National Science Foundation, and the U.S. Department of Energy's Office of Science are examples of national agencies that have become major recipients of federal spending in science. The federal government greatly expanded its scientific funding of programs following World War II. However, over the following decades other budget priorities tended to override federal spending on science. Thus, scientists in numerous disciplines have had to compete for ever-shrinking federal funding for their research. In addition, federal spending decisions can be highly political, meaning that different scientific programs come into and fall out of favor depending on the political priorities of the time.

In March 1965 Don E. Kash wrote about these concerns in "Is Good Science Good Politics" (*Bulletin of the Atomic Scientists*). He stated, "As every interested observer of the American space program is well aware, the Apollo project was determined primarily by political considerations. There have been rumblings of discontent in the scientific community since the project was first enunciated by the President. The essence of the criticism is that the scientific return does not justify the investment." Kash noted that scientific opposition to the project had been "somewhat subdued" because scientists recognized that the endeavor would likely advance engineering and technological knowledge.

In 1991 a coalition including some of the nation's most prestigious science associations (the American Chemical Society, the American Geophysical Union, and the American Physical Society) publicly criticized NASA's costly plans for the space station *Freedom* (later named the *International Space Station* [*ISS*]). In "Coalition of Scientists Decries Space Station; Letter to Senators Questions Potential Cost" (*Washington Post*, July 10, 1991), Curt Suplee notes that the scientists were concerned about the "excessive cost" of the project, which was projected to total about $30 billion. They warned that funding for the station would threaten the "vitality" of other scientific research programs.

The debate became particularly heated in 2004 after President George W. Bush (1946–) outlined his new space exploration goals for the nation, which included crewed missions to the moon and Mars. Meanwhile, the Mars Exploration Rovers were roaming the surface of Mars looking for evidence of past or current water on the planet. Joseph B. Verrengia notes in "Mars Critics Wonder If Billions Aren't Better Spent Elsewhere" (Associated Press, March 8, 2004) that Amitai Etzioni (1929–), a sociologist at George Washington University and a longtime critic of the U.S. space program, believes the scientific community should focus more attention on Earth's oceans because of their potential to yield new energy and medical breakthroughs that would benefit humanity. Etzioni also criticizes the money that is spent looking for water on Mars and asks, "So what if there is water up there? What difference does it make to anyone's life? Will it grow any more food? Cure a disease? This doesn't even broaden our horizons." Etzioni believes any crewed space missions should be financed by private investors, not with taxpayers' dollars.

However, Janet Vertesi, a sociologist at Princeton University, has a different view. In "Don't Gut NASA Space Missions" (CNN.com, December 14, 2013), an opinion piece for CNN, Vertesi argues that robust funding for NASA is good for the nation's economy and its endeavors in the fields of science, technology, engineering, and mathematics (STEM). The agency supports many educational programs throughout the country. Vertesi notes, "NASA funding not only expands the frontiers of our knowledge, it also trains the next generation of STEM leaders in our country."

CREWED VERSUS ROBOTIC MISSIONS

In 1964 Etzioni published *The Moon-Doggle: Domestic and International Implications of the Space Race*, which questions the scientific value of putting astronauts on the moon and criticizes NASA for favoring expensive manned missions over cheaper, more productive robotic missions. This complaint has been a common one in the scientific community since the 1960s.

It is extremely expensive to send explorers into space, particularly human ones. Robotic spacecraft can accomplish more for less money, but they lack the glamour of human explorers. Human explorers inspire young people to become astronauts and encourage voters and politicians to keep funding space travel. NASA knows that machines simply do not reap the same public relations benefits as human astronauts.

President Barack Obama (1961–) has set a path for NASA that calls for the development of a new launch system and crew capsule for human spaceflight missions. As explained in Chapter 7, budget crunches during the early 2010s meant cutbacks in robotic programs, including the cancellation of a robotic Mars missions. Critics complain that human exploration continues to be favored over cheaper and more scientifically robust robotic missions.

James Van Allen (1914–2006) was an acclaimed astrophysicist. As described in Chapter 3, he discovered the vast radiation belts that surround Earth. They are now known as the Van Allen radiation belts. In "Bush's New Space Program Criticized over Costs & Nuclear Fears" (January 15, 2004, http://www.democracynow.org/2004/1/15/bushs_new_space_program_criticized_over), Van Allen was asked his opinion about the Bush administration's plan for NASA to send astronauts to the moon and Mars. Van Allen was highly critical of the scientific validity of NASA's human spaceflight program, complaining that the scientific results had been "very meager."

Steven Weinberg (1933–) is an astronomy professor at the University of Texas, Austin, and a winner of the 1979 Nobel Prize in Physics. He states in the editorial "Obama Gets Space Funding Right" (WSJ.com, February 3, 2010) that "the manned space flight program masquerades as science, but it actually crowds out real science at NASA, which is all done on unmanned missions." Weinberg touts the scientific discoveries from NASA's unmanned observatories, such as the *Hubble Space Telescope*, and the Mars Exploration Rovers as examples of the results that can be obtained for far less money than human space exploration.

Etzioni in "NASA Gets an A for Hype, an F for Findings" (USNews.com, July 21, 2014) makes a similar argument complaining, "The cost of sending humans to Mars is estimated at more than $150 billion." Etzioni points out that "most of NASA's scientific yield" has come from robotic explorers like the *Hubble Space Telescope* and the rovers on Mars.

Martin Rees (1942–) is a British astrophysicist who is critical of manned space missions. In "Prof Sir Martin Rees: Human Spaceflight—Is It Worth the Money and Risk?" (Telegraph.co.uk, January 15, 2016), Rees notes the high costs of the space shuttle program and *International Space Station* and states, "It's hard to see by what criteria this is money well spent. The scientific and technical payoff hasn't been negligible, but it's far lower (and immensely less cost-effective) than unmanned missions have achieved."

Despite these criticisms, public opinion polls show that a majority of Americans support crewed space missions. In "5 Facts about Americans' Views on Space Exploration" (July 14, 2015, http://www.pewresearch.org/fact-tank/2015/07/14/5-facts-about-americans-views-on-space-exploration/), Brian Kennedy reports the results from polling conducted by Pew Research in 2014 about the importance of astronauts to the U.S. space program. Kennedy indicates that 59% of respondents agreed that the use of human astronauts is "essential," whereas 39% said human astronauts are "not essential." The same question was posed to members of the American Association for the Advancement of Science (AAAS), a nonprofit science organization. According to Kennedy, the results were mixed with 47% of the members considering human astronauts "essential," and 52% saying they are "not essential."

Has Space Exploration Been Worth the Cost?

The *Apollo 11* moon landing occurred in July 1969. Polls on the U.S. space program are often conducted on or near the anniversaries of that date. For example, major polls were conducted on the 10th, 25th, 30th, and 40th anniversaries of the landing to quiz the public regarding the benefits of the U.S. space program.

In *Majority of Americans Say Space Program Costs Justified* (July 17, 2009, http://www.gallup.com/poll/121736/Majority-Americans-Say-Space-Program-Costs-Justified.aspx), Jeffrey M. Jones of the Gallup Organization presents the results of polling that was conducted in 1979, 1994, 1999, and 2009. Participants were asked whether they believed the space program has benefited the nation enough to justify its costs. The 1979 poll (which was conducted by NBC News and the Associated Press) found that only 41% of respondents considered the benefits worth the costs. (See Figure 9.1.) By 1994 the space program had earned more respect; 47% of those asked believed the space program's benefits justified its costs. That number climbed to 55% in 1999 and to 58% in 2009.

FIGURE 9.1

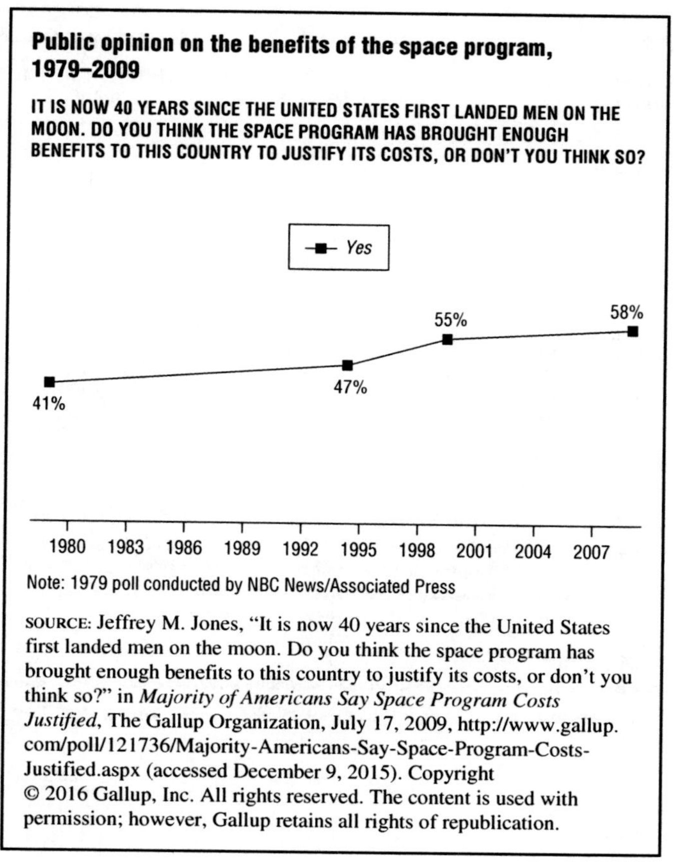

Public opinion on the benefits of the space program, 1979–2009

IT IS NOW 40 YEARS SINCE THE UNITED STATES FIRST LANDED MEN ON THE MOON. DO YOU THINK THE SPACE PROGRAM HAS BROUGHT ENOUGH BENEFITS TO THIS COUNTRY TO JUSTIFY ITS COSTS, OR DON'T YOU THINK SO?

Note: 1979 poll conducted by NBC News/Associated Press

SOURCE: Jeffrey M. Jones, "It is now 40 years since the United States first landed men on the moon. Do you think the space program has brought enough benefits to this country to justify its costs, or don't you think so?" in *Majority of Americans Say Space Program Costs Justified*, The Gallup Organization, July 17, 2009, http://www.gallup.com/poll/121736/Majority-Americans-Say-Space-Program-Costs-Justified.aspx (accessed December 9, 2015). Copyright © 2016 Gallup, Inc. All rights reserved. The content is used with permission; however, Gallup retains all rights of republication.

In the 2011 CBS News and Pew Research Center for the People and the Press polls mentioned earlier, respondents were asked about the financial value of the space shuttle program. CBS News pollsters asked the following question: "Given the costs and risks involved in space exploration, do you think the Space Shuttle program has been a worthwhile program, or not?" Nearly two-thirds (63%) of those asked answered "yes," whereas 31% said "no." Another 6% did not know or did not answer the question. When asked about how they felt about the Space Shuttle Program ending, 48% said they were "disappointed," whereas 16% said they were "pleased." One-third (33%) indicated they "don't care." The remaining 3% did not know or did not answer the question. In a differently worded question, the Pew Research Center for the People and the Press asked participants: "Do you think the space shuttle program has been a good investment for this country or don't you think so?" More than half (55%) said the program has been a good investment, whereas 36% said it has not been a good investment. The remaining 9% were unsure or refused to answer.

In "5 Facts about Americans' Views on Space Exploration," Kennedy describes a Pew Research Center poll conducted during August 2014 in which pollsters asked Americans whether or not the *International Space Station* has been a "good investment." Overall, 64% of respondents considered the space station a "good investment," whereas 29% said it was "not a good investment." Pew asked the same question of members of the AAAS. More than two-thirds (68%) of them deemed the *ISS* a "good investment";31% of AAAS respondents said it was "not a good investment."

AMERICANS RATE NASA'S PERFORMANCE

Gallup annually asks Americans to rate the jobs being done by various federal government agencies. As of February 2016, the most recent polling was conducted during November 2014. As shown in Table 9.2, NASA ranked fourth among the 13 agencies considered, with 50% of respondents saying the space agency was doing an excellent or good job. Table 9.3 shows historical results on this polling question as it relates to NASA. The agency's highest marks were obtained in 1998 when 76% of those asked said that NASA was doing an excellent (26%) or good (50%) job. These results were obtained shortly after John Glenn's (1921–) historic flight aboard the space shuttle *Discovery*.

PRIVATE VERSUS GOVERNMENT ROLE IN SPACE EXPLORATION

Chapter 4 highlights the increasing role that the private sector is playing in the nation's space program. The 2011 CNN/ORC poll addressed this issue by asking poll participants the following question: "In general, do you think the U.S. should rely more on the government or more on private companies to run the country's manned space missions in the future?" More than half (54%) of those asked favored private companies, whereas 38% favored the government. Another 4% said the two sectors should have equal roles, and 2% said "neither." The remaining 2% had no opinion on the subject.

TABLE 9.2

Public opinion on the job being done by selected federal agencies, November 2014

HOW WOULD YOU RATE THE JOB BEING DONE BY—[ITEM A READ FIRST, THEN ITEMS B-I ROTATED]? WOULD YOU SAY IT IS DOING AN EXCELLENT, GOOD, ONLY FAIR, OR POOR JOB?

2014 Nov 11–12 (sorted by "excellent/good")	Excellent/good %	Only fair %	Poor %
The U.S. Postal Service	72	20	8
The Federal Bureau of Investigation, or the FBI	58	27	8
The Centers for Disease Control and Prevention, or the CDC	50	30	16
NASA—the U.S. space agency	50	25	8
The Central Intelligence Agency, or the CIA	49	28	11
The Department of Homeland Security	48	32	16
The Federal Emergency Management Agency, or FEMA	47	31	14
The Food and Drug Administration, or FDA	45	34	19
The Environmental Protection Agency, or EPA	44	32	20
The Secret Service	43	30	16
The Internal Revenue Service, or the IRS	41	29	27
The Federal Reserve Board	38	35	14
The Veterans Administration, or VA	29	29	35

SOURCE: "Q.1. How would you rate the job being done by...? Would you say it is doing an excellent, good, only fair, or poor job?" in *Gallup News Service: Ratings of Government Agencies*, The Gallup Organization, 2014, http://www.gallup.com/file/poll/179525/Govt_Agency_Ratings_I_141120.pdf (accessed December 9, 2015). Copyright © 2016 Gallup, Inc. All rights reserved. The content is used with permission; however, Gallup retains all rights of republication.

TABLE 9.3

Public opinion on the job being done by NASA, selected dates July 1990–November 2014

	Excellent	Good	Only fair	Poor	No opinion
2014 Nov 11–12	13	37	25	8	17
2013 May 20–21	10	32	32	10	17
2009 Jul 10–12	13	45	26	7	10
2007 Sep 14–16	16	40	29	8	6
2006 Jun 23–25	17	40	30	7	6
2005 Aug 5–7	16	44	29	8	3
2005 Jun 24–26	11	42	34	6	7
2003 Sep 8–10	12	38	36	10	4
1999 Dec 9–12	13	40	31	12	4
1999 Jul 13–4	20	44	20	5	11
1998 Nov 20–22	26	50	17	4	3
1998 Jan 30–Feb 1	21	46	21	4	8
1994 Jul 15–17	14	43	29	6	8
1993 Dec 17–19	18	43	30	7	2
1993 Sept 13–15	7	36	35	11	11
1991 May 2–5	16	48	24	6	6
1990 July 19–22	10	36	34	15	5

SOURCE: "Q.1A. NASA—the U.S. Space Agency," in *Gallup News Service: Ratings of Government Agencies*, The Gallup Organization, 2014, http://www.gallup.com/file/poll/179525/Govt_Agency_Ratings_I_141120.pdf (accessed December 9, 2015). Copyright © 2016 Gallup, Inc. All rights reserved. The content is used with permission; however, Gallup retains all rights of republication.

PRACTICAL BENEFITS OF SPACE TRAVEL

Although it is widely acknowledged that space travel has psychological and scientific benefits to society, it is more difficult to point to everyday products that have directly resulted from the nation's space program. Certainly satellites have brought about great changes in telecommunications, navigation, military operations, and weather prediction. All of these developments do affect American lives. The technologies that are associated with space exploration have advanced the fields of robotics, computer programming, and cryogenics (the physics of extremely cold temperatures). In addition, improvements based on NASA technologies have been incorporated into diverse products such as memory foam mattresses, medical imaging devices, eyeglass lenses, golf balls, baby food, pacemakers, and life rafts.

One of the mandates of the 1958 National Aeronautics and Space Act is that the agency and its contractors must publicize any new developments that are significant to commercial industry. NASA accomplishes this through various publications, including a monthly magazine for engineers, managers, and scientists called *NASA Tech Briefs* (http://www.techbriefs.com/tech-briefs) that briefly describes new technologies; *Technical Support Packages*, which describe in detail the technologies that are presented in *NASA Tech Briefs*; and *Spinoff* (https://spinoff.nasa.gov/), an annual publication describing successfully commercialized NASA technology.

According to NASA, in "*Spinoff* Frequently Asked Questions" (May 1, 2011, https://spinoff.nasa.gov/faq.html), a spin-off is "a commercialized product that incorporates NASA technology or expertise." In *Spinoff 2015* (2015, http://spinoff.nasa.gov/Spinoff2015/pdf/Spinoff2015.pdf), NASA describes technologies that have been commercialized by private-sector companies for applications in health and medicine, transportation, public health, consumer goods, energy and environment, information technology, and industrial productivity. (See Figure 9.2.)

In 1988 NASA and the Space Foundation, a private organization, established the Space Technology Hall of Fame (http://www.spacetechhalloffame.org/). Each year a handful of space-based technologies are selected for induction into the hall of fame. Inductees are honored at an annual conference held in Colorado Springs, Colorado, called the National Space Symposium. The two 2015 inductees (http://www.spacefoundation.org/programs/space-technology-hall-fame/inducted-technologies) were technologies devoted to measuring eye movements and protecting buildings and bridges from earthquake vibrations. Previous hall of fame winners familiar to consumers include satellite radio technology and the DirecTV satellite system.

NASA WOOS THE AMERICAN PUBLIC

NASA employs a number of public relations tools that are designed to interest and excite people about space travel. Since its inception, the agency has recognized that public support is crucial to fostering a successful long-term space program.

Television

Throughout the space age NASA has used television as a publicity tool to try to spark greater interest in the space program. Television proved to be one of the greatest public relations tools of the Apollo Program. In 1968 the *Apollo 7* astronauts conducted the first live television interview from space. All the remaining Apollo flights carried television cameras. The worldwide television audience for the *Apollo 11* moon landing in 1969 was estimated at half a billion people.

In 1999 Gallup quizzed Americans about their recollections of the first manned moon landing. Frank Newport of the Gallup Organization indicates in *Landing a Man on the Moon: The Public's View* (July 20, 1999, http://www.gallup.com/poll/3712/landing-man-moon-publics-view.aspx) that 76% of people aged 35 years and older claimed to have watched the event on television as it happened.

NASA TV. NASA operates its own television network called NASA TV (NTV; http://www.nasa.gov/multimedia/nasatv/index.html). NTV broadcasts via satellite and cable and is streamed over the Internet. It features live coverage of NASA activities and missions, video of events for the news media, and educational programming for teachers and students.

As of February 2016, NTV included three channels: the public channel, the media channel, and the education channel. The public channel features live events and mission coverage. The media channel provides similar programming and includes video files. The education channel is geared toward teachers and students and shows two-hour blocks of educational material that is suitable for use in the classroom.

Internet

NASA makes full use of Internet resources for public relations purposes. The agency operates informational websites, publicizes its activities through social networking sites such as Twitter and Facebook, publishes blogs, issues podcasts, and posts NTV videos on YouTube.

NASA's main website (http://www.nasa.gov/) includes thousands of mission photographs and millions of documents that are related to the nation's space endeavors—past, current, and planned. The website provides detailed information about NASA facilities, programs, and missions. There are a variety of multimedia features, including interactive displays, video and audio downloads, and images of Earth and space captured by NASA spacecraft. Furthermore, it provides access to historical archives that include documentation dating back to the earliest days of space travel. According to NASA, the website is visited millions of times each day.

SIGHTING OPPORTUNITIES. One of the ways that NASA tries to engage public interest in space travel is by posting sighting opportunities for its satellites, particularly the *ISS*. NASA's website instructs people how and where to look in the nighttime sky to see the space station as it is passing overhead. Figure 9.3 shows the general paths that the *ISS* takes as it orbits around Earth. Figure 9.4 shows a set of instructions for viewing the *ISS* at a particular location, assuming that skies are clear enough.

This listing identifies the exact date and time at which the *ISS* should become visible to observers on the ground and how long it will remain visible. (See Figure 9.4.) It also gives information about the location of the *ISS* in the sky based on direction (north, south, east, or west) and angle of elevation compared with the horizon. A spacecraft flying directly overhead would be at 90-degree maximum elevation.

In the example shown in Figure 9.4, the *ISS* will appear in the west-southwest direction approximately 10 degrees above the horizon. It will then climb to a maximum elevation of 66 degrees above the horizon and travel out of sight heading toward the northeast. It should disappear from view about 31 degrees above the horizon.

FIGURE 9.2

Earth-based benefits derived from NASA's space programs, 2015

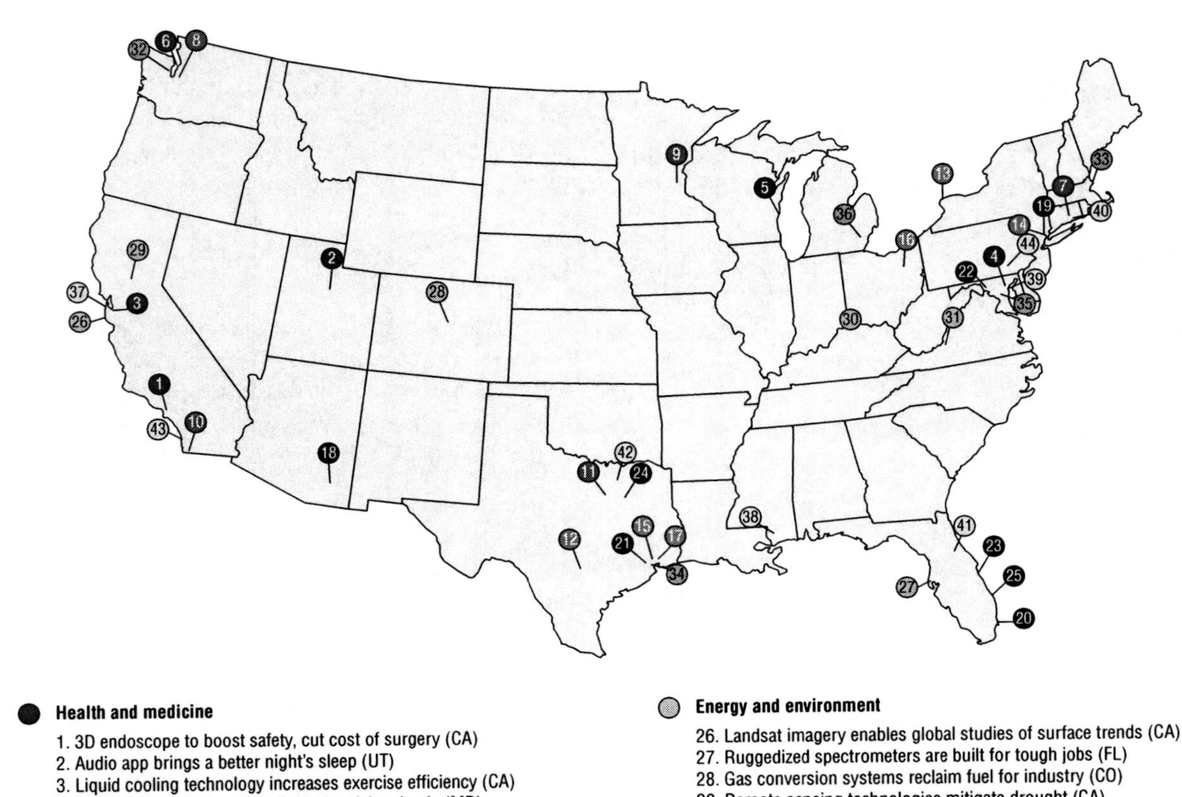

Health and medicine

1. 3D endoscope to boost safety, cut cost of surgery (CA)
2. Audio app brings a better night's sleep (UT)
3. Liquid cooling technology increases exercise efficiency (CA)
4. Algae-derived dietary ingredients nourish animals (MD)
5. Space grant research launches rehabilitation chair (WI)
6. Vision trainer teaches focusing Techniquest at home (WA)

Transportation

7. Aircraft geared architecture reduces fuel cost and noise (CT)
8. Ubiquitous supercritical wing design cuts billions in fuel costs (WA)
9. Flight controller software protects lightweight flexible aircraft (MN)
10. Cabin pressure monitors notify pilots to save lives (CA)
11. Ionospheric mapping software ensures accuracy of pilots' GPS (TX)

Public safety

12. Water mapping technology rebuilds lives in arid regions (TX)
13. Shock absorbers save structures and lives during earthquakes (NY)
14. Software facilities sharing of water quality data worldwide (NY)
15. Underwater adhesives retrofit pipelines with advanced sensors (TX)
16. Laser imaging video camera sees through fire, fog, smoke (OH)
17. 3D lasers increase efficiency, safety of moving machines (TX)

Consumer goods

18. Air revitalization system enables excursions to the stratosphere (AZ)
19. Magnetic fluids deliver better speaker sound quality (NY)
20. Bioreactor yields extracts for skin cream (FL)
21. Private astronaut training prepares commercial crews of tomorrow (TX)
22. Activity monitors help users get optimum sun exposure (MD)
23. LEDs illuminate bulbs for better sleep, wake cycles (FL)
24. Charged particles kill pathogens and round up dust (TX)
25. Balance devices train golfers for a consistent swing (FL)

Energy and environment

26. Landsat imagery enables global studies of surface trends (CA)
27. Ruggedized spectrometers are built for tough jobs (FL)
28. Gas conversion systems reclaim fuel for industry (CO)
29. Remote sensing technologies mitigate drought (CA)
30. Satellite data inform forecasts of crop growth (KY)
31. Probes measure gases for environmental research (VA)

Information technology

32. Cloud computing technologies facilitate Earth research (WA)
33. Software cuts homebuilding costs, increases energy efficiency (MA)
34. Portable planetariums teach science (TX)
35. Schedule analysis software saves time for project planners (MD)
36. Sound modeling simplifies vehicle noise management (MI)

Industrial productivity

37. Custom 3D printers revolutionize space supply chain (CA)
38. Improved calibration shows images' true colors (MS)
39. Micromachined parts advance medicine, astrophysics, and more (MD)
40. Metalworking techniques unlock a unique alloy (CT)
41. Low-cost sensors deliver nanometer-accurate measurements (FL)
42. Electrical monitoring devices save on time and cost (TX)
43. Dry lubricant smooths the way for space travel, industry (CA)
44. Compact vapor chamber cools critical components (PA)

SOURCE: Adapted from "NASA Spinoff Technology across the Nation," in *Spinoff 2015*, National Aeronautics and Space Administration, 2015, http://spinoff.nasa.gov/Spinoff2015/pdf/Spinoff2015.pdf (accessed December 9, 2015)

NASA explains in *Sighting Opportunities* (November 30, 2011, http://spaceflight.nasa.gov/realdata/sightings/help.html) that a spacecraft looks like "a steady white pinpoint of light moving slowly across the sky." Viewers are urged to observe spacecraft with the naked eye or through binoculars. The speed at which spacecraft move makes telescope viewing impractical.

NASA's website provides links to sighting data for hundreds of cities around the world. People at locations not

FIGURE 9.3

Orbits of the *International Space Station*

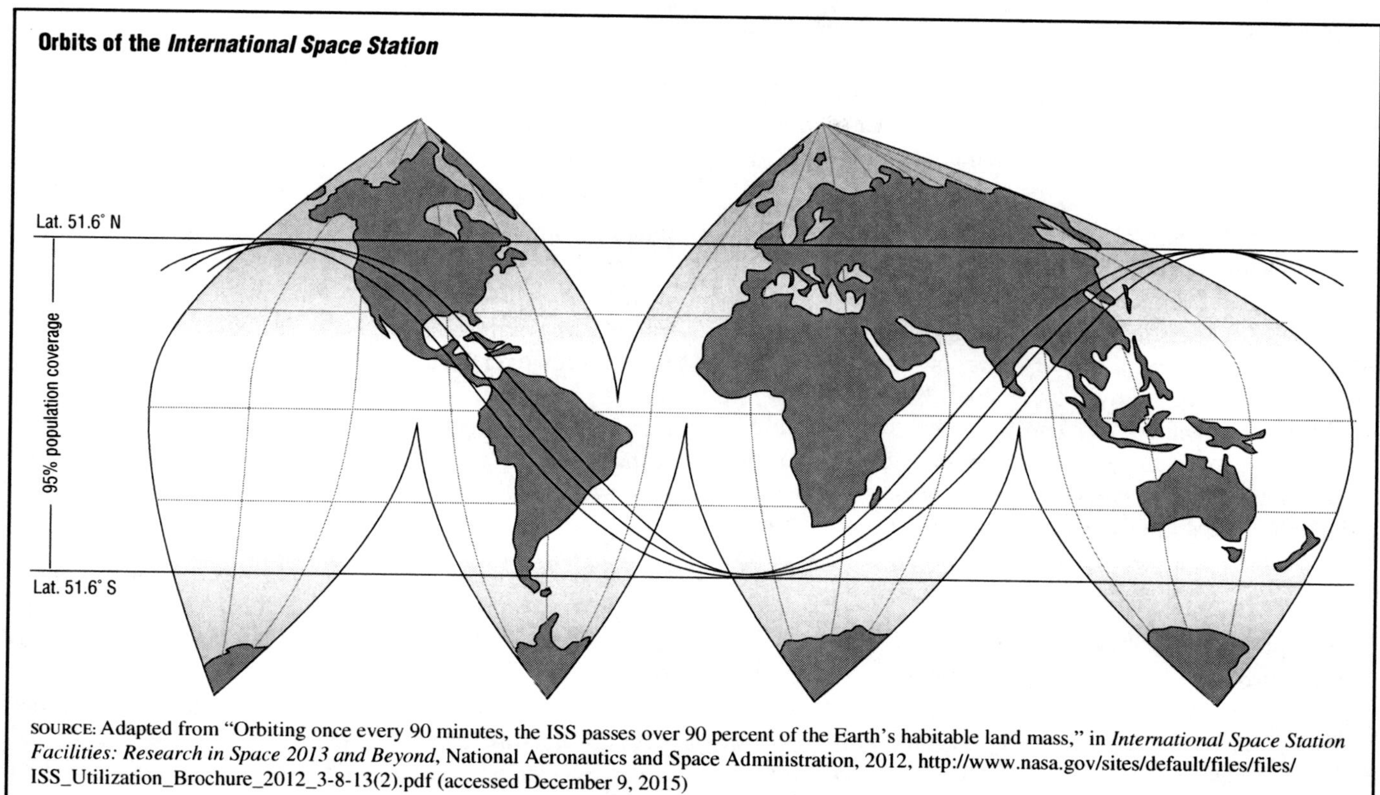

SOURCE: Adapted from "Orbiting once every 90 minutes, the ISS passes over 90 percent of the Earth's habitable land mass," in *International Space Station Facilities: Research in Space 2013 and Beyond*, National Aeronautics and Space Administration, 2012, http://www.nasa.gov/sites/default/files/files/ISS_Utilization_Brochure_2012_3-8-13(2).pdf (accessed December 9, 2015)

FIGURE 9.4

Interpreting sighting data for the *International Space Station*

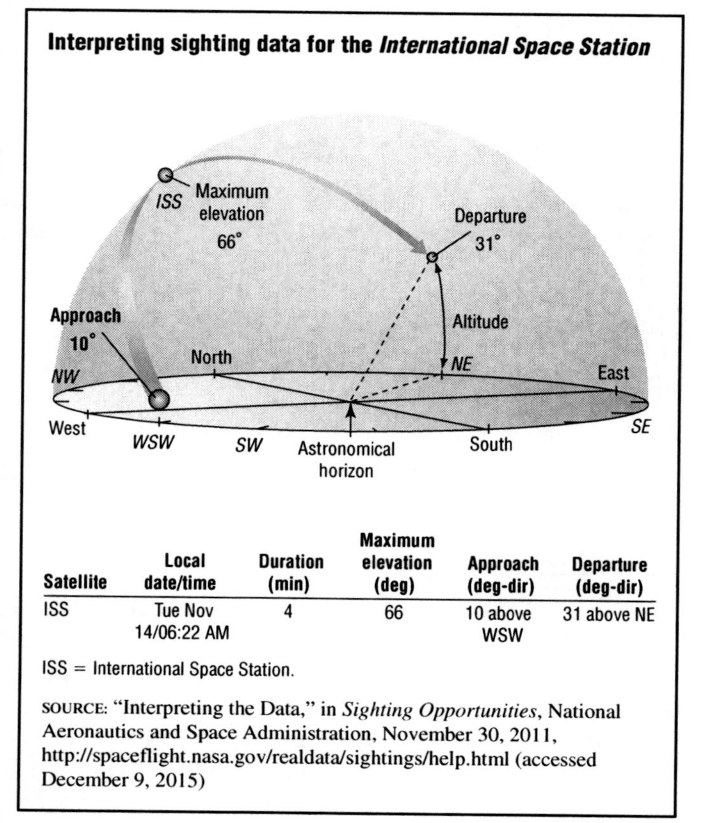

Satellite	Local date/time	Duration (min)	Maximum elevation (deg)	Approach (deg-dir)	Departure (deg-dir)
ISS	Tue Nov 14/06:22 AM	4	66	10 above WSW	31 above NE

ISS = International Space Station.

SOURCE: "Interpreting the Data," in *Sighting Opportunities*, National Aeronautics and Space Administration, November 30, 2011, http://spaceflight.nasa.gov/realdata/sightings/help.html (accessed December 9, 2015)

listed can use an applet (a small application program) called SkyWatch to enter their latitude and longitude and receive viewing information for numerous orbiting satellites.

Ham Radio

Ham radios, also called amateur radios, are defined as radios that make use of radio wave frequencies for purposes of noncommercial, wireless communications. They vary in signal strength and capability. The strongest radios can reach operators on the other side of the world by bouncing signals off the upper atmosphere or using satellites.

In November 1983 the astronaut Owen K. Garriott (1930–) carried a small ham radio with him aboard the space shuttle *Columbia*. During his spare time he used the radio to contact fellow ham operators around the world. This was the first of more than 24 shuttle missions that carried ham radio equipment so astronauts could communicate with their families and other ham operators worldwide and perform interviews for schoolchildren. The program was called the Space Amateur Radio Experiment (SAREX). The Soviet space agency operated a similar ham radio program for cosmonauts aboard the *Mir* space station.

In September 2000 the crew of the space shuttle *Atlantis* carried a ham radio to the *ISS* for use by Expedition crews. The SAREX program was given the new name of Amateur Radio on the International Space Station. Under this program, *ISS* crew members can communicate with ham radio operators all over the world. One aspect of the program includes ham radio stations set up in schools, so students can interview *ISS* crew members.

Educational Programs

NASA operates an extensive student education program that is designed to encourage young people to pursue studies in science, technology, engineering, and mathematics (STEM) and careers in aeronautics and space science. As shown in Table 2.7 in Chapter 2, NASA was allocated $119 million for the program in FY 2015 and was expected to receive $88.9 million for FY 2016. NASA prides itself on its educational programs and the partnerships it establishes with schools, museums, libraries, and science centers throughout the nation to reach as many young people as possible.

Chapter 5 describes the role of the Center for the Advancement of Science in Space, a nonprofit organization that manages the ISS National Laboratory for NASA. The U.S. Government Accountability Office (GAO) indicates in *International Space Station: Measurable Performance Targets and Documentation Needed to Better Assess Management of National Laboratory* (April 2015, http://www.gao.gov/assets/670/669851.pdf) that the NASA Authorization Act of 2010 requires NASA to conduct "scientific outreach and education activities designed to ensure effective utilization of ISS research capabilities." According to the GAO, the education program operated by the Center for the Advancement of Science in Space focuses on promoting the space station as a STEM learning platform.

EDUCATOR RESOURCE CENTERS. Educator Resource Centers (formerly Teacher Resource Centers) are offices maintained at NASA facilities throughout the country. The centers serve as libraries that loan educational materials including lesson plans, audiotapes and videotapes, slides, and miscellaneous print publications to educators.

Art Program

In 1962 the NASA administrator James Edwin Webb (1906–1992) established the NASA Art Program to encourage and collect works of art about aeronautics and space. As of February 2016, the NASA art collection included more than 2,000 works of art in a variety of media, including paintings, drawings, poems, and songs. NASA has donated many of its art works to the National Air and Space Museum in Washington, D.C. In addition, the pieces are displayed at art galleries and museums throughout the country. NASA centers, particularly the Kennedy Space Center in Florida, also display the art works in their visitor areas. Several famous artists have participated in the program, such as Annie Leibovitz (1949–), Norman Rockwell (1894–1978), Andy Warhol (1928–1987), William Wegman (1943–), and Jamie Wyeth (1946–).

Astronauts

Astronauts have always been NASA's greatest public relations agents. The early astronauts became instant heroes during the 1950s and 1960s. They were flooded with fan mail and held up by the media as sterling role models of what was great and daring about the United States. However, after the first moon landing in 1969, public interest in the space program began to fade. The astronauts of later decades were still admired and respected, but they were not treated to the same level of hero worship as their predecessors.

During the early 1980s NASA decided to include a new type of astronaut on space shuttle flights to catch the public's attention. The agency began the Teacher in Space program. NASA hoped that sending a teacher into space would excite the nation's schoolchildren and foster goodwill toward the space program. The schoolteacher Christa McAuliffe (1948–1986) was selected and trained for a mission aboard the space shuttle *Challenger*. She was killed with the other six crew members in 1986 when the shuttle exploded soon after liftoff.

NASA's public relations experiment turned into a nightmare. The *Challenger* catastrophe brought harsh criticism of the agency, which was found to have serious management and safety problems. The loss seemed even more poignant to the public because a teacher, an everyday kind of person, had been one of the victims. NASA decided that space travel was not routine enough to risk the lives of private citizens as goodwill ambassadors.

In 1998 NASA relented somewhat and allowed Glenn, a former Mercury astronaut, to ride aboard the space shuttle *Discovery*. At the time, Glenn was a 77-year-old senator from Ohio. NASA said the mission would reveal new knowledge about the effects of weightlessness and bone loss in older people. Critics complained that it was nothing more than a publicity stunt. Whatever the motivation, the event did greatly improve NASA's image. The public was entranced by the idea of an old hero traveling back into space.

Tourist Attractions

Many NASA facilities have become popular tourist attractions. This is particularly true for centers that are associated with the Apollo Program and the Space Shuttle Program. Most NASA facilities operate their own visitor centers for which admission is free. The Johnson Space Center (Houston, Texas), the Kennedy Space Center (Cape Canaveral, Florida), and the Marshall Space

Flight Center (Huntsville, Alabama) have privately operated tourist centers that charge a fee for admittance.

Contests and Gimmicks

One relatively new way that NASA engages the public in space travel is by holding spacecraft-naming contests. During the 1990s NASA held contests that chose the names for the Mars Pathfinder mission's *Sojourner* rover and the *Magellan* spacecraft.

Other contests have resulted in the names "Chandra" and "Spitzer" being selected for two of NASA's great observatories, and provided the names "Ebb" and "Flow" for the Gravity Recovery and Interior Laboratory mission, all of which are described in Chapter 6. Likewise, the names for the Mars Exploration Rovers *Spirit* and *Opportunity* and the *Mars Science Laboratory*'s rover *Curiosity* resulted from contests. These spacecraft are described in Chapter 7.

Another public relations device used by NASA is to ask people to submit their name for inclusion on CDs or DVDs that will be carried by spacecraft. Numerous NASA missions conducted since the 1990s have included electronic disks carrying the names of millions of people.

Between 2003 and 2006 NASA took suggestions from the public regarding which sites on Mars should be imaged by the Mars orbiter camera on the spacecraft *Mars Global Surveyor*, which was then in orbit around the planet. The spacecraft was lost in November 2006. As of February 2016, the *Mars Reconnaissance Orbiter* was in orbit around Mars. It includes a device called the High Resolution Imaging Science Experiment that takes high-resolution images of the planet's surface. The imager was designed by NASA's partner in the mission, the University of Arizona's Lunar and Planetary Laboratory. In 2010 they introduced HiWish (http://www.uahirise.org/hiwish/), an online application through which the public can suggest specific sites on the surface of Mars to be imaged.

A MOON HOAX?

One of the most offbeat conspiracy theories of the space age is that the U.S. government faked the Apollo moon landings. In 2001 the Fox television network broadcast the program *Conspiracy Theory: Did We Land on the Moon?* Guests on the show claimed the Apollo Program never actually put an astronaut on the moon but faked the lunar landing for television cameras. As of February 2016, the theory continued to be supported on various websites.

Advocates of the hoax theory rely on several key points to support their position. Chief among these are:

- NASA's moon photographs do not show stars in the background behind the astronauts.
- The American flag supposedly planted on the moon by *Apollo 11* astronauts is rippling in a breeze, although there is no atmosphere on the moon.
- There is no blast crater beneath the lunar lander.
- Humans could not have survived exposure to the intense radiation of the Van Allen belts lying between Earth and the moon.

In general, NASA ignores the hoax claims and does not address them publicly. However, the NASA website does include one article of rebuttal: "The Great Moon Hoax" (February 23, 2001, http://science.nasa.gov/headlines/y2001/ast23feb_2.htm). In it Tony Phillips, a NASA science editor, addresses questions about moon photographs and the rippling flag. He points out that the exposure on the moon cameras had to be adjusted to tone down the dazzling brightness of the astronauts' sunlit spacesuits. This caused the background stars to be too faint to appear in the photographs. The rippling flag is explained by the wire inserts that were built into the fabric and by the twisting motion the astronauts used to push the flagpole into the lunar ground.

Phillips notes that moon rocks are the best evidence that astronauts visited the moon. The astronauts brought back 841 pounds (381 kg) of moon rocks, and these rocks have been investigated by researchers from all over the world. Moon rocks differ greatly in mineral and water content from any rocks found on Earth. They also contain isotopes that were created by long-term exposure to high-energy cosmic rays on the lunar surface.

In "Ask an Astrophysicist" (December 1, 2005, http://imagine.gsfc.nasa.gov/ask_astro/space_travel.html), Laura Whitlock addresses the hoax issue concerning the Van Allen radiation belts, which are regions of highly energized ionized particles that are trapped within the geomagnetic fields surrounding Earth. Whitlock explains that early NASA researchers were also worried about the radiation belts. Scientists at the Oak Ridge National Laboratory (ORNL) in Oak Ridge, Tennessee, devised experiments in which bacteria and blood cells were sent aboard unmanned probes into space and returned to Earth. Animal experiments were also performed. The ORNL used the resulting data to design special radiation shields for the Apollo spacecraft. The shields used materials that were left over from nuclear testing performed during the 1950s. Whitlock also notes that the Apollo spacecraft traveled so fast that the astronauts were exposed to Van Allen radiation for only a short time.

In a survey that was conducted in July 1999, the Gallup Organization asked poll participants their views about a possible moon-landing hoax. Gallup notes in *Did Men Really Land on the Moon?* (February 15, 2001, http://www.gallup.com/poll/1993/Did-Men-Really-Land-Moon.aspx) that the vast majority (89%) of those surveyed did not believe the government staged the Apollo moon landing. Only 6% said the landing was a hoax. Another 5% had no opinion.

IMPORTANT NAMES AND ADDRESSES

American Institute of Aeronautics and Astronautics
12700 Sunrise Valley Dr., Ste. 200
Reston, VA 20191-5807
(703) 264-7500
1-800-639-2422
FAX: (703) 264-7551
URL: http://www.aiaa.org/

Ames Research Center
Moffett Field, CA 94035
(650) 604-5000
URL: http://www.nasa.gov/centers/ames/home/index.html

Armstrong Flight Research Center
PO Box 273
Edwards, CA 93524
(661) 276-3311
URL: http://www.nasa.gov/centers/armstrong/home/index.html

Blue Origin
21218 76th Ave. S.
Kent, WA 98032
(253) 437-9300
URL: https://www.blueorigin.com

Canadian Space Agency
John H. Chapman Space Centre
6767 Route de l'Aéroport
Saint-Hubert, Quebec, Canada J3Y 8Y9
(450) 926-4800
FAX: (450) 926-4352
URL: http://www.asc-csa.gc.ca/eng/default.asp

China Aerospace Science and Technology Corporation
16 Fucheng Rd.
Beijing, Haidian District, China 100048
(011-86-10) 68767492
FAX: (011-86-10) 68372291
E-mail: casc@spacechina.com
URL: http://english.spacechina.com/

Commercial Spaceflight Federation
500 New Jersey Ave., Ste. 400
Washington, DC 20001
(202) 715-2924
URL: http://www.commercialspaceflight.org/

European Space Agency
8-10 rue Mario Nikis
75738 Paris, France Cedex 15
(011-33-1) 5369-7654
FAX: (011-33-1) 153697-560
URL: http://www.esa.int/ESA

Federal Aviation Administration
Office of Commercial Space Transportation
800 Independence Ave. SW
Washington, DC 20591
1-866-835-5322
URL: http://www.faa.gov/go/ast

Fédération Aéronautique Internationale
Maison du Sport International
Av. de Rhodanie 54
Lausanne, Switzerland CH-1007
(011-41-21) 345-1070
FAX: (011-41-21) 345-1077
URL: http://www.fai.org/

Glenn Research Center
21000 Brookpark Rd.
Cleveland, OH 44135
(216) 433-4000
URL: http://www.nasa.gov/centers/glenn/home/index.html

Goddard Space Flight Center
Public Inquiries, Mail Code 130
Greenbelt, MD 20771
(301) 286-2000
URL: http://www.nasa.gov/centers/goddard/home/index.html

Indian Space Research Organisation
Antariksh Bhavan, New BEL Rd.
Bangalore, India 560 231
(011-91-80) 23415275
FAX: (011-91-80) 23412253
E-mail: dir.ppr@isro.gov.in
URL: http://www.isro.org/

International Astronomical Union
98-bis Blvd. Arago
F-75014 Paris, France
(011-33-1) 43-25-83-58
FAX: (011-33-1) 43-25-26-16
E-mail: iau@iap.fr
URL: http://www.iau.org/

International Launch Services
1875 Explorer St., Ste. 700
Reston, VA 20190
(571) 633-7400
FAX: (571) 633-7500
E-mail: contactus@ilslaunch.com
URL: http://www.ilslaunch.com

Japan Aerospace Exploration Agency
7-44-1 Jindaiji Higashi-Machi, Chofu-shi
Tokyo, Japan 182-8522
(011-81-422) 40-3000
URL: http://www.jaxa.jp/index_e.html

Jet Propulsion Laboratory
4800 Oak Grove Dr.
Pasadena, CA 91109
(818) 354-4321
URL: http://www.jpl.nasa.gov/

John F. Kennedy Space Center
Kennedy Space Center, FL 32899
(321) 867-5000
URL: http://www.nasa.gov/centers/kennedy/home/index.html

Langley Research Center
8 Lindbergh Way
Hampton, VA 23681-2199
(757) 864-1000
URL: http://www.nasa.gov/centers/langley/home/index.html

Lyndon B. Johnson Space Center
2101 NASA Road 1
Houston, TX 77058
(281) 483-0123

URL: http://www.nasa.gov/centers/johnson/home/index.html

Marshall Space Flight Center
Bldg. 4200
4200 Rideout Rd.
Huntsville, AL 35808
(256) 544-6840
URL: http://www.nasa.gov/centers/marshall/home/index.html

National Academy of Sciences
500 Fifth St. NW
Washington, DC 20001
(202) 334-2000
URL: http://www.nasonline.org/

National Aeronautics and Space Administration
300 E. St. SW, Ste. 5R30
Washington, DC 20546
(202) 358-0001
FAX: (202) 358-4338
URL: http://www.nasa.gov/home/

National Oceanic and Atmospheric Administration
1401 Constitution Ave. NW, Rm. 5128
Washington, DC 20230
(202) 482-3436
URL: http://www.noaa.gov/

National Science Foundation
4201 Wilson Blvd.
Arlington, VA 22230
(703) 292-5111
1-800-877-8339
URL: http://www.nsf.gov/

Office for Outer Space Affairs
United Nations Office at Vienna
Vienna International Centre
Wagramerstrasse 5
A-1220 Vienna, Austria
(011-43-1) 260-60-4950
FAX: (011-43-1) 1-260-60-5830
URL: http://www.unoosa.org/

Planetary Society
60 S. Los Robles Ave.
Pasadena, CA 91101
(626) 793-5100
FAX: (626) 793-5528
E-mail: tps@planetary.org
URL: http://www.planetary.org/

Roscosmos State Corporation
Schepkina St. 42
Moscow, Russia 107996
(011-7-495) 688-90-63
FAX: (011-7-495) 975-44-67
URL: http://en.federalspace.ru/

Scaled Composites
Hangar 78 Airport
1624 Flight Line
Mojave, CA 93501
(661) 824-4541
FAX: (661) 824-4174
URL: http://www.scaled.com/

Space Adventures
8245 Boone Blvd, Ste. 570
Vienna, VA 22182
(703) 524-7172
FAX: (703) 524-7176
URL: http://www.spaceadventures.com

Space Telescope Science Institute
3700 San Martin Dr.
Baltimore, MD 21218
(410) 338-4700
URL: http://www.stsci.edu/portal/

Space Weather Prediction Center
W/NP9
325 Broadway
Boulder, CO 80305
URL: http://www.swpc.noaa.gov/

Stennis Space Center
Stennis Space Center, MS 39529
(228) 688-3333
URL: http://www.nasa.gov/centers/stennis/home/index.html

U.S. Geological Survey Astrogeology Science Center
2255 N. Gemini Dr.
Flagstaff, AZ 86001
(928) 556-7042
FAX: (928) 556-7229
URL: http://astrogeology.usgs.gov/

U.S. Government Accountability Office
441 G St. NW
Washington, DC 20548
(202) 512-3000
E-mail: contact@gao.gov
URL: http://www.gao.gov/

U.S. Strategic Command
901 SAC Blvd., Ste. 1A1
Offutt Air Force Base, NE 68113-6020
(402) 294-4130
FAX: (402) 294-4892
E-mail: pa@stratcom.mil
URL: http://www.stratcom.mil/

Virgin Galactic
E-mail: virgingalactic@virgingalactic.com
URL: http://www.virgingalactic.com

Wallops Flight Facility
Chincoteague Island, VA 23336
(757) 824-1240
URL: http://www.nasa.gov/centers/wallops/home/index.html

White Sands Test Facility
Bldg. 120, 12600 NASA Rd.
Las Cruces, NM 88012
(575) 524-5521
FAX: (575) 524-5798
URL: http://www.nasa.gov/centers/wstf/home/index.html

X Prize Foundation
800 Corporate Pointe, Ste. 350
Culver City, CA 90230
(310) 741-4880
FAX: (310) 741-4974
URL: http://www.xprize.org

RESOURCES

The National Aeronautics and Space Administration (NASA) is the premier resource for information about the U.S. space program. NASA headquarters operates an informative website (http://www.nasa.gov/). There are links to all the facilities that are operated by NASA around the country. Many of these links were consulted for this book. In addition, NASA websites provide publications that discuss in detail many of the agency's programs and missions. Administrative documents, such as budgets and performance reports, are also available and provided much valuable information for this book.

NASA's History Program Office (http://www.hq.nasa.gov/office/pao/History/) maintains an extensive collection of historical documents. This collection includes the complete text of books that were written for NASA about space activities of previous decades.

NASA publishes a press kit for each major mission of the robotic space program. These press kits contain key information on mission objectives, spacecraft design, and science experiments. Another important NASA series is *NASA Facts*. This series provides data about missions and space science and biographies of key historical figures to the U.S. space program.

Other government agencies and organizations were also consulted. The National Oceanic and Atmospheric Administration's Space Environment Center publishes a series of educational papers called "Space Environment Topics" that were an excellent resource. The Congressional Research Service (CRS) is the public policy research arm of Congress. CRS publications discuss the space programs of the United States and other nations in terms of public policy and related factors. The U.S. Government Accountability Office (GAO) is the investigative arm of Congress. GAO publications provided timely analysis of various space programs, including those of the U.S. military.

Other government agencies with online publications about space activities include the Federal Aviation Administration, the U.S. Army at Redstone Arsenal, the U.S. Air Force Space Command, the U.S. Naval Observatory, the National Science Foundation, the National Academy of Sciences, the U.S. Department of Energy's Los Alamos National Laboratory, and the U.S. Geological Survey's Astrogeology Science Center.

Information about international space programs and missions was obtained from the websites of the Canadian Space Agency, the China Aerospace Science and Technology Corporation, the European Space Agency, the Japan Aerospace Exploration Agency, the Indian Space Research Organisation, and the Roscosmos State Corporation. The United Nations Office for Outer Space Affairs operates a website that describes international space law and treaties.

Numerous private companies and organizations are engaged in space-related enterprises and educational programs and provide very informative websites. Breaking news about space activities is available from online news services, including *CNN*, *National Geographic News*, Space.com, and *Spaceflight Now*.

The Smithsonian Institution operates the National Air and Space Museum in Washington, D.C. The museum's website (http://www.nasm.si.edu) provides information about the history of airplane flight and space flight. Throughout the space age, the Gallup Organization has conducted several polls and surveys about American attitudes regarding programs and missions. Other pollsters consulted include CBS News, the University of Chicago's National Opinion Research Center, and the Pew Research Center for the People and the Press.

INDEX

Page references in italics refer to photographs. References with the letter t *following them indicate the presence of a table. The letter* f *indicates a figure. If more than one table or figure appears on a particular page, the exact item number for the table or figure being referenced is provided.*

A

C

D

E

F

G

H

I

J

K

L

N

O

P

S

T

U

V

W

X

Y

Z

CPSIA information can be obtained
at www.ICGtesting.com
Printed in the USA
FFOW03n1701250916
27899FF